Humic Substances in Soil and Crop Sciences: Selected Readings

Humic Substances in Soil and Crop Sciences: Selected Readings

Proceedings of a symposium cosponsored by the International Humic Substances Society; Divisions S-3, S-2, S-4, S-5, and S-7 of the Soil Science Society of America; Divisions A-1 and A-5 of the American Society of Agronomy; and Divisions C-2 and C-3 of the Crop Science Society of America, Chicago, Illinois, 2 December 1985.

Editors

P. MacCarthy, C. E. Clapp, R. L. Malcolm, and P. R. Bloom

Organizing Committee

R. L. Malcolm   C. E. Clapp

Editorial Committee

P. MacCarthy   C. E. Clapp
R. L. Malcolm   P. R. Bloom

Editor-in-Chief ASA

G. H. Heichel

Editor-in-Chief SSSA

D. E. Kissel

Managing Editor

R. C. Dinauer

American Society of Agronomy, Inc.
Soi Science Society of America, Inc.
Madison, Wisconsin, USA

1990

Cover Design: Patricia J. Scullion
Cover Photo: Courtesy of Soil Conservation Service, USDA

•

American Society of Agronomy, Inc.
Soil Science Society of America, Inc.
677 South Segoe Road, Madison, WI 53711 USA

Library of Congress Cataloging-in-Publication Data

Humic substances in soil and crop sciences : selected readings :
    proceedings of a symposium cosponsored by the International
    Humic Substances Society . . . in Chicago, Illinois, December 2,
    1985 . . . / editors, P. MacCarthy . . . [et al.].
        p.   cm.
    Includes bibliographical references.
    Includes index.
    ISBN 0-89118-104-0
    1. Humus—Congresses.   2. Soils—Humic acid content—
Congresses.   3. Crops and soils—Congresses.   I. MacCarthy, Patrick.
II. International Humic Substances Society.
S592.8.H86   1990
631.4'17—dc20   90-37639
                                                                     CIP

Printed in the United States of America

CONTENTS

FOREWORD

Organic matter is a very important constituent of soil, serving as a storehouse of available plant nutrients for higher plants and as nutrient and energy sources for soil biota. Organic matter also increases the ion exchange capacity, pH buffering capacity, and water-holding capacity of soil, helps protect the soil against surface crusting and erosion, and improves soil tilth. Despite its obvious importance, the nature of soil organic matter, and of humic substances in particular, is not well understood.

Research interest in humic substances has expanded significantly in recent years, especially with the advent of new methods such as ^{13}C nuclear magnetic resonance spectroscopy. An increasing awareness of the relationships of humic substances to environmental quality has resulted in increased support for research to better understand this important soil constituent. Greater numbers of publications by scientists of many disciplines also attest to the increased research interest in humic substances.

Humic Substances in Soil and Crop Sciences reviews some of the methodology to characterize humic substances in soil and summarizes information presented by experienced scientists on the effects of these materials on soils and on plant growth, as related to nitrogen fertility. The influence of sewage sludge on soil humic substances and interaction of pesticides with humic substances are also discussed. The importance of humic substances to environmental effects in soil management practices is emphasized.

The Soil Science Society of America is pleased to have cosponsored the symposium "Humic Substances Research Related to Soil and Crop Science" at the annual meeting on 2 December 1985 and this resulting publication. This symposium was cosponsored by Divisions 3, 2, 4, 5, and 7 of the SSSA, Divisions 1 and 5 of the American Society of Agronomy, Divisions 2 and 3 of the Crop Science Society of America, and the International Humic Substances Society.

On behalf of the Soil Science Society of America, we thank the authors, editors, and members of organizing committee.

J. J. MORTVEDT, *past president*
Soil Science Society of America

PREFACE

Humic Substances in Soil and Crop Sciences comprises the contributions of renowned scientists from around the world working in the field of humic substances research. Chapters 2 through 9 in this book were invited contributions presented at a symposium cosponsored by the International Humic Substances Society and divisions of the American Society of Agronomy, Soil Science Society of America, and Crop Science Society of America in Chicago, Illinois on 2 December 1985.

The objective of the symposium, and the resulting book, was to focus attention on the importance of humic substances in soil and crop sciences. The other major constituent of soils, mineral matter, has been more extensively studied in the past and is better understood than the humus fraction. The difficulty in the study of humic substances arises primarily from the extraordinary complexity of these materials, which cannot be crystallized or separated into pure components. This problem has severely limited the fundamental understanding of these substances. As a result, soil organic matter in general and humic substances in particular, have frequently not received the attention in the soil science curriculum, which they properly deserve.

In recent years, however, there has been a growing awareness of, and interest in, humic substances by soil scientists as well as by researchers in many other disciplines. This fact is evidenced by the relatively large number of books that have been published on the subject in the past decade, as well as by numerous publications regularly appearing on this subject in a wide range of journals. Interestingly, while humic substances research has historically been largely the domain of the soil scientist, the awakening of widespread interest in this field in recent years is due primarily to the recognition of important environmental consequences resulting from the presence of these materials in natural waters. At the present time, the study of humic substances is truly an interdisciplinary effort.

The chapters in this book focus on the effects of humic substances in soils and on plant growth. Environmental consequences of humic substances are also addressed. Comparisons between humic substances from soils, streams, and groundwaters are made, and organic matter from soils is considered in addition to extracted humic substances, in order to provide a broader perspective on the subject. Despite the fact that it is not possible to represent the structure of humic substances in discrete molecular terms, it is, nevertheless, possible to understand many of the properties and effects of humic substances in terms of what we do know about these materials.

This book is not presented as a comprehensive treatise on the subject, but rather as *selected readings* that address some aspects of the influence of humic substances on soils and plant growth. It is our hope that these writings will stimulate the reader to delve deeper into the subject and seek answers to the many questions that remain unanswered in the following pages.

The editors are grateful to the authors for their presentations at the symposium and for their contribution to this book. Financial support for the symposium was generously provided by the International Humic Substances

Society and the Soil Science Society of America. The assistance of the many
reviewers in improving the quality of the book is gratefully acknowledged.
We thank Debbie Taylor for extensive typing of the manuscripts, and appreciate the cooperation and help of the ASA Headquarters staff in producing this publication.

The understanding, support, and encouragement of our wives Helen
MacCarthy, Betty Clapp, Mollie Malcolm, and Meg Layese (Bloom), were
indispensable in successfully bringing this book to fruition.

PATRICK MACCARTHY, *co-editor*
Colorado School of Mines, Golden, CO

C. EDWARD CLAPP, *co-editor*
USDA-ARS, University of Minnesota, St. Paul, MN

RONALD L. MALCOLM, *co-editor*
U.S. Geological Survey, Water Resources Division, Denver, CO

PAUL R. BLOOM, *co-editor*
University of Minnesota, St. Paul, MN

CONTRIBUTORS

Tsila Aviad Msc. Agr., Department of Soil and Water Sciences, Faculty of Agriculture, The Hebrew University of Jerusalem, P.O. Box 12, Rehovot 76100, Israel

Paul R. Bloom Professor of Soil Science, Department of Soil Science, University of Minnesota, 1991 Upper Buford Circle, St. Paul, MN 55108

Stephen A. Boyd Associate Professor of Soil Chemistry, Department of Crop and Soil Sciences, Michigan State University, East Lansing, MI 48824-1325

Yona Chen Professor of Soil Chemistry, Department of Soil and Water Sciences, Faculty of Agriculture, The Hebrew University of Jerusalem, P.O. Box 12, Rehovot 76100, Israel

Cary T. Chiou Research Chemist, U.S. Geological Survey, Denver Federal Center, Box 24046, MS408, Denver, CO 80225

C. Edward Clapp Research Chemist and Professor of Soil Biochemistry, USDA-ARS, Department of Soil Science, University of Minnesota, 1991 Upper Buford Circle, St. Paul, MN 55108

Ludwig Haumaier Research Scientist, Institute of Soil Science, University of Bayreuth, P.O. Box 10 12 51, D 8580 Bayreuth, Federal Republic of Germany

Xin-Tao He Research Associate, Department of Agronomy, University of Illinois, Urbana, IL 61801

Reinhold Hempfling Research Scientist, Department of Trace Analysis, Dambachtal 20, D6200 Wiesbaden, Federal Republic of Germany. Formerly with the Institute of Soil Science, University of Bayreuth, Bayreuth, Federal Republic of Germany

Patrick MacCarthy Professor of Chemistry, Department of Chemistry and Geochemistry, Colorado School of Mines, Golden, CO 80401

Ronald L. Malcolm Research Geologist, U.S. Geological Survey, Water Resources Division, Denver Federal Center, Box 25046, MS408, Denver, CO 80225

James P. Martin Professor of Soil Science, Emeritus, Department of Soil and Environmental Sciences, University of California, Riverside, CA 92521 (Deceased 19 December 1989)

M. Schnitzer Principle Research Scientist, Land Resource Research Centre, Agriculture Canada, Central Experimental Farm, Ottawa, Ontario, Canada K1A 0C6

Lee E. Sommers Professor of Agronomy, Department of Agronomy, Colorado State University, Fort Collins, CO 80523

F. J. Stevenson   Professor of Soil Chemistry, Department of Agronomy, University of Illinois, Urbana, IL 61801

Diane E. Stott   Soil Microbiologist, USDA-ARS, National Soil Erosion Laboratory, Purdue University, West Lafayette, IN 47907

Michael A. Wilson   Leader, Coal Chemistry, CSIRO Institute of Earth Resources, P.O. Box 136, North Ryde, New South Wales 2113, Australia

Wolfgang Zech   Professor of Soil Science, Institute of Soil Science, University of Bayreuth, P.O. Box 10 12 51, D8580 Bayreuth, Federal Republic of Germany

Conversion Factors for SI and non-SI Units

Conversion Factors for SI and non-SI Units

To convert Column 1 into Column 2, multiply by	Column 1 SI Unit	Column 2 non-SI Unit	To convert Column 2 into Column 1, multiply by
Length			
0.621	kilometer, km (10^3 m)	mile, mi	1.609
1.094	meter, m	yard, yd	0.914
3.28	meter, m	foot, ft	0.304
1.0	micrometer, μm (10^{-6} m)	micron, μ	1.0
3.94×10^{-2}	millimeter, mm (10^{-3} m)	inch, in	25.4
10	nanometer, nm (10^{-9} m)	Angstrom, Å	0.1
Area			
2.47	hectare, ha	acre	0.405
247	square kilometer, km^2 (10^3 m)^2	acre	4.05×10^{-3}
0.386	square kilometer, km^2 (10^3 m)^2	square mile, mi^2	2.590
2.47×10^{-4}	square meter, m^2	acre	4.05×10^3
10.76	square meter, m^2	square foot, ft^2	9.29×10^{-2}
1.55×10^{-3}	square millimeter, mm^2 (10^{-3} m)^2	square inch, in^2	645
Volume			
9.73×10^{-3}	cubic meter, m^3	acre-inch	102.8
35.3	cubic meter, m^3	cubic foot, ft^3	2.83×10^{-2}
6.10×10^4	cubic meter, m^3	cubic inch, in^3	1.64×10^{-5}
2.84×10^{-2}	liter, L (10^{-3} m^3)	bushel, bu	35.24
1.057	liter, L (10^{-3} m^3)	quart (liquid), qt	0.946
3.53×10^{-2}	liter, L (10^{-3} m^3)	cubic foot, ft^3	28.3
0.265	liter, L (10^{-3} m^3)	gallon	3.78
33.78	liter, L (10^{-3} m^3)	ounce (fluid), oz	2.96×10^{-2}
2.11	liter, L (10^{-3} m^3)	pint (fluid), pt	0.473

Mass

Multiply	SI Unit (Column 1)	non-SI Unit (Column 2)	Multiply
2.20×10^{-3}	gram, g (10^{-3} kg)	pound, lb	454
3.52×10^{-2}	gram, g (10^{-3} kg)	ounce (avdp), oz	28.4
2.205	kilogram, kg	pound, lb	0.454
0.01	kilogram, kg	quintal (metric), q	100
1.10×10^{-3}	kilogram, kg	ton (2000 lb), ton	907
1.102	megagram, Mg (tonne)	ton (U.S.), ton	0.907
1.102	tonne, t	ton (U.S.), ton	0.907

Yield and Rate

Multiply	SI Unit (Column 1)	non-SI Unit (Column 2)	Multiply
0.893	kilogram per hectare, kg ha^{-1}	pound per acre, lb acre^{-1}	1.12
7.77×10^{-2}	kilogram per cubic meter, kg m^{-3}	pound per bushel, bu^{-1}	12.87
1.49×10^{-2}	kilogram per hectare, kg ha^{-1}	bushel per acre, 60 lb	67.19
1.59×10^{-2}	kilogram per hectare, kg ha^{-1}	bushel per acre, 56 lb	62.71
1.86×10^{-2}	kilogram per hectare, kg ha^{-1}	bushel per acre, 48 lb	53.75
0.107	liter per hectare, L ha^{-1}	gallon per acre	9.35
893	tonnes per hectare, t ha^{-1}	pound per acre, lb acre^{-1}	1.12×10^{-3}
893	megagram per hectare, Mg ha^{-1}	pound per acre, lb acre^{-1}	1.12×10^{-3}
0.446	megagram per hectare, Mg ha^{-1}	ton (2000 lb) per acre, ton acre^{-1}	2.24
2.24	meter per second, m s^{-1}	mile per hour	0.447

Specific Surface

Multiply	SI Unit (Column 1)	non-SI Unit (Column 2)	Multiply
10	square meter per kilogram, m^2 kg^{-1}	square centimeter per gram, cm^2 g^{-1}	0.1
1000	square meter per kilogram, m^2 kg^{-1}	square millimeter per gram, mm^2 g^{-1}	0.001

Pressure

Multiply	SI Unit (Column 1)	non-SI Unit (Column 2)	Multiply
9.90	megapascal, MPa (10^6 Pa)	atmosphere	0.101
10	megapascal, MPa (10^6 Pa)	bar	0.1
1.00	megagram per cubic meter, Mg m^{-3}	gram per cubic centimeter, g cm^{-3}	1.00
2.09×10^{-2}	pascal, Pa	pound per square foot, lb ft^{-2}	47.9
1.45×10^{-4}	pascal, Pa	pound per square inch, lb in^{-2}	6.90×10^3

(continued on next page)

Conversion Factors for SI and non-SI Units

To convert Column 1 into Column 2, multiply by	Column 1 SI Unit	Column 2 non-SI Unit	To convert Column 2 into Column 1, multiply by
Temperature			
$1.00\ (K - 273)$	Kelvin, K	Celsius, °C	$1.00\ (°C + 273)$
$(9/5\ °C) + 32$	Celsius, °C	Fahrenheit, °F	$5/9\ (°F - 32)$
Energy, Work, Quantity of Heat			
9.52×10^{-4}	joule, J	British thermal unit, Btu	1.05×10^{3}
0.239	joule, J	calorie, cal	4.19
10^{7}	joule, J	erg	10^{-7}
0.735	joule, J	foot-pound	1.36
2.387×10^{-5}	joule per square meter, J m^{-2}	calorie per square centimeter (langley)	4.19×10^{4}
10^{5}	newton, N	dyne	10^{-5}
1.43×10^{-3}	watt per square meter, W m^{-2}	calorie per square centimeter minute (irradiance), cal cm^{-2} min^{-1}	698
Transpiration and Photosynthesis			
3.60×10^{-2}	milligram per square meter second, mg m^{-2} s^{-1}	gram per square decimeter hour, g dm^{-2} h^{-1}	27.8
5.56×10^{-3}	milligram (H$_2$O) per square meter second, mg m^{-2} s^{-1}	micromole (H$_2$O) per square centimeter second, μmol cm^{-2} s^{-1}	180
10^{-4}	milligram per square meter second, mg m^{-2} s^{-1}	milligram per square centimeter second, mg cm^{-2} s^{-1}	10^{4}
35.97	milligram per square meter second, mg m^{-2} s^{-1}	milligram per square decimeter hour, mg dm^{-2} h^{-1}	2.78×10^{-2}
Plane Angle			
57.3	radian, rad	degrees (angle), °	1.75×10^{-2}

Electrical Conductivity, Electricity, and Magnetism

To convert Column 1 into Column 2, multiply by	Column 1 SI Unit	Column 2 non-SI Unit	To convert Column 2 into Column 1, multiply by
0.1	siemen per meter, S m^{-1}	millimho per centimeter, mmho cm^{-1}	10
10^{-4}	tesla, T	gauss, G	10^4

Water Measurement

To convert Column 1 into Column 2, multiply by	Column 1 SI Unit	Column 2 non-SI Unit	To convert Column 2 into Column 1, multiply by
102.8	cubic meter, m^3	acre-inches, acre-in	9.73 × 10^{-3}
101.9	cubic meter per hour, m^3 h^{-1}	cubic feet per second, ft^3 s^{-1}	9.81 × 10^{-3}
0.227	cubic meter per hour, m^3 h^{-1}	U.S. gallons per minute, gal min^{-1}	4.40
0.123	hectare-meters, ha-m	acre-feet, acre-ft	8.11
1.03 × 10^{-2}	hectare-meters, ha-m	acre-inches, acre-in	97.28
12.33	hectare-centimeters, ha-cm	acre-feet, acre-ft	8.1 × 10^{-2}

Concentrations

To convert Column 1 into Column 2, multiply by	Column 1 SI Unit	Column 2 non-SI Unit	To convert Column 2 into Column 1, multiply by
1	centimole per kilogram, cmol kg^{-1} (ion exchange capacity)	milliequivalents per 100 grams, meq 100 g^{-1}	1
10	gram per kilogram, g kg^{-1}	percent, %	0.1
1	milligram per kilogram, mg kg^{-1}	parts per million, ppm	1

Radioactivity

To convert Column 1 into Column 2, multiply by	Column 1 SI Unit	Column 2 non-SI Unit	To convert Column 2 into Column 1, multiply by
3.7 × 10^{10}	becquerel, Bq	curie, Ci	2.7 × 10^{-11}
37	becquerel per kilogram, Bq kg^{-1}	picocurie per gram, pCi g^{-1}	2.7 × 10^{-2}
0.01	gray, Gy (absorbed dose)	rad, rd	100
0.01	sievert, Sv (equivalent dose)	rem (roentgen equivalent man)	100

Plant Nutrient Conversion

To convert Column 1 into Column 2, multiply by	Elemental	Oxide	To convert Column 2 into Column 1, multiply by
0.437	P	P$_2$O$_5$	2.29
0.830	K	K$_2$O	1.20
0.715	Ca	CaO	1.39
0.602	Mg	MgO	1.66

Chapter 1

An Introduction to Soil Humic Substances

P. MacCARTHY, *Colorado School of Mines, Golden Colorado*

R. L. MALCOLM, *U.S. Geological Survey, Denver, Colorado*

C. E. CLAPP, *Agricultural Research Service, USDA,*
University of Minnesota, St. Paul, Minnesota

P. R. BLOOM, *University of Minnesota, St. Paul, Minnesota*

ABSTRACT

The term *humic substances* refers to an operationally defined, heterogeneous mixture of naturally occurring organic materials. These substances are ubiquitous in nature and arise from the decay of plant and animal residues in the environment. Humic substances are generally classified into humic acid, fulvic acid, and humin on the basis of their solubility in water as a function of pH. It has long been recognized that humic substances have many beneficial effects on soils and consequently on plant growth. Some of these effects have been well documented, but others, such as the alleged direct effect of humic substances on plant growth, are still a matter of controversy. The inherent complexity of humic substances and the inability of researchers to separate them into pure components imposes restrictions on the meaning of the term structure when applied to these materials; we must be satisfied with measuring the average characteristics of these substances. Fortunately, this is sufficient for understanding most of the properties of these materials and for explaining their effects in the soil. In addition to the inherent difficulty of working with a complex mixture, other factors that have contributed to the difficulty of studying these materials have been the inconsistent use of terminology and the previous lack of standard materials for comparison purposes. The recent establishment of standard humic and fulvic acids by the International Humic Substances Society remedies the latter problem.

Agriculturalists since ancient times have recognized significant benefits of soil organic matter to crop productivity. These benefits have been the subject of controversy for centuries and some are still debated today. The following list includes many of the recognized benefits of soil organic matter (Stevenson, 1982):

1. It serves as a slow-release source of N, P, and S, for plant nutrition and microbial growth.
2. It possesses considerable water-holding capacity, and thereby helps to maintain the water regime of the soil.
3. It acts as a buffer against changes in pH of the soil.
4. Its dark color contributes to absorption of energy from the sun and heating of the soil.
5. It acts as a "cement" for holding clay and silt particles together, thus contributing to the crumb structure of the soil, and to resistance against soil erosion.
6. It binds micronutrient metal ions in the soil that otherwise might be leached out of surface soils.
7. Organic constituents in the humic substances may act as plant-growth stimulants.

Some of these effects are discussed in the following chapters of this book. Many of the benefits of soil organic matter have been well documented scientifically, but some effects are so intimately associated with other soil factors that it is difficult to ascribe them uniquely to the organic matter. In fact, soil is a complex, multicomponent system of interacting materials, and the properties of soil result from the net effect of all these interactions. An ultimate understanding of soil properties and behavior would take all of these materials and interactions into account. Because of the complexity of the system, however, we are not yet able to study the "organism" as a whole, and must dissect it into its component parts for individual study. The resulting data must then be integrated in an attempt to provide a picture of the whole soil. Even the study of the separate components is not an easy task because of their individual complexity. The situation is further complicated by a common occurrence in the study of environmental phenomena; that is, the isolated components may not be the same as they were in the original soil. For example, the isolation procedure may alter the form of the materials or may produce artifacts.

The relative importance of various materials in soil for plant growth may vary significantly with cropping and management practices, climate, soil

type, and other factors, and can only be evaluated on a s
organic matter and mineral components of soil share
functions in common; for example, both fractions ha
pacity, water-holding ability, and pH buffering propert
ganic matter represents only about 2 to 6% by weight (on
a typical agricultural soil, it contributes about 50% of the catio.
water-holding, and pH buffering capacities to the soil. A typical ca.
change capacity for soil organic matter is 1 to 2 mol kg^{-1}; the mineral h.
tion has a typical capacity of 0.05 to 0.3 mol kg^{-1}, but is present in much
greater amounts. It should be recognized, of course, that the measured or
practical cation exchange capacity is a pH-dependent quantity. The chapters in this book are directed primarily toward the humic substances of soil,
but in order to address this topic in context it is necessary at times to deal
with soil organic matter in general, and in some cases with whole soil.

1-1 OBJECTIVES OF BOOK

This book is not intended to be a comprehensive treatise on the subject
of soil humic substances; rather, it is presented in the form of *selected readings*
in the field with the purpose of giving the reader a "feel" for the variety
of ways in which humic substances are important in soil and crop sciences.
The objectives of the book are:

1. To present a broad overview of humic substances and to describe
   their importance in soil and crop sciences.
2. To compare soil humic substances to humic substances from waters
   and other sources.

Among the topics covered in this book are: the nature of humic substances from soil, surface water, and groundwater environments, and the
similarities and differences among them (chapter 2, Malcolm); the origin and
formation of soil humic substances (chapter 3, Stott & Martin); selected
methods for characterizing humic substances (chapter 4, Schnitzer); the nature
of nitrogen in humic substances as related to soil fertility (chapter 5, Stevenson & He); sorption of nonionic compounds and pesticides by organic matter and minerals (chapter 6, Chiou); effects of humic substances on plant
growth (chapter 7, Chen & Aviad); ecological aspects of soil organic matter
in tropical soils (chapter 8, Zech et al.); sewage sludge humic and fulvic acids
(chapter 9, Boyd & Sommers); and the use of NMR spectroscopy for studying whole soils (chapter 10, Wilson). Chapter 11 (MacCarthy et al.) summarizes and integrates the material from the preceding chapters, and presents
an ecological overview of the nature of humic substances.

More extensive discussions of the topics addressed in this chapter can
be found in the following references (Kononova, 1966; Greenland & Hayes,
1978; Schnitzer & Khan, 1978; Stevenson, 1982; Aiken et al., 1985b; Frimmel & Christman, 1988; Hayes et al., 1989; Suffet & MacCarthy, 1989).

1-2 HISTORICAL BACKGROUND

The scientific study of humic substances can be traced back more than 200 yr. In 1786, Achard used alkali to extract from peat a material that we now call humic acid (Achard, 1786). Interestingly, Achard's procedure is still the basis of the most common methods for extracting humic substances from soil today. Until the early 1970s most humic studies dealt with materials isolated from soils, although Berzelius in the early 1800s did investigate humic substances isolated from water (Berzelius, 1839). However, from the early 1970s there has been a rapidly growing interest in humic substances in natural waters. One of the contributing factors to this interest in aquatic humic substances was the discovery by Rook (1974) that chlorination of waters leads to the formation of chloroform and other potentially hazardous chlorinated hydrocarbons. This discovery has potentially serious implications because chlorination is one of the most universally used disinfection procedures for municipal waters, and because many chlorinated chemicals are suspected carcinogens. Another factor leading to growth in the study of aquatic humic substances was the development of new technologies for the isolation of reasonable quantities of humic materials from natural waters (Aiken, 1985). The knowledge and techniques being developed from the investigation of aquatic humic materials are being applied to the study of soil humic substances and are leading to progress in this area.

1-3 DEFINITIONS OF TERMS

1-3.1 Soil Organic Matter and Humus

One of the major problems in communicating in the field of humic substances is the lack of precise definitions for unambiguously specifying the various fractions. Unfortunately, the terminology is not used in a consistent manner. For example, the term *humus* is sometimes used synonomously with *soil organic matter,* that is, to denote the organic material in the soil, including humic substances (but exclusive of undecayed plant and animal tissue, their partial decomposition products, and the soil biomass). The term *humus* is sometimes used to represent only the humic substances (Stevenson, 1982). This terminology problem warrants serious attention in the future; but for the present, it is important for the reader to establish how a particular author is using the various terms if one is to understand what the author is saying. An extensive glossary of soil science terms, which will be useful to the readers of this book, has been compiled by the Soil Science Society of America (1987).

The term *soil organic matter* is generally used to represent the organic constituents in the soil, excluding undecayed plant and animal tissue, their partial decomposition products, and the soil biomass (Stevenson, 1982). Thus, this term includes: (i) identifiable high-molecular-weight, organic materials such as polysaccharides and proteins, (ii) simpler substances such as sugars, amino acids, and other small molecules, and (iii) humic substances. It is likely that soil organic matter contains most, if not all, of the organic compounds synthesized by living organisms. Soil organic matter is frequently said to consist of humic substances and nonhumic substances. Nonhumic substances are all those materials that can be placed in one of the categories of discrete compounds such as polysaccharides and sugars, proteins and amino acids, fats, simple organic acids, and so on; humic substances are the other, unidentifiable components. Even this apparently simple distinction, however, is not as clearcut as it might appear. Applying this distinction in practice depends on one's ability to separate the humic from the nonhumic material and to properly identify the discrete substances. The situation is further complicated by the fact that some identifiable substances are covalently linked to humic material and can be released by hydrolysis or by other means. This situation is another example of an instance where the usage of the terminology is not always clear, but in general, a material can be considered to be a humic substance even though it contains covalently bound fragments of identifiable components.

The term *humic substances* is frequently applied to other organic materials of nonspecific nature. For example, extracts from sewage sludge, composts, and manure, that conform to the operational definitions of humic substances are frequently referred to as humic substances. So-called sewage sludge humic substances are discussed in chapter 9 of this book (Boyd & Sommers). The labeling of these materials as humic substances should not mislead one into believing that they are identical to *real* humic substances. Nevertheless, there are similarities in that both classes of materials are composed of complex mixtures of organic substances. Consequently, both materials present similar difficulties to the experimentalist and yield data that are particularly challenging to interpret. Also, sewage sludge and manures may evolve into *real* humic substances with time. Throughout the literature one can find many materials loosely referred to as humic acid; however, researchers should resist the temptation to label diverse materials as ''humic'' just because they are complex, organic mixtures.

1–3.2 Definitions of Humic Substances

Humic substances can be defined operationally as follows (Aiken et al., 1985a):

Humic substances—A category of naturally occurring, biogenic, heterogeneous organic substances that can generally be characterized as being yellow to black in color, of high molecular weight, and refractory.

This statement is really more a description of humic substances rather than a definition, and is typical of the nonspecificity that is prevalent in the study of humic substances. These materials result from the decay of plant and animal residues, and cannot be classified into any of the discrete categories of compounds such as proteins, polysaccharides, and polynucleotides. Humic substances are ubiquitous, and are found in all soils, sediments, and waters. This book focuses on humic substances in soils; however, humic substances in waters are also addressed in order to give a broader perspective to the discussion. Although these materials are known to result from the decay products of biological tissue, the precise biochemical and chemical pathways by which they are formed have not been elucidated. The reader is referred to chapter 3 of this book (Stott & Martin) for a detailed discussion of the formation of humic substances.

Humic substances consist of an extraordinarily complex mixture of organic compounds. Despite the attempts of many researchers over a long period of time no one has yet succeeded in separating humic substances into discrete components. Virtually every separation technique that has been developed by chemists and biochemists has been applied to humic substances. While some of these techniques do succeed in diminishing the degree of heterogeneity of the samples, none of them has come even remotely close to isolating a material that could be called a *pure* humic substance in the classical meaning of the term *pure*. The subject of the extreme heterogeneity of humic substances is addressed again in the summary to chapter 11 of this book (MacCarthy et al.) where this characteristic is considered to be an inherent property of all humic materials.

The inability to define humic substances in specific chemical terms forces us to use a more vague operational definition such as that given above. Over the years, many fractions of humic substances have been isolated and assigned special names. Of these named fractions, only three have stood the test of time as being generally useful, namely, humic acid, fulvic acid, and humin. As with humic substances as a whole, these fractions are operationally defined (Aiken et al., 1985a):

Humic acid—The fraction of humic substances that is not soluble in water under acidic conditions (pH < 2) but is soluble at higher pH values.

Fulvic acid—The fraction of humic substances that is soluble in water under all pH conditions.

Humin—The fraction of humic substances that is not soluble in water at any pH value.

These definitions reflect the methods of isolating the various fractions from the soil. Most studies of humic substances have been conducted on humic and fulvic acids, and the humin fraction has been investigated to a much lesser extent (Rice & MacCarthy, 1988). However, there has been a growing interest in the study of humin in recent years. For a recent review of humin see Hatcher et al. (1985).

1–4 EXTRACTION OF HUMIC SUBSTANCES

1–4.1 Extraction from Soil

Although there are numerous variations of the extraction procedures for isolating humic substances from soil, the essential features embodied in most of these techniques are as follows:

The soil sample is shaken with NaOH solution at a specified temperature and for a specified duration; after centrifugation, the alkaline supernatant solution contains humic and fulvic acids in the salt form. This solution is then decanted and acidified to pH 1.0 to 2.0 with HCl (The exact pH that is used varies somewhat from worker to worker.) The humic acid precipitates, and the fulvic acid, which remains in solution, is decanted after centrifugation. The humic and fulvic acids are then desalted by any of various techniques. The insoluble residue remaining after removal of the alkaline supernatant solution contains the humin, generally mixed with insoluble mineral matter or undecomposed plant material. Any soluble nonhumic material that is mixed with the precipitated humic acid is removed during the subsequent washing with water while desalting the humic acid.

The supernatant solution that remains following acidification of the alkaline extract is more correctly referred to as the *fulvic acid fraction,* rather than as fulvic acid. The fulvic acid fraction contains discrete compounds (such as amino acids and simple sugars) in addition to the humified material (fulvic acid). This distinction is discussed in detail in chapter 2 of this book (Malcolm).

The discrete compounds in the fulvic acid fraction can be separated from the fulvic acid *per se* by passing the fulvic acid fraction through a column of hydrophobic resin. The more hydrophilic compounds (e.g., amino acids, sugars) pass through the column, and the less hydrophilic constituents (fulvic acid) sorb to the resin. The fulvic acid can subsequently be desorbed from the resin with dilute base.

Detailed discussions of the extraction of soil humic substances can be found in references by Stevenson (1982) and Hayes (1985). Techniques for fractionating soil humic substances are described by Swift (1985).

1–4.2 Isolation from Water

Isolating humic substances from waters is a more difficult and time-consuming task, even for those waters that contain a relatively high concentration of humic substances. The development of resin sorption techniques for isolating aquatic humic substances in the early 1970s was a significant advance in this area (Aiken, 1985). The essential features of these techniques are as follows:

The natural water is acidified to approximately pH 2.0 with HCl and is passed through a column containing a resin that is essentially hydrophobic, such as the methyl methacrylate cross-linked polymer, XAD-8. The humic substances sorb to the resin while the more hydrophilic, nonhumic materials pass through the column. Elution of the column at pH 7.0 removes the fulvic acid. Subsequent elution with 0.1 M NaOH desorbs the humic acid. The humic and fulvic acids are then converted to the hydrogen form by passage through a strong acid cation exchanger in the hydrogen form. The isolated humic and fulvic acids are generally freeze-dried in the hydrogen form.

Detailed discussions of the isolation and fractionation of aquatic humic substances are found in the following references (Aiken, 1985; Leenheer, 1985).

1–5 CHARACTERIZATION OF HUMIC SUBSTANCES

Despite inherent difficulties in the study of humic substances and our inability to determine the chemical structure of these materials, a great deal is known about their *composition*. As discussed in chapter 11 of this book (MacCarthy et al.), the question of the absolute structure of humic substances may really be moot. What is the meaning of the term *structure* in the context of a highly complex, heterogeneous mixture? In the ultimate sense it would mean determining the structure of each constituent molecule in the humic assemblage. This is not an achievable goal at the present time. However, we can determine compositional features and other important characteristics of humic substances. Such information affords a basis for understanding many of the properties of humic substances and their functions in soil and water ecosystems.

Humic substances vary in composition, depending on their source, method of extraction, and other parameters. Overall, however, the similarities between different humic substances are more pronounced than their differences. It is, of course, this similarity that allows humic substances to be identified as a class. Furthermore, the measured properties of humic substances are all average properties since these materials are complex mixtures. Among the most common methods for characterizing humic substances are elemental and functional group determinations.

Elemental analysis is one of the more reliable determinations that can be carried out on humic substances, even though it is not without its problems. The elements C and H are most frequently determined, with O generally being obtained by difference. Estimating O content by difference can be a serious source of error in many cases because it assumes that the humic substances are constituted exclusively of C, H, and O. It has been found that C, H, O, N, P, and S generally account for 100% of the composition of humic substances on an ash-free basis, and it is recommended that these six elements be determined where possible. Also, it is recommended that *all* of these elements be determined by direct methods rather than estimating one of them by difference. This procedure allows a cross-check on the analysis by performing a mass balance on the data (MacCarthy & Malcolm, 1989). Average elemental compositions for humic and fulvic acids are given in Table 1–1, and the values for a specific soil are presented in Table 1–2. Generally, humic acid has a greater C content and a smaller O content than fulvic acid. A detailed discussion of the elemental analysis of humic substances is given by Huffman and Stuber (1985), and the implications of elemental composition are discussed by Steelink (1985).

The major functional groups in humic substances are carboxyl, alcohol, phenolic hydroxyl, and carbonyl. Overall, methods for determining functional groups in humic substances are not as reliable as those for measuring the elemental composition. This, again, is a manifestation of the multicomponent nature of humic substances. In the case of elemental determinations, the humic substances are generally subjected to a rigorous degradation procedure in which the material is broken down into simple molecules, such as

Table 1-1. Average elemental composition† of soil humic substances (from Steelink, 1985).‡

Element	Humic acid	Fulvic acid
	%	
Carbon	53.8–58.7	40.7–50.6
Hydrogen	3.2–6.2	3.8–7.0
Oxygen	32.8–38.3	39.7–49.8
Nitrogen	0.8–4.3	0.9–3.3
Sulfur	0.1–1.5	0.1–3.6

† Weight percent on a moisture-free, ash-free basis.
‡ Reprinted with permission from *Humic substances in soil, sediment, and water: Geochemistry, isolation, and characterization.* G.R. Aiken et al. (ed.). Copyright © 1985. Wiley-Interscience, New York.

Table 1-2. Elemental composition† of Sanhedron Al humic and fulvic acids (from Malcolm and MacCarthy, 1986).

Component	C	H	O	N	S	P	Total	Ash
				%				
Humic acid	58.03	3.64	33.59	3.26	0.47	0.10	99.09	1.19
Fulvic acid	48.71	4.36	43.35	2.77	0.81	0.59	100.59	2.25

† Elemental composition expressed as weight percent on a moisture-free, ash-free basis; ash expressed as weight percent.

CO_2 in the case of C determination, or water in the case of H determination. The analysis is thereby transformed into a determination of CO_2 or of water, and so forth. A similar approach does not exist for determination of functional groups in humic substances, and such determinations must be carried out on the intact material.

Because of the heterogeneous nature of humic substances, a specific type of functional group may exist in a range of chemical environments each of which responds somewhat differently to a given measurement probe. An example of this situation occurs with the most common method for determining the acidity of humic substances, that is titration with NaOH. Potentiometric titration of humic substances does not produce a curve with discrete inflection points but generally yields a curve with a gradually sloping profile and no identifiable end-points. This behavior is a reflection of the multicomponent nature of these materials and is not surprising. The weak acidic nature of humic substances is attributed to carboxyl groups, and the very weak acidic behavior is attributed to phenolic groups. The difficulty arises in trying to decide on a cut-off point on the titration curve where consumption of hydroxide at lower pH values is attributed to carboxyl, and consumption of hydroxide at higher pH values is attributed to phenolic groups. In the absence of a discrete end-point such a decision must be made somewhat arbitrarily. Many monoprotic carboxylic acids yield an equivalence point in the pH range of 8 to 9, and for this reason many workers choose an end-point in this region for the titration of humic substances. It is quite possible, however, that some of the carboxyl groups in humic substances are weaker and will not be neu-

Table 1-3. Functional group compositions† of humic and fulvic acids isolated from soils of widely different climatic zones (from Stevenson, 1982).‡

Functional group	Arctic	Cool, temperate Acid soils	Cool, temperate Neutral soils	Subtropical	Tropical	Range
		Humic acids				
Total acidity	560	570–890	620–660	630–770	620–750	560–890
COOH	320	150–570	390–450	420–520	380–450	150–570
Acidic OH	240	320–570	210–250	210–250	220–300	210–570
Weakly acidic + alcoholic OH	490	270–350	240–320	290	20–160	20–490
Quinone C=O	230 }	10–180	450–560	80–150	30–140	10–560
Ketonic C=O	170 }					
OCH₃	40	40	30	30–50	60–80	30–80
		Fulvic acids				
Total acidity	1100	890–1420	--	640–1230	820–1030	640–1420
COOH	880	610–850	--	520–960	720–1120	520–1120
Acidic OH	220	280–570	--	120–270	30–250	30–570
Weakly acidic + alcoholic OH	380	340–460	--	690–950	260–520	260–950
Quinone C=O	200	{170–310	--	{120–260	30–150	{120–420
Ketonic C=O	200				160–270	
OCH₃	60	30–40		80–90	90–120	30–120

† All values expressed in units of meq 100 g^{-1}.
‡ Reprinted with permission from *Humus chemistry: Genesis, composition, reactions.* F.J. Stevenson. Copyright © 1982. Wiley-Interscience, New York.

tralized in that pH range. Accordingly, determination of carboxyl and phenolic hydroxyl groups in humic substances by titrimetry entails a considerable degree of uncertainty.

The acidic nature of humic substances and methods for its determination are discussed by Stevenson (1982) and Perdue (1985). Other methods of functional group determination in humic substances are subject to analogous types of problems. Table 1-3 shows the functional group compositions of humic and fulvic acids from diverse soils. Detailed discussions of the functional group composition of humic substances can be found in the following references (Stevenson, 1982; Perdue, 1985).

Other methods for characterizing humic substances are average molecular weight determination (Wershaw & Aiken, 1985), infrared spectroscopy (MacCarthy & Rice, 1985), and NMR spectroscopy (Wershaw, 1985; Wilson, 1987; see also chapters 2, 4, and 10 in this book). A comprehensive and critical evaluation of what is known about the structural features of humic substances has recently been completed (Hayes et al., 1989).

1-6 STANDARD HUMIC SUBSTANCES

The fact that humic substances are comprised of a complex, unseparable mixture of molecules makes their study particularly difficult compared to that of more discrete materials. Because humic substances are not unique materials but can vary from source to source, and with other factors such as the method of extraction, there is really no unique analytical method for their determination. How could one develop an analytical method for a material that we cannot clearly define? Of course, we can measure the amounts of materials that conform to the operational definitions of the various humic fractions. These materials can be expressed in terms of total mass in each category or in terms of the amount of C present. Such assays are only as good as the isolation and "purification" methods, and are obviously not unique to humic substances.

Another long-standing difficulty in the investigation of humic substances has been the previous lack of a standard material for interlaboratory comparison of experimental data and for calibration of methods. The need for such standards is particularly acute when dealing with nondiscrete materials such as humic substances. This problem has been addressed by the International Humic Substances Society (IHSS) which has established a suite of standard humic substances from soil, peat, leonardite, and surface water (MacCarthy et al., 1986). Researchers can now purchase these materials from the IHSS. (International Humic Substances Society, Standard and Reference Humic Substances, % Department of Chemistry and Geochemistry, Colorado School of Mines, Golden, CO 80401 USA.)

In addition to the standard humic and fulvic acids, the IHSS also maintains a large stockpile of the homogenized soil, peat, and leonardite materials from which the standard humic and fulvic acids were extracted. These materials are of particular value in allowing investigators to compare different extractive methodologies.

REFERENCES

Achard, F.K. 1786. Chemische Untersuchung des Torfs. Crell's Chem. Ann. 2:391–403.

Aiken, G.R. 1985. Isolation and concentration techniques for aquatic humic substances. p. 363–385. In G.R. Aiken et al. (ed.) Humic substances in soil, sediment, and water: Geochemistry, isolation and characterization. Wiley-Interscience, New York.

Aiken, G.R., D.M. McKnight, R.L. Wershaw, and P. MacCarthy. 1985a. An introduction to humic substances in soil, sediment, and water. p. 1–9. In G.R. Aiken et al. (ed.) Humic substances in soil, sediment, and water: Geochemistry, isolation and characterization. Wiley-Interscience, New York.

Aiken, G.R., D.M. McKnight, R.L. Wershaw, and P. MacCarthy (ed.). 1985b. Humic substances in soil, sediment, and water: Geochemistry, isolation, and characterization. Wiley-Interscience, New York.

Berzelius, J.J. 1839. Lehrbuch der Chemie, 3rd ed.; translated by Wohler; Dresden and Leipzig.

Frimmel, F.H., and R.F. Christman (ed.). 1988. Humic substances and their role in the environment. Wiley-Interscience, New York.

Greenland, D.J., and M.H.B. Hayes (ed.). 1978. The chemistry of soil constituents. Wiley-Interscience, New York.

Hatcher, P.G., I.A. Breger, G.E. Maciel, and N.M. Szeverenyi. 1985. Geochemistry of humin. p. 275–302. In G.R. Aiken et al. (ed.) Humic substances in soil, sediment, and water: Geochemistry, isolation and characterization. Wiley-Interscience, New York.

Hayes, M.H.B. 1985. Extraction of humic substances from soil. p. 329–362. In G.R. Aiken et al. (ed.) Humic substances in soil, sediment, and water: Geochemistry, isolation and characterization. Wiley-Interscience, New York.

Hayes, M.H.B., P. MacCarthy, R.L. Malcolm, and R.S. Swift (ed.). 1989. Humic substances II. In search of structure. Wiley-Interscience, Chichester, UK.

Huffman, E.W.D., Jr. and H.A. Stuber. 1985. Analytical methodology for elemental analysis of humic substances. p. 433–455. *In* G.R. Aiken et al. (ed.) Humic substances in soil, sediment, and water: Geochemistry, isolation and characterization. Wiley-Interscience, New York.

Kononova, M.M. 1966. Soil organic matter. Pergamon, Elmsford, New York.

Leenheer, J.A. 1985. Fractionation techniques for aquatic humic substances. p. 409–429. *In* G.R. Aiken et al. (ed.) Humic substances in soil, sediment, and water: Geochemistry, isolation and characterization. Wiley-Interscience, New York.

MacCarthy, P., and J.A. Rice. 1985. Spectroscopic methods (other than NMR) for determining functionality in humic substances. p. 527–559. *In* G.R. Aiken et al. (ed.) Humic substances in soil, sediment, and water: Geochemistry isolation and characterization. Wiley-Interscience, New York.

MacCarthy, P., R.L. Malcolm, M.H.B. Hayes, R.S. Swift, M. Schnitzer, and W.L. Campbell. 1986. Establishment of a collection of standard humic substances. p. 378–379. *In* Vol. 2. Trans Int. Cong. Soil Sci., 13th, 1986. Hamburg, West Germany.

MacCarthy, P., and R.L. Malcolm. 1989. The nature of commercial humic acids. p. 4–13. *In* I.H. Suffet and P. MacCarthy (ed.) Aquatic humic substances: Influence on fate and treatment of pollutants. American Chemical Society, Washington, DC.

Malcolm, R.L., and P. MacCarthy. 1986. Limitations in the use of commercial humic acids in water and soil research. Environ. Sci. Technol. 20:904–911.

Perdue, E.M. 1985. Acidic functional groups of humic substances. p. 493–526. *In* G.R. Aiken et al. (ed.) Humic substances in soil, sediment, and water: Geochemistry, isolation and characterization. Wiley-Interscience, New York.

Rice, J.A., and P. MacCarthy. 1988. Comments on the literature of the humin fraction of humus. Geoderma. 43:65–73.

Rook, J.J. 1974. Formation of haloforms during chlorination of natural waters. Water Treat. Exam. 23:234–243.

Schnitzer, M., and S.U. Khan (ed.). 1978. Soil organic matter. Elsevier, New York.

Soil Science Society of America, Terminology Committee. 1987. Glossary of soil science terms. SSSA, Madison, WI.

Steelink, C.A. 1985. Implications of elemental characteristics of humic substances. p. 457–476. *In* G.R. Aiken et al. (ed.) Humic substances in soil, sediment, and water: Geochemistry, isolation and characterization. Wiley-Interscience, New York.

Stevenson, F.J. 1982. Humus chemistry: Genesis, composition, reactions. Wiley-Interscience, New York.

Suffet, I.H., and P. MacCarthy (ed.). 1989. Aquatic humic substances: Influence on fate and treatment of pollutants. Am. Chem. Soc., Washington, DC.

Swift, R.S. 1985. Fractionation of soil humic substances. p. 387–408. *In* G.R. Aiken et al. (ed.) Humic substances in soil, sediment and water: Geochemistry, isolation, and characterization. Wiley-Interscience, New York.

Wershaw, R.L. 1985. Application of nuclear magnetic resonance spectroscopy to determining functionality in humic substances. p. 561–582. *In* G.R. Aiken et al. (ed.) Humic substances in soil, sediment, and water: Geochemistry, isolation and characterization. Wiley-Interscience, New York.

Wershaw, R.L., and G.R. Aiken. 1985. Molecular size and weight measurements of humic substances. p. 477–492. *In* G.R. Aiken et al. (ed.) Humic substances in soil, sediment, and water: Geochemistry, isolation and characterization. Wiley-Interscience, New York.

Wilson, M.A. 1987. NMR techniques and applications in geochemistry and soil chemistry. Pergamon Press, Oxford.

Chapter 2

Variations Between Humic Substances Isolated from Soils, Stream Waters, and Groundwaters as Revealed by ^{13}C-NMR Spectroscopy

R. L. MALCOLM, *U.S. Geological Survey, Denver, Colorado*

ABSTRACT

Humic substances in soils differ significantly from those in stream waters or groundwaters. Soil humic substances are of higher molecular weight, of greater ^{14}C age, of greater percentage in aromatic C, of more intense color per C atom, and of higher polysaccharide content than their counterparts in streams and groundwaters. CPMAS ^{13}C-NMR spectroscopy is the most useful and definitive characterization tool in demonstrating chemical differences among humic substances from different environments. Most so-called soil fulvic acids are really fulvic acid fractions, because traditional isolation methods do not remove specific compounds solubilized by the alkaline extraction. These unhumified components, such as carbohydrates and low-molecular-weight acids that can account for a large part of the soil fulvic acid fraction, can be removed by treatment with XAD-8 resins. For the samples included in this study, humic substances from all environments have in common (i) an extreme complexity of molecular structure; (ii) an abundance of acidic, ester, and phenolic functional groups; (iii) a predominance of aliphatic character; (iv) an ability to fluoresce; (v) a refractory nature to microbial decay; and (vi) an ability to form complexes with metal ions. Most soil humic substances are comprised primarily of humic acids, whereas fulvic acids constitute >90% of stream humic substances. Humic acids differ significantly from fulvic acids in all these environments, and thus, the humic acid/fulvic acid separation appears to remain a valid characterization technique. Fulvic acids from all types of river waters appear to be remarkably similar, regardless of season, climatic conditions, or vegetation. Fulvic acids in soils vary slightly with soil type, method of extraction, and vegetative cover, but all have an abundance of variously degraded carbohydrate constituents. Groundwater fulvic acids have less intensive color, lower carbohydrate content, higher C content, lower N and O contents, and greater ^{14}C ages than do stream and soil fulvic acids.

Soil humic substances have traditionally been characterized by elemental composition, functional group analysis, protonation character, molecular size and weight, E_4/E_6 values, fluorescence, metal complexation potential, ^{14}C age, oxidative degradation products, and other characteristics (Hayes & Swift, 1978). These same characterizations have been applied to the relatively recent studies of stream-water and groundwater humic substances and have shown variations and similarities between and among humic substances from different sources.

Recent advances in ^{13}C-nuclear magnetic resonance (^{13}C-NMR) spectroscopy have not only shown this technique to be a useful characterization method for humic substances, but also to be the most definitive characterization tool compared to previous characterization methods such as infrared spectroscopy, degradative methods, and others. For relatively ash-free samples of humic substances, the ^{13}C-NMR spectra are semiquantitative to quantitative. The solid-state and liquid-state ^{13}C-NMR spectra of the same relatively ash-free samples often yield nearly the same results in quantitative C chemical shift composition, but a difference of ± 10% in aromaticity between the solid-state and liquid-state spectra of the same sample is also common. The ^{13}C-NMR liquid spectrum will intrinsically have narrower line widths from individual resonances than the corresponding solid-state ^{13}C-NMR spectrum.

The purposes of this chapter are to present solid-state ^{13}C-NMR spectra to show (i) definitive differences between humic and fulvic acids in each environment (soil, surface water, and groundwater); (ii) that both humic and fulvic acids are unique for each environment (soil, surface water, and groundwater); and (iii) differences between soil humic substances extracted and/or isolated by different methods. Traditional characterizations of humic substances will be used to augment the CPMAS ^{13}C-NMR spectroscopy to further establish similarities and differences among humic substances from various sources.

DEFINITIONS OF TERMS

The following terms used in this chapter are defined to enhance clarity and readability:

Fulvic acid fraction—That part of the alkaline soil extract which is also soluble at pH 1. This fraction typically contains a mixture of fulvic acids, simple sugars, oligosaccharides, polysaccharides, low-molecular-weight uncolored acids, HCl, and various other inorganic constituents.

Fulvic acid—The colored, organic, humified part of the fulvic acid fraction which is of nonspecific organic composition.

Cross-polarization (CP) or proton enhanced nuclear induction spectroscopy— The nuclear magnetic resonance technique used to enhance the ^{13}C-NMR spectra by the transfer of net magnetization from the abundant proton spins to the less abundant ^{13}C spins.

Magic angle spinning (MAS)—The solid-state ^{13}C-nuclear magnetic resonance technique used to decrease line broadening in the spectrum by eliminating dipolar ^{13}C-^{1}H interactions and anisotropy effects by rotating the solid sample rapidly at the so-called "magic angle" (54.7°) with respect to the applied magnetic field.

CPMAS—Cross polarization magic angle spinning.

XAD-8 resin—An uncharged but polar macroporous synthetic adsorbent which is a polymer of the methyl ester of methyl acrylic acid. It has limited cross-linkage with a hydrated pore size of approximately 25 μm.

Polyclar AT—A polyvinyl pyrrolidone resin polymer.

fa_1 = percentage aromaticity.

$$= \frac{\text{(Peak area of } ^{13}\text{C-NMR spectrum from 110 to 160 ppm)}}{\text{(Total peak area of } ^{13}\text{C-NMR spectrum from 0 to 230 ppm)}}.$$

fa_2 = percentage aromaticity.

$$= \frac{\text{(Peak area of } ^{13}\text{C-NMR spectrum from 110 to 160 ppm)}}{\text{(Total peak area of } ^{13}\text{C-NMR spectrum from 0 to 230 ppm)} - \text{(Peak area from 160 to 230 ppm)}}.$$

k' = column distribution coefficient.

$$= \frac{\text{mass of humic solute on the resin phase}}{\substack{\text{mass of humic solute in the solution or pore phase} \\ \text{of the column of resin}}}.$$

TMS—Tetramethylsilane.

2–1 MATERIALS AND METHODS

2–1.1 Soil Origins and Extraction Procedures

The six soils used in this study (Table 2–1) were collected and extracted by the author. The soil humic acids and fulvic acid fractions were extracted by a modified NaOH method (Malcolm, 1976), which included pressure filtra-

Table 2-1. Classification, origins, and characteristics of soil samples.

Soil series	Subgroup	Horizon	Geographic location	pH	Vegetation†
Weld	Anidic Paleustolls	Mollic Epipedon	Weld County, eastern CO	7.0	Short-grass prairie (mixed grasses)
Webster	Typic Haplaquolls	Mollic Epipedon	Ames, IA	6.8	Tall-grass prairie (mixed grasses)
Unnamed soil 1 (formerly called Fairbanks)	Typic Cryorthods	A1	Seward, AK	4.8	Mixed deciduous and evergreen forest (sitka spruce and mountain hemlock)
Sanhedrin	Ultic Haploxeralfs	AB	Mendiuna County, CA	6.2	Redwood, Pacific madrone, and ponderosa pine
Unnamed soil 2 (formerly called Seward)	Humic Cryorthods	Bh	Seward, AK	4.5	Mixed evergreen and deciduous forest (sitka spruce and mountain hemlock)
Lakewood	Spodic Quartzips-samments	Bh	Wilmington, NC	5.2	Longleaf pine

† Sitka spruce [*Picea sitchensis* (Bong.) Carr.]; mountain hemlock [*Tsuga mertensiana* (Bong.) Carr.]; redwood [*Sequoia sempervirens* (D. Don) Endl.]; Pacific madrone (*Arbutus menziesii* Pursh); ponderosa pine (*Pinus ponderosa* Laws.); and longleaf pine (*Pinus palustris* Mill.).

tion through a 0.45-μm membrane filter, desalting by dialysis, H saturation by resin exchange, and freeze-drying. These six soils were selected to represent several climatic regions and soil orders within the USA.

2-1.2 Origin and Isolation of Humic Substances from Water

The humic and fulvic acids from surface waters and groundwaters were isolated by the XAD-8 resin technique (Thurman & Malcolm, 1981) with a k' cutoff of 50. Water samples were filtered through a 0.45-μm Ag membrane filter, acidified to pH 2 with concentrated HCl, then adsorbed onto a column of XAD-8 resin. Humic and fulvic acids were desorbed from the XAD-8 resin with dilute NaOH, acidified to pH 1 with concentrated HCl, and then centrifuged to effect a humic acid–fulvic acid separation. Each humic acid and fulvic acid sample was desalted by adsorption onto XAD-8 resin, washed with deionized water, eluted with dilute NaOH, H saturated by resin exchange, and then freeze-dried.

A major portion of the free polysaccharides was removed from the fulvic acid fraction of the Webster and Lakewood soils by XAD-8 resin treatment. A 200-mg sample of each soil fulvic acid fraction was dissolved in 1.0 L of 0.01 M NaOH. The sample was acidified to pH 2.0 with HCl and passed through a 2.0-cm by 25-cm column of XAD-8 resin at a flow rate of 5 bed volumes per hour (the sample flow rate and k' for the isolation of stream humic substances). The column with adsorbed fulvic acid was washed with deionized water until the specific conductance was $<250 \, \mu S \, cm^{-1}$, the fulvic acid was eluted with 0.1 M NaOH, and was H saturated by treatment with ion exchange resin and then freeze-dried.

2-1.3   Chemical Analyses

Dissolved organic carbon (DOC) analyses were conducted on a Beckman 915 Carbon Analyzer or a Technicon DOC Module.[1]

Elemental analyses and ash contents were determined on duplicate samples by Huffman Laboratories, Golden, CO.

Freeze-dried, H-saturated humic and fulvic acids were titrated with dilute standardized NaOH, using a Radiometer automatic titration assembly. The complete titration to a pH of 10 was accomplished within 15 min to limit hydrolysis of esters, especially above pH 8 (Bowles et al., 1989). The titer up to pH 8.0 was attributed to carboxylic acids; twice the titer between pH 8 to pH 10 was attributed to phenolic groups (Bowles et al., 1989).

The XAD-8 resins used in this study were extensively cleaned to prevent contamination from resin bleed. The resin, as received from Rohm and Haas was washed and rinsed daily for 2 wk in 0.1 M NaOH. The <60-mesh (250 μm opening) fines were removed by decantation. The resin was Soxhlet extracted for 2 to 3 d in each of the successive solvents (methanol, acetonitrile, diethyl ether, and methanol). This Soxhlet extraction series was repeated three times before use of the resin.

The CPMAS ^{13}C-NMR spectra were obtained using equipment at the Colorado State University Regional NMR Center, Fort Collins, CO. The ^{13}C-NMR spectrometer operating conditions were: ^{1}H-frequency, 90.1 MHz; ^{13}C-frequency, 22.6 MHz; number of scans, 10 000; spinning rate near 40 kHz; contact time of 1 ms; external standard of hexamethylbenzene; acquisition time of 1024 ms; sweep width of 531.11 ppm; and line broadening of 39.999 Hz.

2-2   RESULTS AND DISCUSSION

2-2.1   Soil Humic Acids

The CPMAS ^{13}C-NMR spectra for six soil humic acids representing acidic and neutral soil subgroups in the USA are given in Fig. 2-1 and 2-2. All six spectra are similar, with five well-resolved and three poorly resolved peaks. The chemical shifts for the five well-resolved peaks are due to unsubstituted aliphatic carbon (0-50 ppm), carbon in C-O of methoxyl groups (50-60 ppm), carbon in all other C-O groups (60-95 ppm), aromatic carbon (110-160 ppm), and carbonyl in carboxyl and ester groups (160-190 ppm). The chemical shifts and assignments for the three poorly resolved peaks are anomeric carbon (95-110 ppm), aromatic carbon in phenolic groups and aro-

---

[1] The use of trade names in this report is for identification purposes only and does not constitute endorsement by the U.S. Geological Survey.

matic ethers (142–160 ppm), and carbonyl carbon in C=O ketonic groups (190–230 ppm).

The distribution of carbon in these soil humic acids as determined by integration of the CPMAS ^{13}C-NMR spectra is given in Table 2–2. These semiquantitative carbon distributions are based on peak area measurements from the spectra presented in Fig. 2–1 and 2–2, which have a relative precision of approximately ±2%. The relative error of the spectra (the spectrometer output) is believed to be ±5% for these relatively ash-free samples (ash contents are given in Table 2–3), because the spectra are within ±5% of the liquid-state ^{13}C-NMR spectra (Malcolm, 1989). These semiquantitative carbon distributions are useful for comparisons among sample spectra, and they give valuable insights concerning general structural relationships among humic substances. Visual interpretations of multipeak spectra can be misleading and deceptive. Even though the method of determining the integrated peak area (manual cut and weigh) may have minor limitations, it has commonly been used to determine aromaticity in humic substances.

Regardless of the intense aromatic carbon peaks in all humic acid spectra in this study, the humic acids are all considerably more aliphatic than aromatic in composition. The fa_1 aromaticity ranged from 25 to 34% with

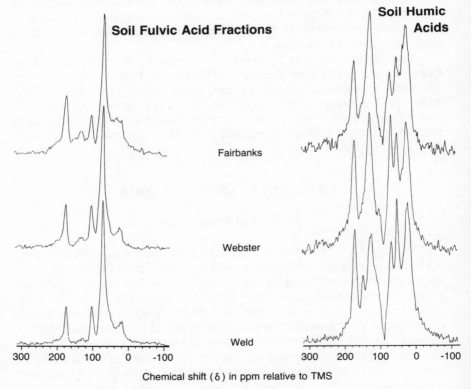

Fig. 2–1. CPMAS ^{13}C-NMR spectra of humic acids and fulvic acid fractions from Weld, Webster, and Fairbanks soils.

an average of about 29%, whereas the more common fa_2 aromaticity (calculated with the total carbonyl carbon region from 160 to 230 ppm excluded from the total spectral area) averaged about 37%. This means that soil humic acids contain three aliphatic carbons for every two aromatic carbons. The fa_2 aromaticity has been inflated by some authors by the inclusion of the 100- to 110-ppm anomeric carbon of carbohydrates into the aromatic carbon region. The inclusion of carbon chemical shifts between 95 to 110 ppm is not believed to be valid, because almost all aromatic resonances appear above 110 ppm and the chemical shift of polysaccharides is known to occur in the 100 to 110 ppm region of the spectra. The 100- to 110-ppm anomeric carbon area may have been included into the aromatic carbon area, because of poor spectral resolution in this region. As shown in the soil humic acid spectra in Fig. 2–1 and 2–2, the broad peak between 100 and 160 ppm usually starts from a valley in the spectrum at about 100 ppm. Even in well-resolved spectra, peak maxima in the anomeric carbon region from 100 to 110 ppm are often not well resolved, varying from a gradual curve with no shoulders to a gradual curve with poorly resolved shoulders. In less-resolved spectra, there was little or no indication of anomeric carbon peaks in the 100- to 110-ppm region; thus the region has been thought to be a continuation of the aromatic carbon chemical shift region of soil humic acids. Whether one

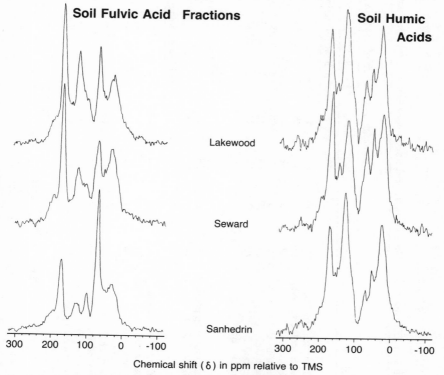

Fig. 2–2. CPMAS ^{13}C-NMR spectra of humic acids and fulvic acid fractions from Sanhedrin, Seward, and Lakewood soils.

Table 2-2. Distribution of C in soil fulvic acid fractions, soil fulvic acids, and soil humic acids determined by solid-state CPMAS ^{13}C-NMR.

Sample	Percentage distribution of C within indicated ppm regions											Degree of aromaticity (in %)	
	0–50	50–61	61–95	50–95	95–110	110–142	142–160	110–160	160–190	190–230	160–230	fa_1	fa_2
Soil fulvic acid fractions													
Weld mollic epipedon	18	6	45	51	10	5	2	6	10	5	15	6	7
Webster mollic epipedon	13	6	43	48	10	8	2	10	12	6	18	10	12
Fairbanks A1	22	6	31	37	8	8	5	13	14	5	19	13	16
Sanhedrin A	22	5	28	33	8	10	3	13	16	8	24	13	17
Seward Bh	23	5	19	24	5	14	5	19	21	8	29	19	27
Lakewood Bh	20	3	17	20	5	20	6	26	22	7	29	26	37
Armadale Bh†	–	–	–	–	–	–	–	28	–	–	23	28	36
Soil fulvic acids													
Elliott mollic epipedon	22	5	16	21	4	21	5	26	24	4	28	26	36
Santa Olalla A†	–	–	–	–	–	–	–	26	–	–	22	26	34
Soil humic acids													
Weld mollic epipedon	30	10	12	22	5	19	6	25	14	4	18	25	30
Webster mollic epipedon	27	8	15	23	4	21	5	27	13	6	19	27	33
Fairbanks A1	30	7	11	18	5	24	6	30	13	5	18	30	36
Sanhedrin A	27	4	8	12	5	27	8	34	16	6	22	34	44
Seward Bh	26	7	12	20	4	20	7	27	18	6	24	27	36
Lakewood Bh	25	5	11	16	4	25	8	33	15	7	23	33	43
Armadale Bh†	–	–	–	–	–	–	–	24	–	–	22	24	30
Elliott mollic epipedon	17	5	12	17	3	36	6	42	18	3	21	42	53
Santa Olalla A†	–	–	–	–	–	–	–	20	–	–	19	20	25

† Estimates from published ^{13}C-NMR spectra.

uses fa_1 or fa_2 aromaticity, soil humic acids are much higher in aromaticity than soil fulvic acids.

The large and intense unsubstituted aliphatic carbon peaks from 0 to 50 ppm, which is only part of the aliphatic carbon region, is equal in area to the aromatic carbon peak. The unsubstituted carbon region is largely methylene carbon with small amounts of methyl and methyne carbon.

Soil humic acids contain approximately 20% C in the C–O region (60–110 ppm) of the spectra, a large portion of which has been assigned to covalently bonded saccharide (carbohydrate) moieties within the humic acid structure. The saccharide moieties of the C–O region of the humic acid spectra cannot be removed by chromatographic separations such as XAD-8 resin chromatography, but they can be partially removed by hydrolysis (Preston & Schnitzer, 1984). All the soil humic acids in this study have a small, intense, well-resolved methoxyl peak (50–60 ppm) that appears to be characteristic for soil humic acids. This methoxyl peak probably originates from lignin oxidation products that are incorporated into soil humic substances.

As expected in humic acids, the carboxyl and ester group carbon peak from 160 to 190 ppm is intense, accounting for approximately 15% of the total carbon. From titration data, approximately two-thirds of the broad peak is attributed to carboxyl carbon and one-third is attributed to ester carbon.

The CPMAS ^{13}C-NMR spectra of soil humic acids in this study are similar to most of the previously published spectra (Hatcher et al., 1981a, b) with the exception that the spectra herein are highly resolved and all exhibit phenolic carbon resonances in the 145- to 160-ppm region. Phenolic carbon resonances are well resolved in the Weld, Seward, and Lakewood soils; spectra for the other three soils have pronounced shoulders. This aromatic substitution region of the CPMAS ^{13}C-NMR spectra (145–160 ppm) may also contain aromatic ethers. Two distinct shoulders in the spectra in this region in some samples may suggest the contribution of both phenolic and ether carbons. Until better resolution can be achieved between phenols and aromatic ether carbon, carboxyl and ester carbons, or until derivatives have been made to unambiguously distinguish these carbon groups, the contribution of phenolic and carboxyl groups to the total acidity of soil humic acids cannot be determined by ^{13}C-NMR spectroscopy.

The spectra of Saiz-Jimenez et al. (1986) provide an exception to the CPMAS ^{13}C-NMR spectra as shown in this chapter. An example of these spectra is shown in Fig. 2–3 for the Santa Olalla soil. The fa_1 aromaticities of their soil humic acids averaged <20% as compared to 30% or higher in this and other studies. The fa_1 aromaticities for their soil fulvic acids were consistently higher than for the soil humic acid counterparts. This reversal in aromaticity between fulvic and humic acids may be attributed to different extractants of soil humic acids or different methods of purification of the extracted humic acids. Saiz-Jimenez et al. (1986) used a mixture of $Na_4P_2O_7$ and NaOH as an extractant for soil humic substances, whereas only NaOH was used in this and other studies. Saiz-Jimenez et al. (1986) also treated their humic acid with a mixture of HCl and HF as a de-ashing treatment; such a treatment was not necessary in this study, because low-ash contents were achieved by filtration through a 0.45-μm membrane filter. These differences in the CPMAS ^{13}C-NMR spectra are even more confusing when the CPMAS ^{13}C-NMR spectrum for the Elliott mollic epipedon soil humic acid as shown in Fig. 2–3 is considered. This soil humic acid is the standard soil humic acid in the International Humic Substances Society's collection of humic and fulvic acids. The Elliott soil was extracted with NaOH and the humic acid treated with HCl and HF for de-ashing. The spectrum is similar to those of the humic acids presented in this study with the major exception that the spectrum lacks a strong unsubstituted aliphatic carbon peak between 0 and 50 ppm. In contrast to the CPMAS ^{13}C-NMR spectrum for the Santa Olalla humic acid, the Elliott humic acid spectrum has an exceptionally strong aromatic carbon peak and an exceptionally weak unsubstituted aliphatic carbon peak. The difference between the Elliott and Santa Olalla humic acid spectra may be due to the difference in extractants or true compositional differences within the soils. These unresolved differences should emphasize the need for soil samples to be treated in a conventional or standard method of extraction before or along with new extractive techniques so that the new techniques can be directly compared with conventional methods.

The elemental compositions of soil fulvic acid fractions, fulvic acids, and humic acids discussed in this chapter are given in Table 2–3. The elemental contents of the Armadale and Santa Olalla soil humic and fulvic acids are taken from the literature (Hatcher et al., 1981a, b; Saiz-Jimenez et al., 1986). The elemental data are typical for soil humic substances (Schnitzer & Khan, 1972; Stevenson, 1982).

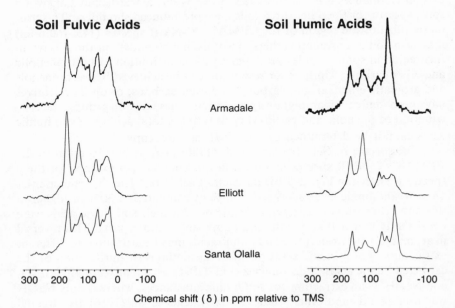

Fig. 2–3. CPMAS ^{13}C-NMR spectra of humic and fulvic acids from Armadale, Elliott, and Santa Olalla soils.

Table 2–3. Elemental composition of selected soil fulvic acid fractions, soil fulvic acids, and soil humic acids.†

Sample	C	H	O	N	S	P	Total	Atomic C/N	Atomic H/C	Percent ash
Soil fulvic acid fractions										
Weld mollic epipedon	44.16	6.05	44.25	3.63	0.62	0.97	99.68	12.16	1.65	2.28
Webster mollic epipedon	53.82	4.45	35.23	3.73	1.14	0.38	98.75	16.83	0.99	1.59
Fairbanks A1	48.59	4.57	41.79	2.67	0.51	0.48	98.61	21.22	1.12	1.69
Sanhedrin A1	48.71	4.36	43.35	2.77	0.81	0.59	100.59	20.51	1.07	2.25
Seward Bh	50.29	3.57	43.67	1.10	0.29	0.01	98.93	53.31	0.85	0.91
Lakewood Bh	51.74	3.53	41.20	1.10	0.65	0.39	98.61	54.85	0.81	1.92
Armadale Bh	50.90	3.60	43.90	1.00	0.60	--	101.40	60.57	0.85	2.00
Soil fulvic acids										
Elliott mollic epipedon	50.48	4.01	42.60	2.68	0.62	0.05	100.44	22.16	0.95	0.79
Santa Olalla A	49.75	4.35	43.18	2.72	--	--	--	21.84	1.05	14.67
Soil humic acids										
Weld mollic epipedon	55.50	5.07	32.76	4.97	0.60	0.16	99.06	13.02	1.10	0.77
Webster mollic epipedon	43.57	5.65	43.87	4.63	0.95	1.24	99.91	10.97	1.55	1.59
Fairbanks A1	57.13	3.74	34.45	3.94	0.39	0.13	99.75	16.91	0.78	0.38
Sanhedrin A1	58.03	3.64	33.59	3.26	0.47	0.10	99.05	20.76	0.75	1.19
Seward Bh	54.43	4.05	36.96	3.38	0.40	0.01	99.23	18.78	0.89	1.89
Lakewood Bh	57.60	3.53	35.07	2.07	0.54	0.10	99.60	32.45	0.73	0.22
Armadale Bh	56.90	5.20	34.90	2.30	0.70	--	--	--	--	3.50
Elliott mollic epipedon	57.99	3.78	33.69	4.18	0.41	0.32	100.37	16.10	0.78	0.90
Santa Olalla A	52.76	5.97	36.90	4.37	--	--	--	14.19	1.36	5.60

† Composition in percentage by weight on a moisture-free, ash-free basis.

2-2.2 Soil Fulvic Acid Fractions Compared to Soil Fulvic Acids

The CPMAS ^{13}C-NMR spectra of the six soil fulvic acid fractions extracted by the author in this study are given in Fig. 2-1 and 2-2. These extracts are designated as fulvic acid fractions because the samples contain varying amounts of saccharides and low-molecular-weight organic acids that were not removed by dialysis. Due to limitations in the isolation procedure, the respective soil fulvic acids are physically admixed with saccharides and specific low-molecular-weight acids.

Saccharides have been shown to be the major constituent admixed with soil fulvic acids by several independent techniques. The first evidence was obtained (Leenheer & Malcolm, 1973) with the Weld and Webster mollic epipedon fulvic acid fractions that were subjected to free-flow electrophoretic separation. A large portion of the fulvic acid fraction of both samples was uncolored and uncharged, and flowed straight down the free-flow electrophoretic curtain. By various tests, this material was shown to consist of simple sugars and easily hydrolyzed polysaccharides. Analysis of the water-soluble fulvic acid fractions also gave intense color reactions with phenoldisulfonic acid, a positive color test for simple sugars. Similar results have been obtained for soil extracts by Hayes et al. (1984). The CPMAS ^{13}C-NMR spectra of soil fulvic acid fractions show intense peaks in the C–O chemical shift region between 60 and 110 ppm, which are characteristic of simple sugars and polysaccharides (Atalla et al., 1980; Earl & VanderHart, 1980). Polysaccharides show an intense peak at about 75 ppm, which is the most intense peak in the ^{13}C-NMR spectra for four of the six soil fulvic acid fractions (Fig. 2-1 and 2-2). The broad peak between 60 and 95 ppm is associated with resonance of carbohydrate carbons 2, 3, and 5; peak centers near 65, 86, and 90 ppm are associated with carbohydrate carbons 6 and 4. The peak near 105 ppm in carbohydrates is associated with the number 1 carbon, the anomeric carbon. The anomeric carbon of carbohydrates near 105 ppm is one of the most intense peaks in four of the six soil fulvic acid fractions and is a sharp shoulder in the other two soil fulvic acid fractions. These six soil fulvic acid fractions were prepared prior to the advent and use of XAD-8 resin technology for desalting and the removal of polysaccharides and low-molecular-weight acids from alkaline soil fulvic acid fractions. Two soil fulvic acid fractions, Webster mollic epipedon and Lakewood Bh, were dissolved in water, then treated by the XAD-8 resin method for removal of saccharides and low-molecular-weight acids. Less than 5% of the C was removed from the Lakewood Bh sample, indicating a small amount of admixed, uncolored simple acids or saccharides in this sample. At the other extreme, slightly more than 50% of the C of the Webster mollic epipedon fulvic acid fraction passed through the XAD-8 resin column. The major portion of the material not retained by the XAD-8 resin column consisted of saccharides.

The CPMAS ^{13}C-NMR spectrum of the Webster soil mollic epipedon fulvic acid is shown in Fig. 2-4. A large portion, but not all, of the saccharide material was removed from the fulvic acid fraction. In a freshly extracted soil fulvic acid fraction that had not been freeze-dried and aged, all of the free saccharides would be expected to be removed from the fulvic acid by the XAD-8 resin method. The large C–O peaks from 50 to 105 ppm in the original fulvic acid fraction were greatly reduced. Peaks for the unsubstituted aliphatic carbon, aromatic carbon, and carboxyl carbon that are typical of soil fulvic acid became prominent in the fulvic acid spectrum.

The soil fulvic acid fractions in this study have six well-resolved peaks and two poorly-resolved peaks in their CPMAS ^{13}C-NMR spectra. Two of the well-resolved peaks at about 75 and 104 ppm are assigned to the C–O chemical shift of polysaccharides as discussed previously. The carboxyl and ester peak is strong and well-resolved in all samples. The unsubstituted aliphatic carbon and aromatic carbon regions are weak but well-resolved in four of the six samples; whereas these two peaks are strong and well-resolved in the two spodic horizons. Peaks in the phenolic carbon region (145–160 ppm), and the C=O ketonic region (190–230 ppm) are poorly-resolved in all spectra for the fulvic acid fractions.

In contrast to soil humic acids that have similar composition and CPMAS ^{13}C-NMR spectra, the soil fulvic acid fractions vary in composition and ^{13}C-NMR spectra. This variation is due primarily to the amount of free saccharides in each fraction. As listed in Table 2–2, the six fulvic acid fractions

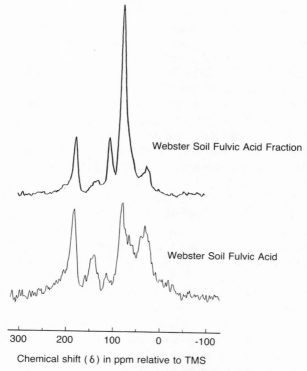

Webster Soil Fulvic Acid Fraction

Webster Soil Fulvic Acid

300   200   100   0   -100

Chemical shift (δ) in ppm relative to TMS

Fig. 2–4. CPMAS ^{13}C-NMR spectra of Webster soil fulvic acid and Webster soil fulvic acid fraction.

show a trend of increasing percentages for the carboxyl, aromatic, and un-substituted aliphatic carbon regions, whereas the percentage for the C–O region that includes the saccharides (the 75- and 104-ppm peaks) decreases. The carboxyl region (160–190 ppm) increases from 10% in the Weld sample to 22% in the Lakewood sample, the aromatic carbon region (110–160 ppm) increases from 6% in the Weld sample to 26% in the Lakewood sample, and the intensity and resolution of the unsubstituted aliphatic region (0–50 ppm) gradually increases from the Weld sample to the Lakewood sample. As these three peaks increase, the contribution of saccharide carbon in the sum of the two parts of the C–O region (60–95 ppm and 95–110 ppm) decreases from 61% in the Weld sample to 25% in the Lakewood sample. These ^{13}C-NMR data agree well with the electrophoretic separations and XAD-8 resin sepa-rations of saccharides from fulvic acid in these fractions. All the data indi-cate that (i) many of the soil fulvic acids as presented in the literature should correctly be designated as soil fulvic acid fractions; (ii) soil saccharides are readily extracted from neutral and acidic soils by NaOH; (iii) free saccharides are not easily removed from fulvic acid fractions by dialysis; and (iv) the inclusion and admixture of saccharides in fulvic acid fractions are common.

Representative CPMAS ^{13}C-NMR spectra of soil fulvic acids remain in question for several reasons. These uncertainties include: (i) the prevalence for the characterization of fulvic acid fractions rather than fulvic acid; (ii) the tendency of most soil researchers to emphasize the humic acid portion of soil and to discard the fulvic acid fraction without investigation; (iii) the diversity of extractants of soil organic matter; and (iv) the possible diversity of fulvic acids among soils. The first two problem areas can be easily and inexpensively resolved by the use of XAD-8 resin methods to isolate fulvic acid. Polyclar AT may also be as effective as XAD-8 resin, but it remains to be tested extensively. In the near future, as these four problems are dealt with systematically, soil fulvic acid will be more rigorously defined.

Presently, the best representative of soil fulvic acid is believed to be the Elliott mollic epipedon fulvic acid. The CPMAS ^{13}C-NMR spectrum of this sample is shown in Fig. 2–3. This sample was prepared by alkaline soil ex-traction and XAD-8 resin treatment for desalting, removal of free saccha-ride, and removal of low-molecular-weight specific acids. The CPMAS ^{13}C-NMR spectra of the Lakewood (Fig. 2–2) soil fulvic acid fraction that has 5% or less of saccharides, and the Armadale (Fig. 2–3) and Seward (Fig. 2–2) soil fulvic acid fractions that also are believed to be low in saccharide content, are very similar to the spectra of the Elliott fulvic acid. After sac-charide removal, the spectrum for the Webster fulvic acid (Fig. 2–4) resem-bles that of the Elliott fulvic acid. The CPMAS ^{13}C-NMR spectrum of the Santa Olalla fulvic acid (Fig. 2–3) with an fa_1 aromaticity of 26%, which is representative of the three soils extracted with a mixed pyrophosphate-alkali extractant, then treated with Polyclar AT, is also very similar to the spectrum of the Elliott fulvic acid.

2–2.3 Stream Fulvic Acids

The CPMAS ^{13}C-NMR spectra representative of stream fulvic acids are shown in Fig. 2–5, and the distribution of carbon from the CPMAS ^{13}C-NMR spectra is given in Table 2–4. The spectra of stream fulvic acids have four major peaks and three minor peaks. The intense unsubstituted aliphatic carbon peak from 0 to 50 ppm dominates the spectra. This broad peak extends into the 50- to 60-ppm chemical shift region and accounts for nearly 40% of the C in stream fulvic acids. The three other well-resolved peaks are the C–O region (50–95 ppm), the aromatic carbon region (110–160 ppm), and the carboxyl and ester region (160–190 ppm). The C–O region of the spectra (50–95 ppm) and the anomeric carbon region (95–110 ppm), which account for approximately 25% of the total C, are believed to be only partially due to covalently bonded saccharides in stream fulvic acids because: (i) the anomeric carbon percentage, which should represent one-fifth of the polysaccharide carbon, is only 2 to 3%; (ii) the presence of free saccharides is unlikely because of isolation by XAD-8 methods; and (iii) only a portion of the C–O region from 60 to 95 ppm can be removed by hydrolysis. The polysaccharide component of the C–O region (60–110 ppm) is probably only

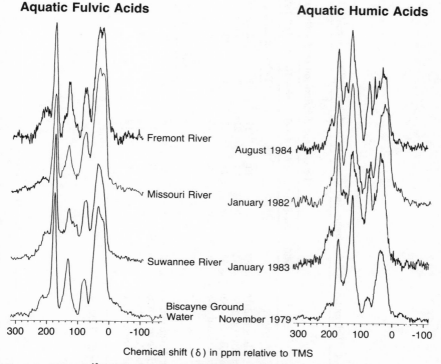

Fig. 2–5. CPMAS ^{13}C-NMR spectra of stream and groundwater fulvic and humic acids from the Fremont River, Missouri River, Suwannee River, and the Biscayne aquifer.

Table 2–4. Distribution of C in stream and groundwater fulvic and humic acids determined by solid-state CPMAS ^{13}C-NMR spectroscopy.

Sample	Percentage distribution of C within indicated ppm regions											Degree of aromaticity (in %)	
	0–50	50–61	61–95	50–95	95–110	110–142	142–160	110–160	160–190	190–230	160–230	fa_1	fa_2
	Fulvic acids												
Fremont River (8/84)†	37	6	11	17	2	12	4	16	17	11	28	16	22
Missouri River (1/82)	41	7	14	21	2	11	3	14	17	5	22	14	18
Suwannee River (1/83)	31	5	14	19	4	13	6	18	18	9	28	18	25
Biscayne Groundwater (11/79)	39	6	9	16	1	13	4	17	19	9	28	17	24
	Humic Acids												
Fremont River (8/84)	23	6	12	18	4	23	9	32	16	7	23	32	42
Missouri River (1/82)	30	5	9	14	3	24	5	29	17	7	24	29	38
Suwannee River (1/83)	23	5	14	19	7	20	10	30	14	8	22	30	38
Biscayne Groundwater (11/79)	26	5	7	12	2	26	10	36	16	7	23	36	47

† Date of collection.

10 to 15%, which is only one-half of the C percentage in this chemical shift region. The fa_1 aromaticity of stream fulvic acids averages approximately 17%. The ester and carboxyl peak from 160 to 190 ppm is believed to be primarily carboxyl in nature, because the 18% C in this peak is equivalent to a carboxyl titer of about 6 millimoles g^{-1}, which is approximately the COOH titer of stream fulvic acids.

The three poorly-resolved peaks result from the anomeric carbon (95–110 ppm), the phenolic carbon (145–160 ppm), and the ketonic carbon (190–230 ppm). As is typical for stream fulvic acids, the anomeric carbon peak is only a weak shoulder in the spectra for two of the three samples and is absent from the third. Stream fulvic acids seldom show a well-resolved phenolic carbon peak, but poorly-resolved shoulders and general smearing in this region (145–160 ppm) indicate that <5% of the C distribution is phenolic in character. The broad and diffuse ketonic region (190–230 ppm) suggests the presence of a variety of aromatic and aliphatic ketones in stream fulvic acids (Leenheer et al., 1987).

The ratio of fulvic to humic acids in stream water is the opposite of that found in soils. Humic acids are three to five times more abundant in soils than are fulvic acids (Stevenson, 1982), whereas fulvic acids in water are 9 to 10 times more abundant than are humic acids (Malcolm, 1985). Humic substances commonly account for 50% of the dissolved organic carbon (DOC) concentrations in stream water, of which 90 to 95% are fulvic acids. These major differences in the proportion of humic and fulvic acids in each environment must be considered in understanding geochemical processes.

The elemental analyses of the stream humic and fulvic acids discussed in this chapter are given in Table 2–5. The elemental analyses given are typical of stream humic substances within the USA.

2–2.4 Stream Humic Acids

Representative CPMAS ^{13}C-NMR spectra for stream humic acids are shown in Fig. 2–5, and the carbon distribution in the CPMAS ^{13}C-NMR spectra is given in Table 2–4. Stream humic acids have five well-resolved and three poorly-resolved peaks in the spectra. The five well-resolved peaks are due to unsubstituted aliphatic carbon (0–50 ppm), methoxyl C–O (50–60 ppm), C–O (60–95 ppm), aromatic carbon (110–160 ppm), and carboxyl and ester (160–190 ppm). The three poorly-resolved peaks are due to anomeric carbon (95–110 ppm), phenolic carbon (145–160 ppm), and ketonic carboxyl (190–230 ppm).

2–2.5 Groundwater Fulvic and Humic Acids

The CPMAS ^{13}C-NMR spectra of humic and fulvic acids in water from the Biscayne aquifer in Miami, FL, are shown in Fig. 2–5. The groundwater humic and fulvic acids have the same peaks in the spectra as stream humic and fulvic acids with the exception that the methoxyl group carbon (50–60 ppm) is not resolved in the groundwater humic acid. The elemental contents

Table 2-5. Elemental composition of selected stream and groundwater humic and fulvic acids.†

Sample	C	H	O	N	S	P	Total	Atomic C/N	Atomic H/C	Percent ash
Fulvic acids										
Fremont River (8/84)‡	56.60	4.74	37.15	1.17	0.92	0.01	100.59	59.00	1.00	2.21
Missouri River (1/82)	55.36	5.14	36.09	1.35	0.77	0.33	99.04	47.82	1.11	0.32
Suwannee River (1/83)	53.60	4.25	40.90	0.70	0.55	0.01	100.0	89.40	0.95	0.82
Biscayne Groundwater (11/79)	55.44	4.17	35.39	1.77	1.06	0.20	98.03	35.54	0.90	0.43
Humic acids										
Fremont River (8/84)	56.59	4.43	35.10	2.59	0.73	0.16	99.60	24.84	0.94	2.83
Missouri River (1/82)	56.97	4.48	34.13	2.36	1.39	0.01	99.34	28.15	0.94	0.89
Suwannee River (1/83)	54.30	4.10	39.20	1.15	0.75	0.01	99.50	56.63	0.91	3.30
Biscayne Groundwater (11/79)	58.28	5.84	30.14	3.39	1.43	0.22	99.30	11.57	0.70	0.10

† Composition in percentage by weight on a moisture-free, ash-free basis.
‡ Date of collection.

of the Biscayne groundwater humic and fulvic acids are given in Table 2–5. The elemental contents of the groundwater humic and fulvic acids are slightly atypical of deep groundwaters because of contamination with soil and surface waters as the aquifer nears the land surface in southern Florida. This mixing of surface waters and groundwaters results in a ^{14}C-age of 800 ± 250 yr for the Biscayne groundwater fulvic acids. This ^{14}C-age is much less than is typical for deep groundwaters. This youthful carbon date is supported by data which found 90.6 ± 2.8% of the C to be modern C. As a result of this contamination, the resulting humic and fulvic acids are lower in C content and higher in O and N contents than is typical for groundwater fulvic and humic acids (Thurman, 1979).

2–3 SUMMARY AND CONCLUSIONS

2–3.1 Comparison of Soil Fulvic Acids with Stream and Groundwater Fulvic Acids

Most of the published data on soil fulvic acids are really based on soil fulvic acid fractions, because the soil fulvic acid fractions contain many constituents other than humified fulvic acid. Because of variation in this mixture within soils, and the different extractants and different methods of isolation, fulvic acids have been attributed as having a vast array of chemical and physical properties. In addition to their diversity, a common property exhibited by fulvic acid fractions is that those fractions from spodic horizons contain small amounts of free saccharides, whereas fractions from other soils contain an abundance of free saccharides. These saccharides are a mixture of simple sugars, oligosaccharides, and polysaccharides. When these soil fulvic acid fractions are compared to fulvic acids that were isolated from streams and groundwaters, there are marked differences due primarily to the major contribution of soil saccharides in the soil fulvic acid fraction. The various methods of isolating stream and groundwater fulvic and humic acids, including the commonly used XAD-8 resin technique, exclude the free saccharides from the isolated fulvic acid. Therefore, soil fulvic acid in the fulvic acid fraction is essentially diluted by unhumified components with properties that are largely uncharacteristic of soil fulvic acid. Soil fulvic acid fractions are polydisperse and have a wide range of molecular weight components (Stevenson, 1982), whereas stream and groundwater fulvic acids approach monodispersity and have a lower molecular weight (800 daltons) (Aiken & Malcolm, 1987). Soil fulvic acid fractions are lower in C and higher in O and N than are stream and groundwater fulvic acids (Table 2–5). The peaks for saccharides dominate the CPMAS ^{13}C-NMR spectra of soil fulvic acids (Fig. 2–1 and 2–2), whereas unsubstituted aliphatic carbons, aromatic carbons, and carboxyl carbons dominate the spectra of stream and groundwater fulvic acids (Fig. 2–3 through 2–5).

Soil fulvic acid can be isolated from the soil fulvic acid fraction by treatment with XAD-8 resin (Malcolm et al., 1979) or Polyclar AT (Saiz-Jimenez

et al., 1986). The resulting fulvic acid should be less polydisperse, higher in
C and N content, and lower in O content due to the loss of the free saccharides. As shown in Fig. 2–4, the CPMAS ^{13}C-NMR spectrum for the XAD-8
purified Webster soil fulvic acid more closely resembles the spectra of stream
fulvic acids than the spectrum of the original Webster soil fulvic acid fraction. The CPMAS ^{13}C-NMR spectra for other soil fulvic acids as prepared
by XAD-8 or Polyclar AT techniques (Fig. 2–3; Saiz-Jimenez et al., 1986)
are similar to the spectra for the purified Webster soil fulvic acid and the
Lakewood Bh and Seward Bh soil fulvic acid fractions that are low (<5%)
in free saccharide substances.

Because few soil fulvic acids have been characterized to date, the
representative characterization data including a CPMAS ^{13}C-NMR spectrum
cannot be presented. From the CPMAS ^{13}C-NMR spectra presented in this
chapter, soil fulvic acids appear to be different from stream or groundwater
fulvic acids in that soil fulvic acids have considerably more unsubstituted
aliphatic carbon, carboxyl carbon, and ketonic carbons in their molecules.
Also, based on ^{14}C-age, soil fulvic acids are several hundred to several thousand years old as compared to the modern ^{14}C-age for stream fulvic acids
(Thurman & Malcolm, 1983; unpublished data of the author).

Stream fulvic acids appear to be different from deep groundwater fulvic acids in CPMAS ^{13}C-NMR spectra, elemental composition, ^{14}C-age,
color intensity per C atom, and saccharide content. Groundwater humic substances have not been studied extensively, but in the few references available
(Thurman, 1979; Malcolm & Thurman, 1981; Thurman, 1985), groundwater
fulvic acids have (i) a bleached color with a 10-fold decrease in color intensity per C atom; (ii) a higher C content and lower O and N contents; and (iii)
by their very nature of being buried for thousands of years, a very old ^{14}C-
age (unpublished data of the author). As shown by the CPMAS ^{13}C-NMR
spectrum (Fig. 2–4) and the carbon distribution percentages determined by
CPMAS ^{13}C-NMR spectroscopy (Table 2–4), the Biscayne groundwater fulvic acid is much higher in unsubstituted aliphatic carbon and aromatic peak
intensity than are the stream fulvic acids.

2–3.2   Comparison of Soil Humic Acids with Stream
and Groundwater Humic Acids

Soil humic acids are different from stream humic acids in molecular size
and weight, N content, ^{14}C-age, and CPMAS ^{13}C-NMR spectra. Stream
humic acids have a molecular size of 0.9 to 1.2 μm and a molecular weight
range between 1500 and 3000 daltons. Soil humic acids have a higher-
molecular-weight range from 1500 to in excess of 50 000 daltons. Soil humic
acids have a strong tendency to aggregate; therefore, they exhibit a broad
range of molecular sizes (Stevenson, 1982). As listed in Tables 2–3 and 2–5,
the N contents of soil humic acids are usually near 4%, which is twice that
of stream humic acids. As stated previously, soil humic acids are usually several hundred to several thousand years old, whereas stream humic acids are
of modern age.

The CPMAS ^{13}C-NMR spectra of soil and stream humic acids are similar with the exception of three peaks—phenolic carbon, methoxyl carbon, and aromatic carbon (Fig. 2–1, 2–2, and 2–3). The methoxyl peaks are much more intense in the spectra of soil humic acids than in those of stream humic acids. There is also a tendency for the aromatic carbon shoulder peaks between 110 and 120 ppm to be more intense in soil humic acids. The intensity of these two peaks may suggest a greater incorporation of lignin degradation products into soil humic acids than into stream humic acids.

In contrast, almost all spectra for stream humic acids show a moderate to strong phenolic peak between 145 and 160 ppm. The phenolic peak of soil humic acids is extremely variable. Soil humic acids studied by the author usually exhibit a weak phenolic peak; sometimes it may be moderate and well resolved as in the Weld, Seward, and Lakewood soils. Other authors have noted either the absence of or merely a slight phenolic carbon peak in the spectra of soil humic acids (Hatcher et al., 1981a).

Groundwater humic substances from deep groundwaters are very different from stream or soil humic substances in color intensity per C atom, ^{14}C-age, elemental composition, and CPMAS ^{13}C-NMR spectra. The ^{14}C-age of groundwater humic acids is usually several thousand years; for example, the Biscayne groundwater humic acid (Fig. 2–5 and Table 2–5) is 1790 ± 140 yr BP. The ages of other groundwater humic substances have been reported by Thurman (1979). Groundwater humic substances are also very bleached in color, have higher C and H contents, and lower O and N contents than their stream and soil counterparts.

The CPMAS ^{13}C-NMR spectra of groundwater humic acids exhibit extreme differences from stream and soil humic substances. The groundwater humic substances (Fig. 2–3) are higher in unsubstituted aliphatic carbon and aromatic carbon, and are lower in C–O carbon and phenolic carbon. With increased time of burial, the labile phenolic groups and the saccharide portions of the molecule of a groundwater humic acid are lost. In the reducing groundwater condition, the O content is lowered, the H content is increased, and the aromatic carbon composition is sharply increased.

The differences between surface-water humic acids and groundwater humic acids are marked for deep groundwater, but due to mixing of more shallow groundwaters with waters percolating from the surface to the subsurface at different rates and magnitudes, humic substances in groundwaters at shallow and moderate depths are expected to exhibit characteristics intermediate between those in streams and deep groundwaters.

2–3.3 Comparison of Fulvic Acids with Humic Acids in Soil, Stream, and Groundwater Environments

With few exceptions, it can be stated that fulvic acids in each respective environment are lower in molecular size and weight, lower in C and N contents, higher in O content, lower in aromaticity, lower in color intensity per C atom, and have different CPMAS ^{13}C-NMR spectra than humic acids in the same environment. Many of these properties are closely associated and are not independent. Color intensity in humic acids has long been associated with the aromatic chromophores. The higher O content of fulvic acids enables them to have a higher carboxyl and ketonic content than humic acids.

The greater diversity and complexity of humic acids compared to fulvic acids is well demonstrated by the greater number of well-resolved carbon regions in the CPMAS ^{13}C-NMR spectra. In some spectra, humic acids have seven well-resolved peaks as compared to five for some fulvic acids. By visual examination, the major differences between the CPMAS ^{13}C-NMR spectra of humic and fulvic acids in each source (soil, stream, or groundwater) can be recognized.

ACKNOWLEDGMENTS

The solid-state ^{13}C-NMR spectra presented in this chapter were obtained using equipment at the Colorado State University Regional NMR Center, funded by National Science Foundation Grant no. CHE-8208821. The special assistance of Dr. Gary Maciel at Colorado State University is appreciated.

REFERENCES

Aiken, G.R., and R.L. Malcolm. 1987. Molecular weight of aquatic fulvic acids by vapor pressure osmometry. Geochim. Cosmochim. Acta 51:2177–2184.

Atalla, R.H., J.C. Gast, D.W. Sindorf, V.J. Bartuska, and G.E. Maciel. 1980. ^{13}C-NMR spectra of cellulose polymorphs. J. Am. Chem. Soc. 102:3249–3251.

Bowles, E.C., R.C. Antweiler, and P. MacCarthy. 1989. Acid-base titration and hydrolysis of Suwannee River fulvic acid. p. 205–230. In R.C. Averett et al. (ed.) Humic substances in the Suwannee River, Georgia; Interactions, properties, and proposed structures. Geol. Surv. Open-File Rep. (U.S.) 87–557.

Earl, W.L., and D.L. VanderHart. 1980. High resolution, magic angle sample spinning ^{13}C-NMR of solid cellulose I. J. Am. Chem. Soc. 102:3251–3252.

Hatcher, P.G., G.E. Maciel, and L.W. Dennis. 1981a. Aliphatic structure of humic acids: A clue to their origin. Org. Geochem. 3:43–48.

Hatcher, P.G., M. Schnitzer, L.W. Dennis, and G.E. Maciel. 1981b. Aromaticity of humic substances in soils. Soil Sci. Soc. Am. J. 45:1089–1094.

Hayes, M.H.B., J.E. Dawson, J.L. Mortensen, and C.E. Clapp. 1984. Electrophoretic characteristics of extracts from sapric histosol soils. p. 31–41. In M.H.B. Hayes and R.S. Swift (ed.) Volunteered papers of the 2nd Conf. Int. Humic Subtances Soc., Univ. of Birmingham Press, Birmingham, England.

Hayes, M.H.B., and R.S. Swift. 1978. The chemistry of soil organic colloids. p. 179–230. In D.J. Greenland and M.H.B. Hayes (ed.) The chemistry of soil constituents. John Wiley & Sons, Chichester, England.

Leenheer, J.A., and R.L. Malcolm. 1973. Fractionation and characterization of natural organic matter from certain rivers and soils by free-flow electrophoresis. Geol. Surv. Water Supply Pap. (U.S.) 1817-E.

Leenheer, J.A., M.A. Wilson, and R.L. Malcolm. 1987. Presence and potential significance of aromatic-ketone groups in aquatic humic substances. Org. Geochem. 11:273–280.

Malcolm, R.L. 1976. Method and importance of obtaining humic and fulvic acids of high purity. J. Res. U.S. Geol. Surv. 4:37–40.

Malcolm, R.L. 1985. The geochemistry of stream fulvic and humic substances. p. 181–209. In G.R. Aiken et al. (ed.) Humic substances in soil, sediment, and water: Geochemistry, isolation, and characterization. Wiley-Interscience, New York.

Malcolm, R.L. 1989. Applications of solid-state ^{13}C-NMR spectroscopy to geochemical studies of humic substances. p. 339–371. *In* M.H.B. Hayes et al. (ed.) Humic substances II: In search of structure. John Wiley & Sons, Chichester, England.

Malcolm, R.L., G.R. Aiken, and E.M. Thurman. 1979. Chromatographic techniques for isolation and purification of aqueous humic substances. p. 161. *In* Agronomy Abstr., ASA, Madison, WI.

Malcolm, R.L., and E.M. Thurman. 1981. Humic substances in ground water—Amount, impact on chlorination, and effect on pollution transport potential. Trans. Am. Geophys. Union 62:284.

Preston, C.M., and M. Schnitzer. 1984. Effects of chemical modifications and extractions on the carbon-13 NMR spectra of humic materials. Soil Sci. Soc. Am. J. 48:305–311.

Saiz-Jimenez, C., B.L. Hawkins, and G.E. Maciel. 1986. Cross-polarization, magic-angle spinning ^{13}C nuclear magnetic resonance spectroscopy of soil humic fractions. Org. Geochem. 9:277–284.

Schnitzer, M., and S.U. Khan. 1972. Humic substances in the environment. Marcel Dekker, New York.

Stevenson, F.J. 1982. Humus chemistry. Wiley-Interscience, New York.

Thurman, E.M. 1979. Isolation, characterization, and geochemical significance of humic substances from groundwater. Ph.D. Thesis. Univ. of Colorado, Boulder (Diss. Abstr. 79-23295).

Thurman, E.M. 1985. Humic substances in groundwater. p. 87–103. *In* G.R. Aiken et al. (ed.) Humic substances in soil, sediment, and water: Geochemistry, isolation, and characterization. Wiley-Interscience, New York.

Thurman, E.M., and R.L. Malcolm. 1981. Preparative isolation of aquatic humic substances. Environ. Sci. Technol. 15:463–466.

Thurman, E.M., and R.L. Malcolm. 1983. Structural study of humic substances: New approaches and methods. p. 1–23. *In* R.F. Christman and E.T. Gjessing (ed.) Aquatic and terrestrial humic materials. Ann Arbor Science, Ann Arbor, MI.

Chapter 3

Synthesis and Degradation of Natural and Synthetic Humic Material in Soils

D. E. STOTT, *Agricultural Research Service, USDA, National Soil Erosion Research Laboratory, Purdue University, West Lafayette, Indiana*

J. P. MARTIN, *University of California, Riverside, CA*

ABSTRACT

Soil humus is composed predominantly of two types of substances, humic substances and polysaccharides. Humic acids are the major extractable component of soil humic substances; fulvic acids are usually a small component of soil; and soil humin is a major nonextractable component. Humic acids are complex macro-molecules consisting of an array of aromatic and aliphatic structures. Soil polysac-charides, nonhumic components, are discussed in this chapter because of their important role in soil structure and aggregation. They consist of a variety of sugar units of both plant and microbial origin. Decomposition of fresh organic residues is discussed in terms of the fate of the constituent carbon. Before residues can be considered a part of the true soil humus, they undergo profound transformation and no longer resemble the original material. An overview of some current hypotheses on the formation of humic acids and soil polysaccharides is presented along with possible mechanisms for their increased resistance to biodegradation. Use of model humic acid-type molecules have aided in elucidating some of the possible ways by which humic acids are synthesized and protected in the soil environment. Informa-tion on the characterization of humic acids and soil polysaccharides that has accumu-lated through the use of various degradative and nondegradative procedures is examined.

Soil humus is a mixture of numerous organic constituents but two types of components, humic acids and polysaccharides, may constitute up to 80% of total extractable humus (Hayes & Swift, 1978; Stevenson, 1982b). Humic acids, dark brown to black in color, make up the larger fraction. They are thought to be complex aromatic macromolecules with amino acids, amino sugars, peptides, aliphatic acids and other aliphatic compounds involved in linkages between the aromatic groups (Bondietti et al., 1972; Flaig et al., 1975; Schnitzer, 1977; Hayes & Swift, 1978; Hatcher et al., 1981; Steven-son, 1985). Phenolic polymers such as lignin (synthesized by plants) and phenolic and other aromatic compounds synthesized by soil microorganisms are important sources of aromatic units for new humus (Haider et al., 1975; Martin & Haider, 1980b). Studies using ^{13}C-NMR techniques have high-lighted the important contribution of complex aliphatic structures to humic substances (Hatcher et al., 1981, 1983). Susceptible aromatic units might be cleaved while the rings are still part of the humic acid molecule, thus reduc-ing the phenolic groups and forming aliphatic acid structures. This process might account for some of the aliphatic units found in humic acids (Craw-ford, 1981; Ellwardt et al., 1981). Soil polysaccharides constitute 10 to 30% of the soil humus (Swincer et al., 1968). Some of the polysaccharides are strongly associated with clay colloids or humic acid macromolecules (Steven-son, 1982b, 1983). Microbially synthesized polysaccharides have, in the past, been considered a major precursor of this humus fraction (Martin & Haider, 1971).

This chapter (i) gives an overview of the incorporation of various fresh organic residues into the natural soil humus; (ii) presents some current hypotheses on the formation of soil humic materials; (iii) describes the char-acterization of humic molecules; and (iv) discusses the biological transfor-mations of newly formed soil humus.

Table 3-1. Decomposition of various organic substrates in a coarse-loamy mixed, thermic Mollic Haploxeralf incubated in the laboratory at $22 \pm 2\,°C$ under continuous aeration and $-33\mathrm{kPa}$ water potential (authors' data).

Substrate	Weeks after adding amendment to soil				
	1	2	4	8	12
	— percentage of applied C evolved as CO_2 —				
Glucose	73	80	82	88	89
Glycine	74	80	83	86	89
Chlorella pyrenoidosa protein	40	61	75	83	84
Pyruvic acid	62	69	75	79	81
Cellulose	27	43	52	70	77
Benzoic acid	55	60	64	69	71
Sudan grass (*Sorghum sudanense* L.)	34	47	54	62	64
Cornstalks (*Zea mays* L.)	30	38	47	53	59
Chitosan	20	35	45	53	56
Mucor rouxii cells	26	36	44	50	54
Ferulic acid	29	34	38	43	46
Coumaryl alcohol	15	26	31	36	45
Prune wood (*Prunus* spp.)	12	25	33	40	45
Cow manure	10	17	26	31	34
Hendersonula toruloidea cells	13	21	28	35	39
4,5 Dichlorocatechol	16	20	26	31	34
Catechol	10	13	17	20	23
Douglas fir wood (*Pseudotsuga menziesii* L.)	2	3	5	9	15
H. toruloidea melanin	3	5	8	11	13
C. pyrenoidosa protein linked into model H.A.	5	6	9	10	10
Peat moss	1	2	3	4	8
Soil humic acid	<1	<1	<1	1	1

3-1 DECOMPOSITION AND STABILIZATION OF ORGANIC RESIDUES IN HUMUS

Carbon is continuously being bound into organic compounds, mainly by photosynthetic processes. This organic C must later be released so that it may be used by future generations of plants, and indirectly, by animals. This process is accomplished through the biodegradation of organic residues by microorganisms in soil and water. About 60 to 80% of the C from most plant residues is returned to the atmosphere as CO_2 after 1 yr (Jenkinson, 1971; Sauerbeck & Gonzalez, 1977; Paul & Van Veen, 1978). From 5 to 15% of the C remaining in the soil occurs in the microbial biomass, while the remainder is partially stabilized in the form of new humus (Kassim et al., 1981; Stott et al., 1983a, b).

Various constituents of organic residues decompose at different rates (Table 3-1). Use of uniformly or specifically ^{14}C-labeled organic substrates has enabled researchers to more precisely follow the decomposition, transformation, and incorporation into the humus of carbonaceous materials added to the soil.

3-1.1 Plant Residues

The rapid decomposition of plant residues upon incorporation into soil has been the subject of several investigations. Decomposition of ^{14}C-labeled mature wheat straw (*Triticum aestivum* L.) and other plant residues was studied in small field plots in Germany and Costa Rica (Sauerbeck & Gonzalez, 1977). After 1 yr, about $65 \pm 5\%$ of the added C was lost as CO_2 from the temperate German soils and $70 \pm 5\%$ from the tropical Costa Rican soils. Field crop residue decomposition studies in England indicated that about 69% of the C from ryegrass (*Lolium perenne* L.) roots and tops, wheat straw, and cornstalks (*Zea mays* L.) was lost as CO_2 during 1 yr (Jenkinson, 1971). In Australian soils, the decomposition of ^{14}C-labeled medic (*Medicago littoralis*) resulted in the loss of >50% of the added C in 4 wk (Ladd et al., 1981). After 8 wk, the total loss of labeled medic C increased to about 70%, and to 75% after 1 yr. Nyhan (1975) followed the decomposition of ^{14}C-labeled blue grama grass tops and roots in a Colorado field soil. In 412 d, 54 to 57% of the herbage C and 26 to 37% of the root C was lost. When fresh unlabeled blue grama (*Bouteloua gracilis* L.) residues were added to soil containing partially degraded labeled plant material, there was no effect on rates of ^{14}C loss as compared to soil with no added fresh residues.

Laboratory decomposition studies have been criticized for not being representative of what occurs under field conditions. Over a period of sever-

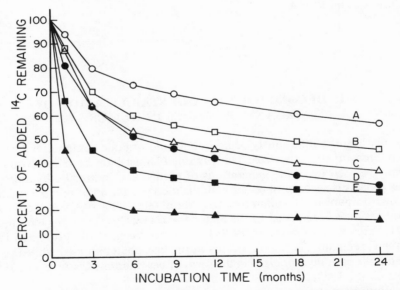

Fig. 3–1. Decomposition of substrates: (A) ring and 2-(side chain) cornstalk lignin carbon (B) 3-(side chain) cornstalk lignin carbon (C) 1-(side chain) cornstalk lignin carbon, (D) $-OCH_3$ cornstalk lignin carbon (E) whole wheat straw and cornstalk carbon, (F) wheat straw polysaccharide carbon in a coarse-loamy, mixed, thermic, Mollic Haploxeralf. Incubated under continuous aeration, $22 \pm 2\,°C$ and -33 kPa water potential (Martin et al., 1980; Stott et al., 1983a).

al months, however, it has been found that C losses from laboratory tests are essentially the same as those from field studies in temperate zones (Jenkinson, 1971; Sauerbeck & Gonzalez, 1977; Ladd et al., 1981). In continuously aerated laboratory incubation studies, 67 to 69% of ^{14}C-labeled wheat straw or cornstalk carbons were lost as CO_2 after 1 yr (Fig. 3-1; Martin et al., 1980; Stott et al., 1983a). These incubations were carried out under conditions that were favorable for microbial activity in temperate soils, i.e., 22 $\pm$ 2 °C and about -33 kPa water potential (60% water-holding capacity for the soils used). In the field, there are times when temperature and water potential conditions are not optimal for rapid decomposition. Most of the decomposition of available C substrates occurs, however, during a relatively short time period. When environmental factors become favorable, biodegradation proceeds rapidly until a more stable residue of new humus remains. Thus, over 1 yr, laboratory and field test data would be comparable.

3-1.2 Readily Available Substrates

Readily available, water-soluble substrates such as sugars, amino acids, pyrimidines, and aliphatic acids are metabolized within a few hours or days after being added to soil (Table 3-1). While C from these sources is rapidly evolved as CO_2, initially 40 to 60% of the C may be transformed into microbial biomass and associated metabolic products (Wagner, 1975; Paul & Van Veen, 1978). These microbial cells and products are also subject to decomposition and transformation. After 1 yr, 70 to 90% of the original substrate C will have evolved as CO_2. Of the C remaining in the soil, 10 to 20% of the original substrate C will be present in the microbial biomass, and the remainder will be incorporated into new humus (Table 3-2; Martin et al., 1980; Stott et al., 1983a, b). With time, the amount of original substrate C found in the microbial biomass will decline and that in humus will increase (Kassim et al., 1982). In most soils, 2 to 4% of the total organic C will be part of the microbial biomass (Jenkinson & Powlson, 1976; Anderson & Domsch, 1978).

After a 1-yr incubation, about 20% of the C from readily biodegradable compounds is associated with soil humus. Some of this C is in the form of aromatic molecules. After soil amended with labeled glucose or wheat straw was incubated for 6 wk, ^{14}C activity was found in 16 phenolic compounds released upon reductive degradation of the soil humic acids (Martin et al., 1974). Up to 80% of the residual C from these substrates is present in the form of polysaccharides and peptides, however, and is released as sugar and amino acid units upon acid hydrolysis (Jenkinson, 1971; Oades & Wagner, 1971; Martin et al., 1980). This result is expected because the major portion of the metabolized C not initially lost as CO_2 would be used for the synthesis of microbial cell components (Stott et al., 1983a).

Simple phenolic compounds and other aromatic molecules may be present in plant and microbial tissues, and are released during biodegradation (Flaig et al., 1975; Linhares & Martin, 1979b; Kassim et al., 1981). In pure culture studies, most of these phenolic compounds can be readily de-

Table 3-2. Distribution of ^{14}C activity after incubation of various ^{14}C-labeled organic compounds in a coarse-loamy, mixed thermic Mollic Haploxeralf (Martin et al., 1980; Stott et al., 1983a, b).

Labeled substrates	Incuba-tion period (months)	% ^{14}C lost as ^{14}C	Estimated % of residual ^{14}C in:		% of humus ^{14}C found in:		% of humus ^{14}C lost upon acid hydrolysis
			Biomass	Humus	HA†	FA	
UL‡ wheat straw	6	65	9.1	90.9	26	31	62
	12	69	8.2	91.8	26	31	56
	24	74	4.8	95.2	36	14	48
UL wheat straw polysaccharide	6	80	13.9	86.1	24	ND§	65
	12	83	9.5	90.5	17	39	73
	24	84	7.8	92.2	22	28	58
Ring wheat straw lignin	6	11	0.5	99.4	29	32	18
	12	16	0.2	99.8	36	30	5
2-side chain C wheat straw lignin	6	8	0.6	99.4	34	31	20
	12	13	0.6	99.4	38	25	5
UL glucose	6	78	19.0	81.0	20	24	74
	12	81	16.0	84.0	22	29	78
UL fungal melanin	6	11	0.1	99.9	71	ND	8
	12	13	0.7	99.3	68	ND	5
Ring catechol	12	40	3.6	96.4	ND	ND	31
Ring 4,5-dichloro catechol	12	54	6.9	93.1	ND	ND	31
Ring catechol linked into model HA	12	12	2.1	97.9	ND	ND	21
Ring 4,5-dichloro catechol linked into model HA	12	22	2.1	97.9	ND	ND	43

† HA = humic acid, FA = fulvic acid.   § ND = not determined.
‡ UL = uniformly labeled.

graded by several species of soil microorganisms (Dagley, 1971). In soil, however, biodegradation and stabilization are related to a specific compound's reactivity in respect to oxidative polymerization reactions (Haider & Martin, 1975; Martin & Haider, 1976, 1979a; Kassim et al., 1981). For example, about 70% of the C from benzoic acid evolves as CO_2 after 90 d, while over 70% of the catechol C is stabilized in the soil humus (Tables 3-1 and 3-2). Even after 1 yr, 60% of catechol C remains in soil. About 3% of the residual catechol C will be found in the microbial biomass and the rest will be in the soil humus (Table 3-2). Incorporation of the more reactive phenols into humus involves intact benzene rings. If the ring were cleaved before being incorporated into a humic acid molecule, the resulting aliphatic acid would decompose readily, with up to 80% of its C lost as CO_2.

3-1.3 Polysaccharides, Proteins, and Lipids

Sixty percent more of the C in most organic residues returned to the soil consists of cellulose and other polysaccharides. In addition, fresh residues can contain 6% or more protein. While polysaccharides and proteins decompose rapidly, their rate of biodegradation is slower than simple sugars and acids (Table 3-1). After 6 to 12 mo, 70 to 85% of the C from polysaccharides or proteins will evolve as CO_2 (Sauerbeck & Gonzalez, 1977; Martin et al., 1980). Up to 16% of the residual C from these substrates will be found in the soil microbial biomass (Table 3-2; Stott et al., 1983a, b). The small decrease in C lost as CO_2, as compared to the structurally simpler substrates, indicates that some of the C, either in its original form or as partially degraded units, is stabilized by incorporation into new humus (Bondietti et al., 1972; Verma et al., 1975). Proteins and polysaccharides containing amino acid or amino sugar units could be incorporated into humic acid macromolecules via nucleophilic addition to quinones or by strong H bonding (Bondietti et al., 1972; Martin et al., 1978). In addition, some polysaccharides are located within dense soil aggregates, and thus may be further protected from biodegradation (Tisdall & Oades, 1982). Also, highly branched molecules may be more recalcitrant (resistant to microbial degradation) than straight chain molecules (Alexander, 1977).

Lipids account for 1 to 6% of the organic matter in mineral soils and up to 20% of that in organic soils (Braids & Miller, 1975). Certain plant and microbial tissues also contain appreciable amounts of lipids. Kowalenko and McKercher (1971) proposed that the primary source of soil phospholipids is from the microbial biomass rather than from plant tissues. The work of Wagner and Muzorewa (1977) indicates that a majority of extractable soil lipids are of microbial origin and might be incorporated into soil humus without undergoing extensive degradative transformation. The ether-extractable portion (which contains the lipid fraction) of 30 soil fungi varies from 5 to 40% of the total dry weight (J.P. Martin, 1978, unpublished data). The ether-extractable portion and the original mycelium degrade at about the same rate in soil, indicating that lipid material is readily decomposed. Variation in decomposition rates between the mycelia of different fungal species is related to melanin content rather than lipid content.

3-1.4 Lignin

Advances in radioisotope methodology used to label plants and model lignins has enhanced our knowledge of the biodegradation and transformation of lignin in soil (Hackett et al., 1977; Haider et al., 1977a; Kirk et al., 1977; Martin & Haider, 1977; Crawford, 1981).

Two studies (Martin & Haider, 1979b; Martin et al., 1980) on lignin biodegradation in soil used ^{14}C to label various portions of the coumaryl or coniferyl alcohol component of the lignin molecule. After 1 yr, 30% of the C from the ring or the second C of this side chain from either the coumaryl or coniferyl alcohol components was lost as CO_2; after 2 yr, 43% was lost (Fig. 3-1). In comparison, the same studies showed that 47 to 55% of the C from the methoxyl group or the first or third C of the side chain was evolved as CO_2 after 1 yr, and 55 to 69% evolved after 2 yr (Fig. 3-1). Drying, rewetting, and additions of readily biodegradable organic substrates exert little effect on the degradation rate of lignin (Martin & Haider, 1979a; Haider & Martin, 1981).

While the majority of residual carbons from readily available substrates is associated with peptides and polysaccharides (Wagner & Mutakar, 1968; Jenkinson, 1971), the primary portion of residual lignin carbons is associated with aromatic molecules (Table 3-2; Martin et al., 1980; Stott et al., 1983a). Biomass estimates also support this conclusion: after 6 to 12 mo, 5 to 15% of the residual carbons from readily available substrates are present in soil microbial biomass compared to 0.2 to 1% of residual lignin carbons (Martin et al., 1980; Kassim et al., 1981; Stott et al., 1983a). If most of the aromatic rings in lignin had been cleaved by microbial enzymes with release of aliphatic acids, a major portion of the ring carbons would have been utilized for synthesis of cell components, and the metabolized lignin C should be present in soil humus in the form of peptides and polysaccharides. Actually the very small amount of lignin C found in biomass is less than would be expected on the basis of total C evolved as CO_2, suggesting that lignin may be partially degraded by processes of co-metabolism.

3-1.5 Microbial Melanins

Certain soil microorganisms, especially fungi and actinomycetes, synthesize dark-colored phenolic macromolecules called *melanins*. The melanins are formed in culture media or are present in cells and spores (Reisinger & Kilbertus, 1974). Generally melanins are very resistant to biodegradation (Table 3-1; Linhares & Martin, 1978). A loss of 5 to 21% of the C from *Aspergillus glaucus* and *Hendersonula toruloidea* melanins was noted over 1 yr from a variety of Californian and Chilean soils (Martin et al., 1982; Stott et al., 1983a). Like lignin, the majority of the residual melanin C was recovered from the humic acid fraction and very little was incorporated in the microbial biomass (Table 3-2).

3-2 FORMATION OF HUMIC SUBSTANCES

3-2.1 Humic Acid

Humic acids appear to be complex macromolecules of aromatic units with linked amino acids, peptides, amino sugars, aliphatic acids and other organic constituents (Kononova, 1966; Bondietti et al., 1972; Flaig et al., 1975; Stevenson, 1982b). The aromatic rings are of the di- and triphenol type, contain both free OH groups and quinone double linkages, and are bridged by O, S, and N atoms, as well as by NH, CH_2, and other groups (Stevenson, 1985).

The ability of numerous phenolic and related compounds to undergo enzymatic and autoxidative polymerization reactions is probably of importance in the formation of humic acids. During the early stages of microbial decomposition, simple phenolic compounds, portions of lignin and melanin molecules, and other phenolic polymers are transformed by β-oxidation of side chains, addition of hydroxyl groups, oxidation of methyl groups, and decarboxylation. Some of these compounds, especially *o*-dihydroxy and tri-hydroxy phenols, are highly reactive and may undergo autoxidation to form macromolecules when the pH is 6 or greater (Haider et al., 1975). These and other less reactive compounds, such as coniferyl alcohol, ferulic acid, orcinol, and vanillic acid, are readily oxidized by peroxidases and phenolases synthesized by soil organisms (Ladd & Butler, 1975; Skujins, 1976; Martin & Haider, 1980a; Suflita & Bollag, 1981; Burns, 1982). Radicals are formed by the oxidation of the phenols and are stabilized through linkage to form dimers and, upon further oxidation, quinones. Less reactive phenols, such as *p*-hydroxybenzoic and 2,5-dihydroxybenzoic acids, other aromatic compounds, and molecules with –SH or free amino groups, such as amino acids, peptides, and amino sugar polysaccharides may be linked into the developing humic polymers through nucleophilic addition to the quinones (Mayaudon et al., 1973; Flaig et al., 1975; Martin & Haider, 1980b). Condensed ring aromatic structures may be linked into humic acid molecules (Mathur, 1971).

Lignin has been considered a major source of aromatic structural units for humus formation in many soils (Hurst & Burges, 1967; Flaig et al., 1975; Martin & Haider, 1980b; Stevenson, 1982b). Decomposition studies with labeled lignin indicate that a large portion of lignin C is incorporated into soil humic acids (see previous section). Degradative analysis of soil humic substances shows, however, that lignin undergoes profound changes during humification (Martin & Haider, 1980b). Such changes include marked increases in exchange acidity and N moieties such as peptides, an increase in COOH groups, a decrease in methoxyl groups, and wide differences in the compounds recovered by various degradative procedures and pyrolysis (Maximov et al., 1977; Meuzelaar et al., 1977; Schnitzer, 1977; Saiz-Jiminez et al., 1979).

During lignin biodegradation, numerous simple phenolic substances are released in small amounts (Flaig et al., 1975), but portions of lignin molecules at all stages of decomposition could undergo oxidation reactions and be linked into humic acid molecules (Ladd & Butler, 1975; Martin & Haider, 1980b). Once linked, some of the exposed rings may be cleaved via microbial action, leaving a web of highly branched aliphatic structures (Crawford, 1981; Ellwardt et al., 1981).

It has been suggested since the early part of this century that melanins, secondary metabolites formed by microbes during carbohydrate catabolism, may be similar to soil humic acids (Martin & Haider, 1971). Studies have shown that some melanins are similar to soil humic acids in elemental composition, exchange acidity, amalgam reductive degradation products, resistance to microbial decomposition, structures released upon pyrolysis and oxidative degradation procedures, low polysaccharide content, and ^{13}C-NMR spectra (Kang & Fellbeck, 1965; Haider et al., 1975; Saiz-Jiminez et al., 1975; Schnitzer & Neyroud, 1975; Hayes & Swift, 1978; Linhares & Martin, 1978, 1979a; Lüdemann et al., 1982). Because of these similarities and inasmuch as melanic fungi may constitute the major portion of active fungal hyphae in soil (Waid, 1960), microbial melanins probably contribute to the formation of resistant soil humus.

Table 3-3. Some phenolic compounds isolated from the culture media of some common soil fungi (Haider & Martin, 1970; Martin et al., 1974; Martin & Haider, 1980b).

Acids	Toluenes
Caffeic	p-Cresol
Cresorsellinic	m-Cresol
2,4-Dihydroxybenzoic	2,4-Dihydroxy
2,5-Dihydroxybenzoic	2,6-Dihydroxy
2,6-Dihydroxybenzoic	3,5-Dihydroxy
3,4-Dihydroxybenzoic	4-Methyl-2,6-dihydroxy
3,5-Dihydroxybenzoic	2,3,5-Trihydroxy
3,5-Dihydroxy-4-methylbenzoic	2,3,6-Trihydroxy
Ferulic†	2,4,5-Trihydroxy
Gallic	2,4,6-Trihydroxy
p-Hydroxybenzoic	3,4,5-Trihydroxy
m-Hydroxybenzoic	
p-Hydroxycinnamic	Others
6-Methylsalicylic	
Orsellinic	Catechol
Protocatechuic	5-Methylpyrogallol
Salicylic	Phloroglucinol
Syringic†	Pyrogallol
2,3,4-Trihydroxybenzoic	Resorcinol
2,4,6-Trihydroxybenzoic	
Vanillic†	

† Present when fungi were grown in media containing plant residues.

Studies indicate that there is a huge variation in the number and kinds of structural units in fungal melanins. Some melanins consist mostly of orsellinic acid derived phenols, some contain a combination of phenols derived from p-hydroxycinnamic and orsellinic acids, and others have a variety of anthraquinones and phenols (Martin et al., 1972; Haider et al., 1975; Martin & Haider, 1980). *Hendersonula toruloidea* synthesized over 40 phenolic and aromatic compounds, and *Eurotium echinulatum* over 50. Most of these aromatics can be found as consistent units of fungal melanins (Table 3-3; Haider et al., 1975; Saiz-Jiminez et al., 1975). Microbial cytoplasmic and cell wall components may be linked into these polymers via enzymatic or autoxidative reactions (Nelson et al., 1979). As with lignin, melanins and other microbially synthesized materials undergo decomposition and transformation before becoming part of the true soil humus.

3-2.2 Model Humic Acid Molecules

Many model humic acid-type polymers, including specifically ^{14}C-labeled preparations, have been made using phenolase and peroxidase enzymes as catalysts, or by autoxidative polymerization reactions using highly reactive compounds such as hydroxy-phenols. Phenolic compounds that have been identified as either lignin-derived or microbially synthesized have been polymerized in reaction mixtures containing from 10 to 25 different substances (Bondietti et al., 1972; Martin et al., 1972a; Verma et al., 1975; Martin & Haider, 1976; Haider et al., 1977a; Suflita & Bollag, 1981; Brannon & Sommers, 1985a). Such model polymers are very resistant to degradation in soil. After 3 to 6 mo, only 2 to 21% of the added polymer C was evolved as CO_2. Generally, the more complex the mixture of reactive compounds, the more stable the resulting humic acid-type macromolecule.

Haider et al. (1965) showed that in the presence of a phenoloxidase, amino acids and phenols would form cross-linkages and that the terminal amino acid linked to the phenol was stable to acid hydrolysis. Ladd and Butler (1966) made simple model humic acid-type molecules from either p-benzo-quinone or catechol with a N source of either NH_4^+ salts, amino acids, or proteins. Total weight and amino-N distribution of the protein–phenol molecules were similar to soil humic acids. Covalent linkage of these compounds provides much greater resistance to biodegradation. Liu et al. (1985) studied the enzymatically catalyzed cross-coupling reactions between amino acid esters and phenols. Linkage was achieved through a covalent bonding of the amino-N and an aromatic C, forming a quinone. Additional polymerization resulted in ring formation through N to form a heterocyclic compound.

Using model polymers, Brannon and Sommers (1985a) have reported one mechanism for stabilizing organic P in soil. The stabilization involves the incorporation of organic compounds containing both amine and phosphate ester functional groups into humic acid macromolecules via oxidative polymerization. They estimated that about 80% of the added organic P was stabilized and protected against microbial attack, whereas the remaining portion was readily accessible and therefore subject to rapid biodegradation (Brannon & Sommers, 1985b).

Changes in resistance to biodegradation of various organic compounds in soil humus have been studied through the use of model humic acid-type molecules. Intimate association of most readily available substrates with preformed model humic acids does not provide much protection against biodegradation as compared to model humic acids with those same substrates incorporated during formation (Martin et al., 1978). Notable exceptions are proteins or amino acids having free amino or sulfhydryl groups. Probably the protection is related to strong H bonding between the substrates and humic acids rather than linkage into previously formed polymers.

Organic molecules differ in their reactivity and ability to link into a humic acid molecule as it is being formed. Proteins and components of microbial cell walls and cytoplasm link readily and are stabilized against degradation (Table 3–4). Added in free form, 65 to 75% of the C from these molecules will be lost as CO_2 after 12 wk (Verma & Martin, 1976). Only 10 to 40% of the C will be evolved as CO_2 when the substrates are first linked into model humic acids before being added to soil. Certain individual amino acids, such as glycine and lysine, form linkages with the model humic acids less readily, but once stabilized, are very resistant to degradation (Martin & Haider, 1980a). Substitution on the benzene ring changes the reactivity of phenols. Chlorine substituents on a chlorocatechol molecule apparently interfere with the cross linkages in the model polymer, thus allowing the substituted rings to be more susceptible to microbial attack (Table 3–4, Stott et al., 1983b).

Table 3-4. Amount of various compounds linking into model polymers in relation to their decomposition, free or linked into model polymers, in a Mollic Haploxeralf soil after 12 wk (Verma & Martin, 1976; Martin & Haider, 1980a; Stott et al., 1983b).

Labeled substrates[†]	Percentage linked into polymer	$\%\,^{14}C$ lost as CO_2	
		Free	Linked
UL *Chlorella* protein	99	74	26
UL *Nostoc muscorum* cell walls	99	67	39
UL *Nostoc muscorum* cytoplasm	98	74	12
Ring catechol	96	27	6
Ring 4-chlorocatechol	93	32	10
Ring 4,5-dichlorocatechol	80	36	12
Ring ferulic acid	75	57	7
UL cysteine	63	72	13
UL lysine	34	78	17
UL glucosamine	28	72	19
UL cytosine	4	70	ND
Ring 2,4-D	0	70	ND

[†] UL = Uniformly labeled. Ring = ring portion of molecule labeled.

The decomposition of catechol that has been incorporated into model humic acids is similar to that of whole model polymers and soil humic acids (Tables 3-1 and 3-4).

Model humic acid-type molecules provide a useful tool in determining if stabilization in humus plays an important role in the fate of pesticides and other xenobiotics added to the soil (Martin & Stott, 1981; Stott et al., 1983b). More work is needed to determine how aliphatic units are linked into or associated with humic acids so as to become resistant to biodegradation.

3-2.3 Polysaccharides

Soil polysaccharides should not be considered as humic substances. After extraction, soil humus can be separated into the traditional "humic acid", "fulvic acid" and "humin" fractions. Most of the soil polysaccharides are recovered from the fulvic acid and humic fractions (Greenland & Oades, 1975; Linhares & Martin, 1979a; Saiz-Jiminez et al., 1979; Hatcher et al., 1985). Most amino sugar polysaccharides remain in the humin fraction (Guchert et al., 1977), indicating that most polysaccharide or humic molecules containing significant amounts of amino sugar units do not precipitate upon acidification of the NaOH extract or are not soluble under the conditions of extraction. Some metal complexes of polysaccharides containing uronic acid units are not soluble in the NaOH solution (Martin et al., 1966). It is also likely that some metal–clay–polysaccharide complexes are not soluble. Stevenson (1982b) considers the soil polysaccharides as coprecipitates rather than a type of true humic substance [see discussions by Hatcher et al. (1985) and by Malcolm in chapter 2 of this book]. The soil polysaccharide fraction, after the removal of most contaminants, has little or no color. Despite questions of definition, soil polysaccharides form an important fraction of the soil humus, influencing soil aggregation, structure, and stability (Hayes,

1980; Foster, 1981; Tisdall & Oades, 1982; Emerson et al., 1986). Thus, we have included the soil polysaccharide fraction in our discussion of humus.

Collectively, soil polysaccharides constitute the second most abundant component of soil humus (Martin & Haider, 1971; Foster, 1981). Various investigators have determined that polysaccharides make up from 5 to 30% of the soil humus (Greenland & Oades, 1975; Cheshire, 1977). As methods for quantifying this fraction usually give low values, the actual percentages may be higher (Gupta & Sowden, 1965; Swincer et al., 1968). For most soils, 1 to 5% of the polysaccharides are associated with the humic acid fraction (Ivarson & Sowden, 1962; Oades, 1967; Butler & Ladd, 1971; Lowe, 1978; Linhares & Martin, 1979a); however, up to 10% association with humic acids has been reported in allophanic soils (Griffith & Schnitzer, 1975).

In the past, it has been hypothesized that a major portion of the soil polysaccharide fraction originates through microbial synthesis (Swincer et al., 1968; Cheshire, 1977). Also, soil polysaccharides contain amino sugar units, which are synthesized by many microorganisms, but usually are not present in plant polysaccharides (Stevenson, 1957; Bremner, 1967; Benzing-Purdie, 1981). Stevenson (1983) found amino sugars in both the humic acid and polysaccharide fractions of a silt loam. However, Cheshire and Hayes (1990), in a recent review, have concluded that microbial synthesis is only one of the processes in the complicated formation of soil polysaccharides. Cheshire and co-workers have determined that the pentose sugar units found in soil polysaccharides are primarily derived from plant tissues (Cheshire & Hayes, 1990).

Several factors are involved in polysaccharide stabilization in soil humus. Microbial and plant polysaccharides are both subject to decomposition, but rates vary with specific polymers (Martin, 1971). While plant and microbial polysaccharides contain one to three and occasionally up to five types of monomer units, soil polysaccharides contain 10 or more major monomer units with many other types present in smaller concentrations (Sowden & Ivarson 1962; Oades, 1967; Swincer et al., 1968; Martin & Haider, 1971). Plant and microbial polysaccharides, at any stage of decomposition, may recombine to form more resistant structures (Martin, 1971). Highly branched molecules may be more resistant to biodegradation than straight chain structures. Also, uronic acids comprise about 10% of the polysaccharide fraction (Greenland & Oades, 1975). These, along with molecules with *cis*-hydroxyls or phosphoric acid esters, form salts or complexes with di- or trivalent metal ions (Martin et al., 1966; Martin, 1971). The resulting complexes may have a greatly reduced susceptibility to biodegradation. Polysaccharides containing amino sugar or amino acid units may also be stabilized in humic acid macromolecules via nucleophilic addition to quinones or by strong H bonding (Bondietti et al., 1972; Martin et al., 1978). Also, polysaccharides may form complexes with clay minerals through metal ion-clay linkages which increase resistance to degradation (Greenland, 1965; Harris et al., 1966; Guckert et al., 1977; Stevenson, 1979, 1982b). Polysaccharides may be further protected when they are located within dense soil aggregates (Adu & Oades, 1978, Tisdall & Oades, 1982).

3-3 CHARACTERIZATION OF NATURAL AND SYNTHETIC HUMIC MATERIALS

Since the bulk of the soil organic matter consists of the relatively large humic acids and polysaccharide molecules, scientists have been interested in determining the structures of these substances. Degradative analyses were the primary means of gathering structural information up through the 1970s. These methods include acid hydrolysis, oxidative degradation, reductive degradation, pyrolysis and several others (Schnitzer, 1977). Results from these procedures must be interpreted with care. The breakdown products recovered may not be directly derived from the actual molecular structure, but rather may be a result of chemical alterations that can occur during the degradative reactions (Martin et al., 1974; Maximov et al., 1977; Hayes & Swift, 1978; Mathur & Schnitzer, 1978). For this reason it is important to understand the chemical reactions that might occur in the system that is being studied. These reactions have been reviewed in detail (Hayes & Swift, 1978; Hayes & Himes, 1986). In the last two decades, there has been an increased use of nondegradative spectroscopic methods (Hayes & Himes, 1986).

3-3.1 Elemental Analysis

The major elements found in humic acid molecules are C and O. Carbon content ranges from 50 to 60% and O content from 30 to 35% (Stevenson, 1982b). Nitrogen, H, and S contents have been reported as 1 to 6%, 4 to 6%, and 0 to 2%, respectively (Hayes & Swift, 1978; Stevenson, 1982b).

3-3.2 Acid Hydrolysis

Simple acid hydrolysis will release sugars from polysaccharides and amino acids from peptides. The values reported in the literature are probably low because some of the amino acids and monosaccharides are destroyed during the hydrolysis. Also, the terminal amino acid of a peptide chain when linked to a phenolic unit is not readily cleaved by hydrolysis (Haider et al., 1965).

Upon 6 M HCl hydrolysis, up to 50% or more of the N in humic acid macromolecules can be recovered as amino acid N. The amino acids thus recovered include all those found in an average microbial or plant protein (Bremner, 1967; Sowden et al., 1976). Thus, peptides or proteins represent a major form of humic acid N, indicating that about 6 to 20% of the humic acid macromolecule consists of peptide structures.

The amino acid distribution patterns of four Brazilian soils show that aspartic acid, glutamic acid, glycine, and alanine are the dominant amino acids (Table 3-5; Coelho et al., 1985). Melanins synthesized by two fungal species showed similar distribution patterns. The variations noted between the soil and fungal macromolecules are more quantitative than qualitative in nature and the similarities are greater than any differences. This supports the hypothesis that fungal melanins play a role in humic acid formation. The

Table 3-5. Distribution† of amino acids in humic acids from Brazilian soils and melanins from fungi grown on glucose media with sodium nitrate as the N source (Coelho et al.,1985).

Amino acid	Red yellow podzol (forest)	Red yellow podzol (sorghum)	Dusky red latosol (legumes)	Red yellow latasol (tropical savanna)	Dark red latasol (tropical savanna)	*Hendersonula toruloidea* melanin	*Aspergillus glaucus* melanin
Acidic							
Aspartic acid	12.3	10.6	14.6	12.6	14.4	10.8	10.3
Glutamic acid	9.6	8.3	8.5	9.5	9.0	7.9	10.3
Basic							
Arginine	1.8	2.2	1.7	1.7	1.7	2.2	2.4
Histidine	4.2		2.6	1.5	2.2	2.6	1.6
Lysine	3.8	3.7	3.7	3.7	4.2	2.6	3.1
Neutral							
Alanine	8.1	8.6	10.6	10.3	9.8	9.8	8.5
Glycine	13.1	12.2	14.0	11.9	11.0	10.8	11.0
Isoleucine	2.8	3.3	3.1	3.3	3.1	3.6	3.6
Leucine	3.7	5.6	5.4	5.5	5.3	6.3	5.1
Phenylalanine	3.3	3.0	1.9	2.7	2.5	2.2	2.7
Proline	3.7	4.5	4.5	3.9	2.5	6.6	6.9
Threonine	3.9	4.5	5.8	5.3	4.0	7.2	6.7
Tyrosine	2.5	3.9	2.5	3.0	3.1	1.1	2.5
Serine	3.4	4.3	4.9	5.0	4.3	7.5	7.3
Valine	4.6	5.7	4.9	4.9	4.7	5.7	5.3
Sulfur-containing							
Cysteic acid	0	0.6	0	0.6	0.6	0.2	0.2
Half cystine	0.6	0	0	0	0	0.7	0.1
Methionine	0.4	0.8	0.6	0.9	1.0	0.4	0.2

† Relative molar distribution = α-amino-N of each amino acid × 100/total amino and imino acid-N.

source of N and the C-to-N (C/N) ratio, will affect the distribution of N in fungal melanins (Bondietti, 1972). The smaller the C/N ratio, the greater the total N content of the melanins and a greater proportion of the N will be in structures that are readily hydrolyzed by 6 M HCl (Table 3–6). At all C/N ratios, melanins from fungi grown in the presence of a peptone-N source contained two to three times more N than those from a KNO_3-N source.

Mild acid hydrolysis procedures are used to estimate quantities of polysaccharides in the soil and amounts associated with humic acid molecules (Greenland & Oades, 1975). The major sugar units released by hydrolysis of soil polysaccharides include: glucose, galactose, mannose, arabinose, xylose, fucose, rhamnose, glucuronic acid, galacturonic acid, glucosamine and galactosamine (Greenland & Oades, 1975; Cheshire, 1977; Hayes & Swift, 1978).

3–3.3 Oxidative Degradation

Oxidative degradation of polymers involves the methylation and oxidation of polymers with permanganate, persulfate, or cupric oxide. Using the

Table 3-6. Nitrogen distribution in fungal melanins as affected by the C/N ratio and N source in the growth medium (Bondietti, 1972).

	N source					
	Peptone			KNO$_3$		
C/N ratio	Total N content	Acid soluble N[†]	Amino acid N[‡]	Total N content	Acid soluble N	Amino acid N
	%					
	Hendersonula toruloidea					
5	11	70	32	7	65	42
10	10	72	32	7	73	42
50	6	66	34	5	63	38
100	3	64	20	3	55	44
	Stachybotrys chararum					
5	8	76	38	5	55	29
10	6	72	39	5	53	28
50	5	56	39	3	50	32
100	4	47	43	3	54	38

[†] Portion of melanin N solubilized upon 6 M HCl hydrolysis.
[‡] Portion of melanin N released as amino-N upon acid hydrolysis.

KMnO$_4$ and the alkaline CuO methods, Schnitzer (1977) found that the major degradation products released from humic acids were a series of di-, tri-, tetra-, penta-, and hexabenzene carboxylic acids, phenols, and fatty acids. It is likely that most of these are formed by oxidative splitting of covalent C–O bonds and cleavage of susceptible rings with the side chains being oxidized or converted to COOH groups (Maximov et al., 1977; Hayes & Swift, 1978). Similar fragments are formed upon oxidation of model humic acid polymers made from simple phenolic compounds (Mathur & Schnitzer, 1978). The major degradation products of lignins are veratric acid, several dimethoxydicarboxy benzenes, biphenyls, and diphenyl ethers (Griffith & Schnitzer, 1975). The relatively mild persulfate oxidation method has been used on fungal melanins and soil humic acids (F. Martin et al., 1981, 1983). While the products released from the soil humic acids showed a dominance of phenolic and aromatic acids, aliphatic acids (especially n-C$_{16}$ and n-C$_{18}$ acids) accounted for 70 to 80% of the compounds released from fungal melanins.

3–3.4 Amalgam Reductive Degradation

Sodium amalgam (an alloy of Na and Hg) reductive degradation of humic acids yielded 15 to 35% ether-soluble substances and a large number of simple phenolic compounds. These include both microbial- and lignin-related substances (Burges et al., 1964; Piper & Posner, 1972; Martin et al., 1974; Hayes & Swift, 1978). The yield of phenolic substances corresponded to 3 to 32% by weight of the starting soil organic matter (Piper & Posner, 1972; Martin et al., 1974). Yields of aromatic components in model humic acids ranged from 4 to 23%, with the lowest yields from polymers incorporating o-dihydroxyphenols and the highest from those made with resorcinol-type phenols only (Martin et al., 1974). The incorporation of amino acids, peptides, or amino sugars reduced yields. Degradation of fungal polymers yielded 4 to 16% of the starting material and lignin-type phenols were present only when the fungi were grown on plant residues (Table 3-3).

Tests with pure aromatic compounds have shown that many are partially destroyed by the reduction process, and thus the actual amounts of phenolic compounds in the polymers are likely to be higher than indicated by this method. The simple phenols released from model humic acids by sodium amalgam degradation have been shown to be the same as those used to synthesize the macromolecules, or to be slightly altered compounds formed by decarboxylation or migration of an OH group to the other benzene ring of a ring-to-ring ether linkage (Martin et al., 1974).

3-3.5 Pyrolysis–Mass Spectrometry

Pyrolysis–mass spectra of a soil humic acid and a fungal melanin are shown in Fig. 3-2. Also shown are two plant polymers, cellulose and lignin, which together constitute 70 to 80% of the plant biomass. The cellulose spectrum shows prominent peaks for acetic acid, furan, furfural, furfuryl alcohol, and their methylated derivatives (Haider et al., 1977b). Lignin fragments were larger than those from cellulose, and the signals indicate the presence of phenol and derivatives of p-coumaryl, coniferyl, guaiacyl, and syringyl alcohols, all of which are structural units of lignins. Pyrolysis of soil humic acids and fungal melanins yields numerous organic fragments (Meuzelaar et al., 1977). The most prominent fragments in the spectra contained S and N and were related to proteins, benzenes, and phenols. Portions that were less prominent, although still significant, were related to polysaccharides and other aliphatic structures. Spectral peaks that are typical for lignin were either very small or absent from the humic acid and melanin spectra (Martin et al., 1977; Meuzelaar et al., 1977). The identified low boiling-point compounds from fungal melanins were mostly derived from peptide, carbohydrate and other aliphatic materials in the polymer structures (Saiz-Jiminez et al., 1979). Many compounds identified from pyrolysis-mass spectra of whole soils are related to polysaccharides (Saiz-Jiminez et al., 1979).

3-3.6 Nuclear Magnetic Resonance Spectrometry

Improvements in techniques for obtaining ^{13}C NMR spectra of humic substances and lignins has increased the value of this approach for structural studies of these substances (Hatcher et al., 1983). Most spectra show peaks or shoulders corresponding to aliphatic, peptide, aromatic, polysaccharide, and carboxylic acid structures (Gonzalez-Vila et al., 1976; Hatcher et al., 1981; see also chapter 2 (Malcolm) of this book). Fungal melanins, such as those from Epicoccum nigrum and Hendersonula toruloidea, give spectra with chemical shifts similar to those of soil humic acids (Lüdemann et al., 1982). Most aquatic humic acids show weaker signals in the aromatic regions

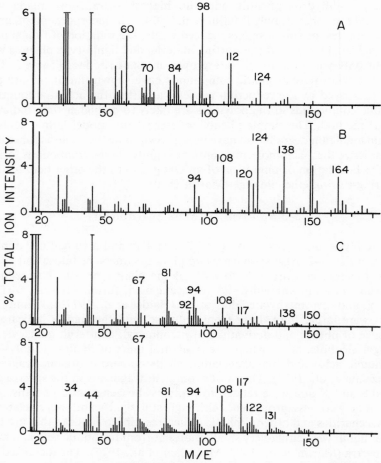

Fig. 3–2. Spectra obtained by curie-point pyrolysis techniques used in conjunction with low-voltage electron impact ionization mass spectrometry: (A) cellulose, (B) lignin from wheat straw, (C) humic acid from a Mollisol, and (D) *Hendersonula toruloidea* melanin (Martin and Haider, 1980b).

(Hatcher et al., 1983; Malcolm, 1985). For both melanins and humic acids, weak peaks or shoulders have been noted in the 150-ppm region suggesting a low phenolic group content (Hatcher et al., 1981; Lüdemann et al., 1982). In a comparison of several model, fungal, and soil humic acid molecules, Hatcher and Martin (1983 unpublished) found that the signal at 150 ppm apparently was directly related to the quantity of phenolic groups in the molecules. The model humic acid with hydroxybenzoic acids, hydroxyphenols, and toluenes plus protein produced a spectrum most like those from soil humic acids and fungal melanins (P.G. Hatcher, 1983, personal communication). The phenol signal was still stronger than for soil humic acids suggesting that other stuctural units present in natural humic molecules reduce the concentration (w/w) of phenolic groups.

3-3.7 Electron Spin Resonance Spectrometry

Electron spin resonance spectrometry has shown that fungal melanins differ from soil humic acids in that melanins have a higher concentration of free radicals (Riffaldi & Schnitzer, 1972; Saiz-Jiminez & Shafizadeh, 1985). This indicates that the fungal melanins are more reactive than the relatively stable humic acids.

3-4 BIOLOGICAL TRANSFORMATIONS OF SOIL HUMUS

Although soil humus decomposes at a rate of 2 to 5% yr^{-1}, studies with ^{14}C-labeled organic substrates have shown that new humus decomposes faster than old humus (Jenkinson, 1971; Sauerbeck & Gonzalez, 1977; Paul & Van Veen, 1978). Based on a 10-yr field study, the mean residence time of new humus formed under ^{14}C-labeled ryegrass was calculated as 26 yr (Jenkinson & Rayner, 1977). Sauerbeck and Gonzalez (1977) estimated from a long-term field trial with ^{14}C-labeled wheat straw that it would take 25 yr for 99% of the added C to be converted to CO_2. In a 2-yr incubation study, Martin et al. (1980) calculated that it would take about 33 yr to lose 99% of the C from wheat straw and 30 yr to lose that amount from the polysaccharide fraction of wheat straw. Also 44 yr would be required for a 99% loss of C from cornstalk lignin. These time spans are considered short. If the lignin C remaining in the new humus continued to decompose at the same rate that occurred during the last 90 d of a 2-yr period, calculations show that it would take 60 yr to lose 99% of the added C.

The mean age of the C in soil can be estimated using radioisotopic dating. Simonson (1959) calculated the mean age of humus C in five pasture soils in the midwestern USA as 200 to 400 yr. Paul et al. (1964) reported half-lives of 1700 to 2200 yr for humus in some Canadian soils. The estimated mean residence times for humus in allophanic soils were 2000 to 5000 yr (Wada & Aomine, 1975). Jenkinson and Rayner (1977) determined the mean age of the humus in some plots at the Rothamsted Experimental Station as 1240 yr.

As noted by Paul et al. (1964) and Campbell (1978), the soil fraction hydrolyzed with 6 M HCl was younger than the nonhydrolyzed fraction. This would be expected since major portions of new humus are in structures that are hydrolyzed by acid, mostly peptides and polysaccharides, and are synthesized in abundance by soil microorganisms from the carbohydrates and proteins readily available in plant residues (Jenkinson, 1971; Oades & Wagner, 1971; Martin et al., 1980; Stott et al., 1983a). Even though they will be stabilized by linkage through free amino or sulfhydryl groups, peptides and polysaccharides will still decompose faster than the aromatic portions of the new humus molecules (Table 3-4); Bondietti et al., 1972; Verma et al., 1975).

Perhaps a major cause for the greater stability of older humus is that it has formed clay–humus or clay–humus–metal ion complexes, affording greater resistance to biodegradation. Newer humus is less likely to be stabi-

Table 3–7. Decomposition of some ^{14}C-labeled organic substrates in allophanic and nonallophanic soils after 12 wk (Martin et al., 1982; Zunino et al., 1982).

Soil series†	% Allophane	% ^{14}C lost as CO_2						
		Glucose	Cellulose	Wheat straw	Chlorella protein	H. toruloidea cells	H. toruloidea melanin	Cornstalk lignin
Steinbeck loam	0	78	70	53	60	34	3	13
Greenfield sandy loam	0	77	78	54	67	31	5	18
Lo Aguirre sandy loam	0	76	74	53	60	30	ND‡	ND
Lo Aguirre + allophanic concentrate	16	58	49	34	40	19	ND	ND
Corte Alto topsoil 2–20 cm	15	58	42	30	35	20	<1	ND
Puerto Octay topsoil 2–20 cm	26	52	23	32	38	14	<1	3

† Steinbeck loam (fine-loamy, mixed mesic Ultic Haplustalf); Greenfield sandy loam (coarse-loamy, mixed thermic Typic Haploxeralf). Current taxonomic information is not available for the Lo Aguirre, Corte Alto, and Puerto Octay soils that are located in Chile; however, the Corte Alto and Puerto Octay soils are derived from volcanic ash (Andepts) and are high in allophane and organic carbon (8.9 and 8.1% organic C, respectively.)

‡ ND = not determined.

lized in this manner since many of the binding sites on the clay micelles would already be occupied. Studies of soil development and humus accumulation on young landscapes supports this concept (Stevenson, 1982a).

Some clays and primary minerals have been reported to act as catalysts for humic acid formation (Filip et al., 1977; Shindo & Huang, 1983; Wang et al., 1985). Respirometer studies indicated that oxidative polymerization of hydroxyphenol mixtures was not enhanced by the presence of montmorillonite, kaolinite or vermiculite (Martin & Haider, 1986). In similar studies, muscovite (Filip et al., 1977) and calcium-saturated illite (Wang et al., 1985) catalyzed model humic acid formation. The presence of clays such as montmorillonite and kaolinite has been shown to accelerate the formation of melanin indirectly by increasing growth and biomass production of fungi (Bondietti et al., 1971; Filip et al., 1972). Allophanic clays appear to be especially active in stabilizing organic C. Loss of C as CO_2 from readily available substrates such as glucose, cellulose, and protein is reduced by about 25% in the presence of allophane (Table 3-7). Biodegradation of more recalcitrant molecules such as melanin and lignin is also reduced (Martin et al., 1982; Zunino et al., 1982).

Some of the older humus might be protected inside dense aggregates, while more recently formed humus would be deposited on aggregate surfaces (Adu & Oades, 1978). One reason for the accelerated decomposition of organic matter in cultivated soils is related to the disruption and breaking apart of these aggregates, exposing more surface area to microbial attack. In addition, there may be a small portion of the new humus which is extremely resistant or, through transformations, becomes highly resistant to biodegradation. After a time, this fraction could increase and eventually constitute a major part of the humus.

3-5 CONCLUSIONS

Although 55 to 70% of all plant and animal residue C returned to the soil is released into the atmosphere as CO_2 after 1 yr, there are differential rates of decomposition among different constituents. In general, the more C lost as CO_2 from a given substrate, the more of the residual C that will be found in the soil microbial biomass (on a relative basis), and less will be stabilized in the soil humus. Most of the residual C from readily available substrates will be present in new humus as peptides and polysaccharides. Lignin and many fungal melanins, which decompose slowly, are among the many sources of aromatic units for soil humic acids. Their structures are partially degraded, transformed, and recombined through enzymatic and autoxidative reactions to form stable humic macromolecules.

Humic acids appear to be complex molecular structures consisting of hydroxyphenols, hydroxybenzoic acids, toluenes, and other aromatic moieties. Also included in the newly-formed humic acid molecules are peptides, amino acids, amino sugars, fatty acids, microbial cell wall and cytoplasmic fractions, and other aliphatic structures which comprise the majority of the C in humic acids. Microbial cleavage of rings of the aromatic units linked into humic acids can leave a network of highly branched aliphatic structures. This might be the source of some of the aliphatic units, which are evident by ^{13}C NMR spectroscopy methods.

Microbially synthesized polysaccharides appear to be the major precursors of the soil polysaccharide fraction of humus. Recent studies, however, are highlighting the importance of plant polysaccharides. The microbial and plant polysaccharides are subject to partial degradation, transformation, and repolymerization in much the same way as humic acids, resulting in structures that are more resistant to decomposition.

New humus decomposes faster than older humus. The mean residence time of new humus C has been calculated to be 25 to 35 yr. Older humus may have a mean residence time of hundreds to thousands of years. Complexing of humic and polysaccharide molecules with metal ions and clays micelles may greatly increase their resistance to biodegradation. Protection within dense aggregates may also increase their stability.

Work with model humic acids has given us valuable insights into the structure and activity of soil humic acids. Along with ^{14}C-labeled organic substrates, model humic acid molecules have enabled researchers to study the mechanisms by which various compounds can be stabilized in the soil environment to form humus.

REFERENCES

Adu, J.K., and J.M. Oades. 1978. Physical factors influencing decomposition of organic materials in soil aggregates. Soil Biol. Biochem. 10:109–115.

Alexander, M. 1977. Introduction to soil microbiology. 2nd ed. John Wiley & Sons, New York.

Anderson, J.P.E., and K.H. Domsch. 1978. Mineralization of bacteria and fungi in chloroform-fumigated soils. Soil Biol. Biochem. 10:207–213.

Benzing-Purdie, L. 1981. Glucosamine and galactosamine distribution in soil as determined by gas-liquid chromatography of soil hydrolysates: Effect of acid strength and cations. Soil Sci. Soc. Am. J. 45:66–70.

Bondietti, E. 1972. Stabilization of nitrogen in humic complexes by soil fungi. Ph.D. diss. Univ. of California, Riverside. (Diss. Abstr. 72-31485).

Bondietti, E., J.P. Martin, and K. Haider. 1971. Influence of nitrogen source and clay on growth and phenolic polymer production by Stachybotrys species, Hendersonula toruloidea, and Aspergillus sydowi. Soil Sci. Soc. Am. Proc. 35:917–922.

Bondietti, E., J.P. Martin, and K. Haider. 1972. Stabilization of amino sugar units in humic-type polymers. Soil Sci. Soc. Am. Proc. 36:597–602.

Braids, O.C., and R.H. Miller. 1975. Fats, waxes, and resins in soil. p. 343–368. In J.E. Gieseking (ed.) Soil components. Vol. 1. Organic components. Springer-Verlag, New York.

Brannon, C.A., and L.E. Sommers. 1985a. Preparation and characterization of model humic polymers containing organic phosphorus. Soil Biol. Biochem. 17:213–219.

Brannon, C.A., and L.E. Sommers. 1985b. Stability and mineralization of organic phosphorus incorporated into model humic polymers. Soil Biol. Biochem. 17:221–227.

Bremner, J. 1967. Nitrogenous compounds. p. 32–66. In A.D. McLaren and G.H. Peterson (ed.) Soil biochemistry. Marcel Dekker, New York.

Burges, N.A., H.M. Hurst, and S.B. Walkden. 1964. The phenolic constituents of humic acid and their relation to lignin of the plant cover. Geochim. Cosmochim. Acta. 28:1547–1564.

Burns, R.G. 1982. Enzyme activity in soil: Location and possible role in microbial ecology. Soil Biol. Biochem. 14:423–427.

Butler, J.H.A., and J.N. Ladd. 1971. Importance of molecular weight of humic and fulvic acids in determining their effects on protease activity. Soil Biol. Biochem. 3:295–340.

Campbell, C.A. 1978. Soil organic carbon, nitrogen, and fertility. p. 173–272. *In* M. Schnitzer and S.V. Kahn (ed.) Soil organic matter. Elsevier Sci. Publ., Amsterdam.

Cheshire, M.V. 1977. Origins and stability of soil polysaccharides. J. Soil Sci. 28:1–10.

Cheshire, M.V., and M.H.B. Hayes. 1990. Composition, origins, structures and reactivities of soil polysaccharides. p. 307–336. *In* M.F.L. DeBoodt et al. (ed.) NATO Advanced Research Workshop on soil colloids and their associations in soil aggregates. Plenum Publ., London.

Coelho, R.R.R., L.F. Linhares, and J.P. Martin. 1985. Amino acid distribution in some fungal melanins and of soil humic acids from Brazil. Plant Soil 87:337–346.

Crawford, R.L. 1981. Lignin biodegradation and transformation. John Wiley & Sons, New York.

Dagley, S. 1971. Catabolism of aromatic compounds by microorganisms. Adv. Microb. Physiol. 6:1–46.

Ellwardt, P.C., K. Haider, and L. Ernst. 1981. Untersuchunger des mikrobiellen lignin abbaues durch ^{13}C-NMR-spectroskopie zu specifisch ^{13}C angereicherten DHP-lignin aus coniferylalkohol. Holzforschung 35:103–109.

Emerson, W.W., R.C. Foster, and J.M. Oades. 1986. Organo-mineral complexes in relation to soil aggregation and structure. p. 521–548. *In* P.M. Huang and M. Schnitzer (ed.) Interactions of soil minerals with natural organics and microbes. SSSA Spec. Publ. 17. SSSA, Madison, WI.

Filip, Z., W. Flaig, and E. Rietz. 1977. Oxidation of some phenolic substances as influenced by clay minerals. p. 91–96. *In* Soil Org. Matter Stud., Proc. Symp. (Braunschweig, 6–10 Sept. 1976) Vol. 2. IAEA-FAO, Vienna.

Filip, Z., K. Haider, and J.P. Martin. 1972. Influence of clay minerals on growth and metabolic activity of *Epicoccum nigrum* and *Stachybotrys chartarum*. Soil Biol. Biochem. 4:135–145.

Flaig, W., H. Beutelspacher, and E. Reitz. 1975. Chemical composition and physical properties of humic substances. p. 1–211. *In* J.E. Gieseking (ed.) Soil components, Vol. I. Organic components. Springer-Verlag, New York.

Foster, R.C. 1981. Polysaccharides in soil fabrics. Science 214:665–667.

González-Vila, F.J., H. Lentz, and H.F. Lüdemann. 1976. ^{13}C nuclear magnetic resonance of natural humic substances. Biochem. Biophys. Res. Comm. 72:1063–1070.

Greenland, D.J. 1965. Interactions between clays and organic compounds in soils. Part II. Adsorption of soil compounds and its effect on soil properties. Soil Fert. 28:415–425.

Greenland, D.J., and J.M. Oades. 1975. Saccharides. p. 213–261. *In* J.E. Gieseking (ed.) Soil components. Vol. 1. Organic compounds. Springer-Verlag, New York.

Griffith, S.M., and M. Schnitzer. 1975. The isolation and characterization of stable metal-organic complexes from tropical volcanic soils. Soil Sci. 120:126–127.

Guckert, A., H.H. Tok, and F. Jacquin. 1977. Biodégradation de polysaccharides bactériens adsorbés sur une montmorillonite. p. 403–411. *In* Soil Org. Matter Stud., Proc. Symp. (Braunschweig, 6–10 Sept. 1976) Vol. 1. IAEA-FAO, Vienna.

Gupta, U.C., and F.J. Sowden. 1965. Studies on methods for determination of sugars and uronic acids in soils. Can. J. Soil Sci. 45:237–240.

Hackett, W.F., W.J. Conners, T.K. Kirk, and J.G. Zeikus. 1977. Microbial decomposition of synthetic ^{14}C-labeled lignins in nature: Lignin biodegradation in a variety of natural materials. Appl. Environ. Microbiol. 33:43–51.

Haider, K., L.R. Frederick, and W. Flaig. 1965. Reactions between amino acid compounds and phenols during oxidation. Plant Soil 22:49–64.

Haider, K., and J.P. Martin. 1970. Humic acid-type phenolic polymers from *Aspergillus sydowi* culture medium, *Stachybotrys* spp. cells and autoxidized phenol mixtures. Soil Biol. Biochem. 2:145–156.

Haider, K., and J.P. Martin. 1975. Decomposition of specifically carbon-14 labeled benzoic and cinnamic acid derivatives in soil. Soil Sci. Soc. Am. Proc. 39:657–662.

Haider, K., and J.P. Martin. 1981. Decomposition in soil of specifically ^{14}C-labeled model and cornstalk lignins and coniferyl alcohol over two years as influenced by drying, rewetting, and additions of an available C substrate. Soil Biol. Biochem. 13:447–452.

Haider, K., J.P. Martin, and Z. Filip. 1975. Humus biochemistry. p. 195–244. *In* E.A. Paul and A.D. McLaren (ed.) Soil biochemistry. Vol. 4. Marcel Dekker, New York.

Haider, K., J.P. Martin, and E. Rietz. 1977a. Decomposition in soil ^{14}C-labeled coumaryl alcohols: Free and linked into dehydropolymer and plant lignins and model humic acids. Soil Sci. Soc. Am. J. 41:556–562.

Haider, K., B.R. Nagar, C. Saiz-Jiminez, H.L.C. Meuzelaar, and J.P. Martin. 1977b. Studies on soil humic compounds, fungal melanins and model polymers by pyrolysis mass-spectrometry. p. 213–262. In Soil Org. Matter Stud., Proc. Symp. (Braunschweig, 6–10 Sept. 1976) Vol. 2. IAEA-FAO, Vienna.

Harris, R.F., G. Chesters, and O.N. Allen. 1966. Dynamics of soil aggregation. Adv. Agron. 18:107–169.

Hatcher, P.G., I.A. Breger, L.W. Dennis, and G.E. Maciel. 1983. Solid-state ^{13}C-NMR of sedimentary humic substances: New revelations of their chemical composition. p. 32–81. In R.F. Christman and E.T. Gjessing (ed.) Aquatic and terrestrial humic substances. Ann Arbor Science, Ann Arbor, MI.

Hatcher, P.G., I.A. Breger, G.E. Maciel, and N.M. Szeverenyi. 1985. Geochemistry of humin. p. 275–302. In G.R. Aiken et al. (ed.) Humic substances in soil, sediment, and water. John Wiley & Sons, New York.

Hatcher, P.G., M. Schnitzer, L.W. Dennis, and G.E. Maciel. 1981. Aromaticity of humic substances in soils. Soil Sci. Soc. Am. J. 45:1089–1094.

Hayes, M.H.B. 1980. The role of natural and synthetic polymers in stabilizing soil aggregates. p. 263–295. In R.C.W. Berkeley et al. (ed.) Microbial adhesion to surfaces. Soc. of Chem. Indust. Ellis Horwood Ltd., Publ. Chichester.

Hayes, M.H.B., and F.L. Himes. 1986. Nature and properties of humus-mineral complexes. p. 103–158. In P.M. Huang and M. Schnitzer (ed.), Interactions of soil minerals with natural organics and microbes. SSSA Spec. Publ. 17. SSSA, Madison, WI.

Hayes, M.H.B., and R.S. Swift. 1978. The chemistry of soil organic colloids. p. 179–320. In D.J. Greenland and M.H.B. Hayes (ed.) The chemistry of soil constituents. John Wiley & Sons, New York.

Hurst, H.M., and N.A. Burges. 1967. Lignin and humic acids. p. 260–286. In A.D. McLaren and G.H. Peterson (ed.) Soil biochemistry. Marcel Dekker, New York.

Ivarson, K.C., and F.J. Sowden. 1962. Methods for the analysis of carbohydrate material in soil. I. Colorimetric determinations of uronic acids, hexoses and pentoses. Soil Sci. 94:245–250.

Jenkinson, D.S. 1971. Studies on decomposition of ^{14}C-labeled organic matter in soil. Soil Sci. 111:64–70.

Jenkinson, D.S., and D.S. Powlson. 1976. The effects of biocidal treatments on metabolism in soil. V. A Method for measuring soil biomass. Soil Biol. Biochem. 8:209–213.

Jenkinson, D.S., and J.S. Rayner. 1977. The turnover of soil organic matter in some of the Rothamsted classical experiments. Soil Sci. 123:298–305.

Kang, K.S., and G.T. Felbeck. 1965. A comparison of the alkaline extracts of tissues of Aspergillus niger with humic acids from soils. Soil Sci. 99:175–181.

Kassim, G., J.P. Martin, and K. Haider. 1981. Incorporation of a wide variety of organic substrate carbons into soil biomass as estimated by the fumigation procedure. Soil Sci. Soc. Am. J. 45:1106–1112.

Kassim, G., D.E. Stott, J.P. Martin, and K. Haider. 1982. Stabilization and incorporation into biomass of phenolic and benzenoid carbons during biodegradation in soil. Soil Sci. Soc. Am. J. 46:305–309.

Kirk, T.K., W.J. Conners, and J.G. Zeikus. 1977. Advances in understanding the microbiological degradation of lignin. Adv. Phytopathol. 11:369–394.

Kowalenko, C.G., and R.B. McKercher. 1971. Phospholipid P content of Saskatchewan soils. Soil Biol. Biochem. 3:243–247.

Kononova, M.M. 1966. Soil organic matter. 2nd ed. Pergamon Press, Oxford.

Ladd, J.N., and J.H.A. Butler. 1966. Comparison of some properties of soil humic acids and synthetic phenolic polymers incorporating amino derivatives. Aust. J. Soil Res. 4:41–54.

Ladd, J.N., and J.H.A. Butler. 1975. Humus-enzyme systems and synthetic organic polymer-enzyme analogs. p. 143–194. In E.A. Paul and A.D. McLaren (ed.) Soil biochemistry. Vol. 4. Marcel Dekker, New York.

Ladd, J.N., J.M. Oades, and M. Amato. 1981. Microbial biomass formed from ^{14}C, ^{15}N-labeled plant material decomposing in soils in the field. Soil Biol. Biochem. 13:119–126.

Linhares, L.F., and J.P. Martin. 1978. Decomposition in soil of humic acid-type polymers (melanins) of Eurotium echinulatum, Aspergillus glaucus sp. and other fungi. Soil Sci. Soc. Am. J. 42:738–743.

Linhares, L.F., and J.P. Martin. 1979a. Carbohydrate content of fungal humic acid-type polymers (melanins). Soil Sci. Soc. Am. J. 43:313–318.

Linhares, L.F., and J.P. Martin. 1979b. Decomposition in soil of emodin, chrysophanic acid, and a mixture of anthraquinones synthesized by an *Aspergillus niger* isolate. Soil Sci. Soc. Am. J. 43:940–945.

Liu, S.-Y., A.J. Freyer, R.D. Minard, and J.-M. Bollag. 1985. Enzyme catalyzed complex-formation of amino acid esters and phenolic humus constituents. Soil Sci. Soc. Am. J. 49:337–342.

Lowe, L.E. 1978. Carbohydrates in soil. p. 65–93. *In* M. Schnitzer and S.U. Khan (ed.) Soil organic carbon, nitrogen, and fertility. Elsevier Scientific Publ. Co., Amsterdam.

Lüdemann, H.D., H. Lentz, and J.P. Martin. 1982. Carbon-13 nuclear magnetic resonance spectra of some fungal melanins and humic acids. Soil Sci. Soc. Am. J. 46:957–962.

Malcolm, R.L. 1985. Geochemistry of stream fulvic and humic substances. p. 181–210. *In* G.R. Aiken et al. (ed.) Humic substances in soil, sediment, and water. John Wiley & Sons, New York.

Martin, F., F.J. González-Vila, and J.P. Martin. 1983. The persulfate oxidation of fungal melanins. Soil Sci. Soc. Am. J. 47:1145–1148.

Martin, F., C. Saiz-Jiminez, and A. Cert. 1977. Pyrolysis gas chromatography of soil humic fractions. I. The low-boiling point compounds. Soil Sci. Soc. Am. J. 44:1114–1118.

Martin, F., C. Saiz-Jiminez, and F.J. González-Vila. 1981. The persulfate oxidation of a soil humic acid. Soil Sci. 132:200–204.

Martin, J.P. 1971. Decomposition and binding action of polysaccharides in soil. Soil Biol. Biochem. 3:33–41.

Martin, J.P., J.O. Ervin, and R.A. Shepherd. 1966. Decomposition of the iron aluminum, zinc and copper salts or complexes of some microbial and plant polysaccharides in soil. Soil Sci. Soc. Am. Proc. 30:196–200.

Martin, J.P., and K. Haider. 1971. Microbial activity in relation to soil humus formation. Soil Sci. 111:54–63.

Martin, J.P., and K. Haider. 1976. Decomposition of specifically ^{14}C-labeled ferulic acid: Free and linked into model humic acid-type polymers. Soil Sci. Soc. Am. J. 40:377–380.

Martin, J.P., and K. Haider. 1977. Decomposition in soil of specifically carbon-14-labelled DHP and cornstalk lignins, model humic acid-type polymers, and coniferyl alcohols. p. 23–32. *In* Soil Org. Matter Stud., Proc. Symp. (Braunschweig, 6–10 Sept. 1976) IAEA-FAO, Vienna.

Martin, J.P., and K. Haider. 1979a. Effect of concentration on decomposition of some ^{14}C-labeled phenolic compounds, benzoic acid, glucose, wheat straw, and *Chlorella* protein in soil. Soil Sci. Soc. Am. J. 43:917–920.

Martin, J.P., and K. Haider. 1979b. Biodegradation of ^{14}C-labeled model and cornstalk lignins, phenols, model phenolase humic polymers, and fungal melanins as influenced by a readily available carbon source and soil. Appl. Environ. Microbiol. 38:283–289.

Martin, J.P., and K. Haider. 1980a. A comparison of the use of phenolase and peroxidase for the synthesis of model humic acid-type polymers. Soil Sci. Soc. Am. J. 44:983–988.

Martin, J.P., and K. Haider. 1980b. Microbial degradation and stabilization of ^{14}C-labeled lignins, phenols, and phenolic polymers in relation to soil humus formation. *In* T.K. Kirk et al. (ed.) Lignin biodegradation: Microbiology, chemistry, and potential applications. Vol. I. CRC Press, Boca Raton, FL.

Martin, J.P., and K. Haider. 1986. Influence of mineral colloids on turnover rates of soil organic carbon. p. 283–304. *In* P.M. Huang and M. Schnitzer (ed.) Interactions of soil minerals with natural organics and microbes. SSSA Spec. Publ. 17. SSSA, Madison, WI.

Martin, J.P., and K. Haider, and E. Bondietti. 1972a. Properties of model humic acid synthesized by phenoloxidase and autoxidation of phenols and other compounds formed by soil fungi. p. 171–186. *In* D. Povoledo and H.L. Golterman (ed.) Humic substances: Their structure and function in the biosphere. Proc. Int. Meet. Humic Substances, Nieuwersluis, The Netherlands. Pudoc, Wageningen.

Martin, J.P., and K. Haider, and G. Kassim. 1980. Biodegradation and stabilization after 2 years of specific crop, lignin, and polysaccharide carbons in soils. Soil Sci. Soc. Am. J. 44:1250–1255.

Martin, J.P., and K. Haider, and C. Saiz-Jiminez. 1974. Sodium amalgam reductive degradation of fungal and model phenolic polymers, soil humic acids, and simple phenolic compounds. Soil Sci. Soc. Am. Proc. 38:760–765.

Martin, J.P., and K. Haider, and D. Wolf. 1972b. Synthesis of phenols and phenolic polymers by *Hendersonula toruloidea* in relation to humic acid formation. Soil Sci. Soc. Am. Proc. 36:311–315.

Martin, J.P., A.A. Parsa, and K. Haider. 1978. Influence of intimate association with humic polymers on biodegradation of [^{14}C] labeled organic substrates in soil. Soil Biol. Biochem. 10:483–486.

Martin, J.P., and D.E. Stott. 1981. Microbial transformations of herbicides in soil. Proc. West. Soc. Weed Sci. 34:39–55.

Martin, J.P., H. Zunino, P. Peirano, M. Caiozzi, and K. Haider. 1982. Decomposition of ^{14}C-labeled lignins, model humic acid polymers, and fungal melanins in allophanic soils. Soil Biol. Biochem. 14:289–293.

Mathur, S.P. 1971. Characterization of soil humus through enzymatic degradation. Soil Sci. 111:147–157.

Mathur, S.P., and M. Schnitzer. 1978. A chemical and spectroscopic characterization of some synthetic analogues of humic acids. Soil Sci. Soc. Am. J. 42:591–596.

Mayaudon, J., M. El Halfawi, and M.A. Chalvignac. 1973. Propietes des diphenol oxydases extraites des sol. Soil Biol. Biochem. 5:369–383.

Maximov, O.B., T.V. Shevts, and Y.N. Elkin. 1977. On permanganate oxidation of humic acids. Geoderma 19:63–78.

Meuzelaar, H.L.C., K. Haider, B.R. Nager, and J.P. Martin. 1977. Comparative studies of pyrolysis-mass spectra of melanins, model phenolic polymers, and humic acids. Geoderma 17:239–252.

Nelson, D.W., J.P. Martin, and J.O. Ervin. 1979. Decomposition of microbial cells and components in soil and their stabilization through complexing with model humic acid-type phenolic polymers. Soil Sci. Soc. Am. J. 43:84–88.

Nyhan, J.W. 1975. Decomposition of ^{14}C-labeled plant materials in a grassland soil under field conditions. Soil Sci. Soc. Am. Proc. 39:643–648.

Oades, J.M. 1967. Carbohydrates in some Australian soils. Aust. J. Soil Res. 5:103–115.

Oades, J.M., and G.H. Wagner. 1971. Biosynthesis of sugars in soils incubated with ^{14}C glucose and ^{14}C dextran. Soil Sci. Soc. Am. J. 35:914–917.

Paul, E.A., C.A. Campbell, D.A. Rennie, and K.D. McCullum. 1964. Investigations of the dynamics of soil humus utilizing carbon dating techniques. Trans. Int. Congr. Soil Sci. 8th, 1964. 3:201–208.

Paul, E.A., and J.A. Van Veen. 1978. The use of tracers to determine the dynamic nature of organic matter. In Trans. Int. Congr. Soil Sci., 11th, 1978. 3:61–02.

Piper, T.J., and A.M. Posner. 1972. Sodium amalgam reduction of humic acid II. Application of the method. Soil Biol. Biochem. 4:525–531.

Reisinger, O., and G. Kilbertus. 1974. Biodegradation et humification IV. Microorganismen intevenant dan la decomposition des cellules d'Aureobasidium pollulans. (De Bary) Arnaud. Can. J. Microbiol. 10:299–306.

Riffaldi, R., and M. Schnitzer. 1972. Electron spin resonance spectrometry of humic substances. Soil Sci. Soc. Am. Proc. 36:301–305.

Saiz-Jiminez, C., K. Haider, and J.P. Martin. 1975. Anthraquinones and phenols as intermediates in the formation of dark-colored, humic acid-like pigments by Eurotium echinulatum. Soil Sci. Soc. Am. J. 39:649–653.

Saiz-Jiminez, C., K. Haider, and H.L.C. Muezelaar. 1979. Comparisons of soil organic matter and its fractions by pyrolysis mass spectrometry. Geoderma 22:25–38.

Saiz-Jiminez, C., F. Martin, and A. Cert. 1979. Low boiling-point compounds produced by pyrolysis of fungal melanins and model phenolic polymers. Soil Biol. Biochem. 11:305–309.

Saiz-Jiminez, and F. Shafizadeh. 1985. Electron spin resonance spectrometry of fungal melanins. Soil Sci. 139:319–325.

Sauerbeck, D., and M.A. Gonzalez. 1977. Field decomposition of carbon-14-labeled plant residues in various soils of the Federal Republic of Germany and Costa Rica. p. 159–169. In Soil Org. Matter Stud., Proc. Symp. (Braunschweig, 6–10 Sept. 1976) Vol. 1. IAEA-FAO, Vienna.

Schnitzer, M. 1977. Recent findings on the characterization of humic substances extracted from soil from widely differing climatic zones. p. 117–132. In Soil Org. Matter Stud., Proc. Symp. (Braunschweig, 6–10 Sept. 1976). Vol. 2. IAEA-FAO, Vienna.

Schnitzer, M. and J.A. Neyroud. 1975. Further investigations on the chemistry of fungal "humic acids". Soil Biol. Biochem. 7:365–371.

Shindo, H., and P.M. Huang. 1985. Catalytic polymerization of hydroquinone by primary minerals. Soil Sci. 139:505–511.

Simonson, R.W. 1959. Outline of generalized theory of soil genesis. Soil Sci. Soc. Am. Proc. 23:152–156.

Skujins, J.J. 1976. Extracellular enzymes in soil. CRC Crit. Rev. Microbiol. 6:383–421.

Sowden, F.J., S.M. Griffith, and M. Schnitzer. 1976. The distribution of nitrogen in some highly organic tropical volcanic soils. Soil Biol. Biochem. 8:55–60.

Sowden, F.J., and K.C. Ivarson. 1962. Decomposition of forest litters: 3. Changes in the carbohydrate constituents. Plant Soil 16:389–400.

Stevenson, F.J. 1957. Investigations of amino polysaccharides in soil. Soil Sci. 83:113–122.

Stevenson, F.J. 1979. Lead-organic matter interactions in a mollisol. Soil Biol. Biochem. 11:493–499.

Stevenson, 1982a. Origin and distribution of nitrogen in soil. p. 1–42. In F.J. Stevenson (ed.) Nitrogen in agricultural soils. Agron. Monogr. 22. ASA, Madison, WI.

Stevenson, F.J., 1982b. Humus chemistry. John Wiley & Sons, New York.

Stevenson, F.J. 1983. Isolation and identification of amino sugars in soil. Soil Sci. Soc. Am. J. 47:61–65.

Stevenson,F.J. 1985. Geochemistry of soil humic substances. p. 13–52. In G.R. Aiken et al. (ed.) Humic substances in soil, sediment, and water. John Wiley & Sons, New York.

Stott, D.E., G. Kassim, W.M. Jarrell, J.P. Martin, and K. Haider. 1983a. Stabilization and incorporation into biomass of specific plant carbons during biodegradation in soil. Plant Soil 70:15–26.

Stott, D.E., J.P. Martin. D.D. Focht, and K. Haider. 1983b. Biodegradation, stabilization in humus, and incorporation into soil biomass of 2,4,-D and chlorocatechol carbons. Soil Sci. Soc. Am. J. 47:66–70.

Suflita, J.M., and J.-M. Bollag. 1981. Polymerization of phenolic compounds by a soil-enzyme complex. Soil Sci. Soc. Am. J. 45:297–302.

Swincer, G.D., J.M. Oades, and D.J. Greenland. 1968. Studies on soil polysaccharides. II. The composition and properties of polysaccharides in soils under pasture and under fallow wheat rotation. Aust. J. Soil Res. 6:226–239.

Tisdall, J.M., and J.M. Oades. 1982. Organic matter and water-stable aggregates in soils. J. Soil Sci. 33:141–163.

Verma, L., and J.P. Martin. 1976. Decomposition of algal cells and components and their stabilization through complexing with model humic acid-type phenolic polymers. Soil Biol. Biochem. 8:85–90.

Verma, L., J.P. Martin, and K. Haider. 1975. Decomposition of carbon-14-labeled proteins, peptides, and amino acids; free and complexed with humic polymers. Soil Sci. Soc. Am. J. 39:279–283.

Wada, K., and S. Aomine. 1975. Soil development during the quarternary. Soil Sci. 116:170–177.

Wagner, G.H. 1975. Microbial growth and carbon turnover. p. 269–305. In E.A. Paul and A.D. McLaren (ed.) Soil biochemistry. Vol. 3. Marcel Dekker, New York.

Wagner, G.H., and U.K. Mutakar. 1968. Amino components of soil organic matter formed during humification of ^{14}C glucose. Soil Sci. Soc. Am. Proc. 32:683–686.

Wagner, G.H., and E.I. Muzorewa. 1977. Lipids of microbial origin in soil organic matter. p. 99–104. In Soil Org. Matter Stud., Proc. Symp. (Braunschweig, 6–10 Sept. 1976) Vol. 2. IAEA-FAO, Vienna.

Waid, J.S. 1960. The growth of fungi in soil. p. 54–75. In D. Parkinson and J.S. Waid (ed.) The ecology of soil fungi. Liverpool University Press. Liverpool, UK.

Wang, T.S.C., J.-H. Chen, and W.-M. Hsiang. 1985. Catalytic synthesis of humic acids containing various amino acids and dipeptides. Soil Sci. 140:3–10.

Zunino, H., F. Borie, S. Aguilera, J.P. Martin, and K. Haider. 1982. Decomposition of ^{14}C-labeled glucose, plant and microbial products and phenols in volcanic ash-derived soils of Chile. Soil Biol. Biochem. 14:37–43.

Chapter 4

Selected Methods for the Characterization of Soil Humic Substances

M. SCHNITZER, *Land Resource Research Center, Agriculture Canada, Ottawa, Ontario, Canada*

ABSTRACT

Four relatively new instrumental methods that are currently being used to study humic stances are (i) solution- and solid-state ^{13}C nuclear magnetic resonance (NMR) spectroscopy; (ii) electron spin resonance (ESR) spectroscopy; (iii) pyrolysis–mass spectrometry (Py-MS); and (iv) supercritical gas extraction (SCGE). ^{13}C-NMR spectra provide us with an inventory of the different types of C (paraffinic, aliphatic and aromatic C, and C in CO_2H, ketonic, and quinonoid groups) in soils and humic materials. ESR spectroscopy tells us about the symmetry and coordination of paramagnetic metals in metal–humic acid and mutal–fulvic acid complexes. Pyrolysis–soft ionization–mass spectrometry is a promising technique for identifying major humic components. Supercritical gas extraction is an important method for the relatively mild and efficient extraction of major humic components. These methods can be used separately or in combination and will, during the next decade, generate detailed and specific information on humic substances. This information will allow us to better understand and use these materials in agriculture. Examples illustrating the application of each of these methods are presented in this chapter.

Analytical methods constitute the backbone of chemistry, and without the availability of adequate and reliable analytical methodology, new developments will be curtailed. It is due in part to the lack of suitable methods that soil organic chemistry has lagged behind other specializations in soil science. The recent commercial availability of novel and sophisticated instruments, however, is drastically changing soil organic chemistry and making it possible to significantly advance knowledge in this field. The purpose of this chapter is to describe some of these exciting developments so that readers will become aware of their existence and make use of them.

Applications of the following methods to the analysis of soil humic substances are discussed in this chapter: (i) *Solution-* and *solid-state* ^{13}C nuclear magnetic resonance (NMR) spectroscopy; (ii) electron spin resonance (ESR) spectroscopy; (iii) pyrolysis–mass spectrometry (Py-MS); and (iv) supercritical gas extraction (SCGE).

4–1 ^{13}C NUCLEAR MAGNETIC RESONANCE (NMR) SPECTROSCOPY

Of the methods developed in recent years, ^{13}C-NMR spectroscopy is probably the most useful for the characterization of soil organic matter and its components. A recent monograph by Wilson (1987) presents a detailed discussion of the theory of ^{13}C-NMR and describes applications to humic substances. Figure 4–1, curve (a) shows the *solution-state* ^{13}C-NMR spectrum of a humic acid extracted from the Ah horizon of a Mollisol from central Alberta (Preston & Schnitzer, 1984). The spectrum provides an inventory of the different components of the humic acid. The presence of unsubstituted aliphatic C (i.e. C in straight-chain, branched and cyclic alkanes, alkanoic acids, and other aliphatic components) is indicated by signals in the 0- to 50-ppm region of the spectrum. Carbon in proteinaceous materials (amino acids, peptides, and proteins) shows resonances between 40 and 60 ppm, and C in carbohydrates gives signals between 61 to 105 ppm. Signals between 106 and 165 are due to aromatic C, while those near 155 ppm arise from phenolic C. The strong signal between 170 and 180 ppm comes from C in carboxyl groups, with possibly some overlapping from phenolic, amide, and ester carbons. Thus, the ^{13}C-NMR spectrum indicates the presence in the humic acid of a wide variety of components and structures whose determinations by other methods would either be laborious and time-consuming or not possible at all. In this sense, the method offers unique possibilities.

Curve (b) in Fig. 4–1 shows the *solution-state* ^{13}C-NMR spectrum of the residue from the same humic acid after hydrolysis for 24 h with hot 6M HCl. Most of the resonances in the 40- to 105-ppm region (arising from proteinaceous materials and carbohydrates) are no longer observed because of the hydrolytic removal of these materials by the hot acid. Also, the intensity of the signal between 170 and 180 ppm, due largely to C in carboxyl groups, has been reduced because of partial decarboxylation of the material by the strong acid. On the other hand, intensities of two other major humic acid components, that is, unsubstituted aliphatic C (0 to 50 ppm) and aromatic C (106 to 165 ppm) remain undiminished and can thus be examined in greater detail.

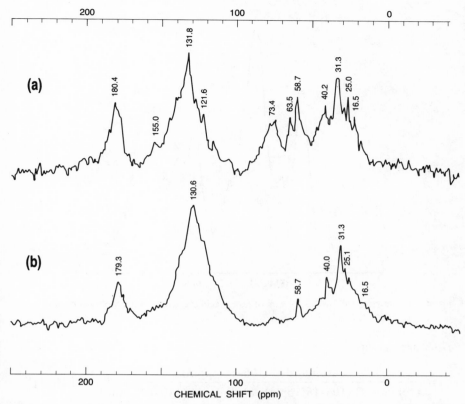

Fig. 4-1. Solution-state ^{13}C-NMR spectra of: (a) humic acid extracted from a Mollisol Ah horizon; (b) same humic acid after hydrolysis with hot, $6M$ HCl.

The *solid-state* spectrum of the same humic acid is shown in Fig. 4–2. The bands in this spectrum are broader and the spectrum contains fewer details than the *solution-state* spectrum of the same humic acid. However, the major types of C in the humic acid are, nonetheless, well separated. Phenolic C shows a distinct signal at 152 ppm and C in carbonyl (C=O) groups produces a resonance near 230 ppm. One of the advantages of *solid-state* ^{13}C-NMR spectroscopy is that whole untreated soils can be analyzed directly; it is not necessary to extract the organic matter and to dissolve the sample. The application of ^{13}C-NMR spectroscopy to whole soil is discussed by Wilson in chapter 10 of this book. A second advantage is that *solid-state* ^{13}C-NMR spectroscopy of humic substances is usually less time-consuming than *solution-state* ^{13}C-NMR of these materials.

One of the important characteristics which can be derived from ^{13}C-NMR spectra of humic materials and soil is the degree of aromaticity. The latter can be calculated in the following manner:

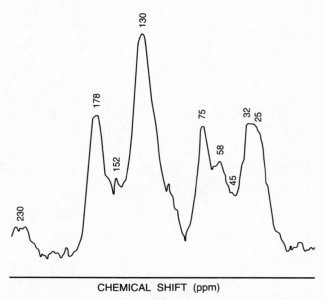

CHEMICAL SHIFT (ppm)

Fig. 4–2. Solid-state ^{13}C-NMR spectrum of humic acid extracted from a Mollison Ah horizon.

f_a % aromaticity

$$= \frac{\text{aromatic C (106–165 ppm)}}{\text{aromatic C (106–165 ppm)} + \text{aliphatic C (0–105 ppm)}} \times 100. \quad [1]$$

Carbon in carboxyl groups is omitted in these calculations because it is not known how many carboxyl groups are bonded to aromatic and how many to aliphatic structures. *Solid-state* ^{13}C-NMR spectra of humic acids extracted from different soils are presented in Fig. 4–3 (Hatcher et al., 1981). Aromaticities of the humic acids in this figure range from 49 to 54%.

Valuable information on the structural features of humic materials can be obtained by combining ^{13}C-NMR spectroscopy with chemical methods. In this manner, effects on the chemical structure of humic substances of different extractants, methylation, hydrolysis, oxidation, and reduction can be evaluated. The type of wide-ranging information that can be obtained from ^{13}C-NMR spectra of humic materials is summarized in Table 4–1, showing that the scope of this method is indeed very impressive.

In Table 4–2, measurements by *solution-state* and *solid-state* ^{13}C-NMR spectroscopy of the principal types of C in 18 humic acids extracted from Natriborolls in Saskatchewan are compared. Aromatic C and degree of aromaticity were computed as indicated in Table 4–1. The data in Table 4–2 show fairly good agreement between the two methods.

As applications of ^{13}C-NMR spectroscopy to humic materials and soil continue to increase rapidly, two major problems have come to the surface. The first concerns the quantitative determination of carbohydrates. Several

Table 4–1. Interpretations of chemical shifts in ^{13}C-NMR spectra of humic substances.

Aliphatic C (alkanes + fatty acids)	0–39 ppm
Protein C, peptide C, amino acid C, methoxyl C	40–60 ppm
Carbohydrate C	61–105 ppm
Aromatic C	106–165 ppm
Phenolic C	150–165 ppm
Carboxylic C	170–190 ppm
Carbonyl C (ketonic)	210–230 ppm
Aliphaticity (%)	$\dfrac{(0–105)\text{ ppm}}{(0–165)\text{ ppm}} \times 100$
Aromaticity (%)	$\dfrac{(106–165)\text{ ppm}}{(0–165)\text{ ppm}} \times 100$

Table 4–2. Comparison of the principal types of carbon in humic materials as determined by *solution-state* and *solid-state* ^{13}C-NMR spectroscopy.

	Weight/weight % by	
Type of carbon	Solution-state ^{13}C-NMR	Solid-state ^{13}C-NMR
Aliphatic C	39.5 ± 6.4	40.9 ± 3.8
Aromatic C (excluding phenolic)	43.3 ± 3.8	35.9 ± 2.6
Carboxyl C	17.2 ± 2.4	23.2 ± 4.4
Aromaticity (including phenolic)	52.5 ± 6.5	46.0 ± 3.0

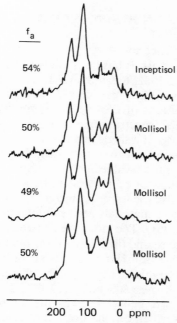

Fig. 4–3. Solid-state ^{13}C-NMR spectra of humic acid extracted from different soils. Aromaticities (f_a), calculated as explained in the text, are shown in the figure (Hatcher et al., 1981).

workers (Sweet & Perdue, 1982; Oades et al., 1987; Schnitzer & Preston, 1987) have noted that chemical monosaccharide analyses account for only a portion of the intensity in the 61- to 105-ppm region normally attributed to carbohydrates. This observation means that ^{13}C-NMR spectroscopy cannot be used for the quantitative analysis of carbohydrates in humic subtances and soils. It is possible that altered and degraded carbohydrates produce signals in this region, but these materials are no longer monosaccharides. Another problem has arisen with phenolic carbons whose signals are often barely noticeable in ^{13}C-NMR spectra of soil humic acids (Hatcher et al., 1981). To throw light on this situation, Preston and Schnitzer (1987) studied the effect of pH on phenolic and carboxyl regions by determining chemical shifts of known phenols, hydroxybenzoic acids, and hydroxybenzene dicarboxylic acids in base and in dimethylsulfoxide solution. They observed that shifts for both carboxyl and phenolic carbons covered wide ranges, and in basic solution, some phenolic carbons occurred at shifts > 170 ppm, a region normally attributed to carboxyl carbons. The range of shifts for phenolic carbons in base solution was 142.6 to 173.3 ppm, and in dimethylsulfoxide solution the range was 132.5 to 164.6 ppm. For carboxyl carbons, the observed range extended from 166.5 to 172.6 ppm in dimethylsulfoxide, and from 176.4 to 180.5 ppm in base solution. Thus, in *solution-state* ^{13}C-NMR spectra of humic acids, signals due to phenolic C may overlap with those of carboxyl C so that quantitative determinations of these functional groups by ^{13}C-NMR need to be checked by other methods.

4–2 ELECTRON SPIN RESONANCE (ESR) SPECTROSCOPY

Electron spin resonance spectroscopy measures free radicals (unpaired electrons) in humic materials. It has been known since the early 1960s (Rex, 1960; Steelink & Tollen, 1962; Steelink, 1964) that humic materials possess free radicals, which most likely participate in a wide variety of reactions. The theory and applications of ESR spectroscopy have been described by Atherton (1973). The ESR spectrum of a typical humic acid is shown in Fig. 4–4. The spectrum consists of a single line devoid of hyperfine splitting. From the spectrum (by comparison with a standard), the number of free radicals per unit weight as well as the g-value and line-width can be calculated. From the magnitude of g-values for humic materials, with most samples ranging between 2.0038 and 2.0042 (Senesi & Schnitzer, 1977), it appears that prominent free radicals in these materials are semiquinones or substituted semiquinones (Blois et al., 1961). The latter are of two types: (i) permanent free radicals with long lifetimes, and (ii) transient free radicals with relatively short lives (several hours). The transient free radicals in humic materials can be generated in relatively large concentrations by chemical reduction, irradiation, or increase of pH (Senesi & Schnitzer, 1977). The permanent free radicals, by contrast, appear to be stabilized by the complex chemical structure of the humic materials which can act both as electron donors and electron acceptors. These oxidation-reduction reactions are reversible and can be assumed to proceed in the following manner:

[2]

[3]

The semiquinones (shown in the center) can be produced either by the reduction of quinones or by the oxidation of phenols. Under alkaline conditions semiquinone anions are formed. It is of special interest that humic substances can participate in oxidation-reduction reactions with transition metal ions and biological systems via free radical intermediates.

So far, however, ESR spectroscopy has contributed little to our understanding of the structural organic chemistry of humic materials. The main reason for this limitation is that it has been difficult to split the signal. To the best of the author's knowledge there are only a few reports in the literature that desribe splitting of the ESR signal of humic materials. Figure 4–5, curve (a), shows the ESR spectrum of a 1.0% (w/w) fulvic acid solution adjusted to pH 7 (Senesi et al., 1977a). This spectrum consists of a single asymmetrical line with a g-value of 2.0039 and a line width of 3.0 G, and represents a free radical concentration of 2.99×10^{17} spins g^{-1}. Ten minutes after the addition of H_2O_2 solution, hyperfine structure in the form of a triplet [Fig.

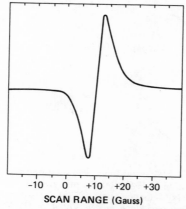

-10 0 +10 +20 +30
SCAN RANGE (Gauss)

Fig. 4–4. ESR spectrum of a typical humic acid (1 gauss = 10^{-4} tesla).

4–5, curve (b)] appears. The best resolution of the triplet is achieved between 10 and 120 min [Fig. 4–5, curve (c)]. The total line width of the split signal is 4.2 G while the g-value is 2.0041. The three lines of the triplet indicate interaction of the unpaired electron with two equivalent protons. The ESR parameters are similar to those of 2,6-dimethoxy-*p*-benzosemiquinone (Fitzpatrick & Steelink, 1969). After 2 h of contact with the H_2O_2, a slow decay of the signal is observed. After 22 h, the signal returns to that of the initial fulvic acid [Fig. 4–5, curve (d)]. When additional H_2O_2 is added a well-resolved triplet reappears [Fig. 4–5, curve (b)]. Thus, only after mild oxidation does ESR spectroscopy provide information on the nature of the free radical in the fulvic acid.

In a recent study, Cheshire et al. (1985) observed hyperfine splitting in ESR solution spectra of humic acids extracted from several horizons of a Spodosol. The upper horizons, relatively rich in lignin, contained humic acids with some spectral characteristics similar to those of lignin. Hyperfine splitting depended on the concentration of alkali (was sharper in $4M$ KOH than in $0.2M$ KOH) and the period of contact. The authors were unable to identify the free radicals in the humic acids and concluded that the radical species in the humic acids examined constituted a more complex mixture than had been implied in earlier reports.

Other applications of ESR spectroscopy are concerned with the characterization of metal–humic acid or metal–fulvic acid complexes (Schnitzer, 1978). Metal–organic binding is illustrated by the spectra in Fig. 4–6. Figures 4–6, curve (a), and 4–6, curve (b) (Senesi et al., 1977b) show ESR spectra of a naturally-occurring iron–fulvic acid complex extracted from an Inceptisol in the West Indies and after additional purification that lowered the iron content from 2.8 to 0.3% (w/w). As a result of purification, the intensity of resonance A (at g = 4.1) is significantly decreased but those of the remaining resonances remain more or less unchanged. However, an addi-

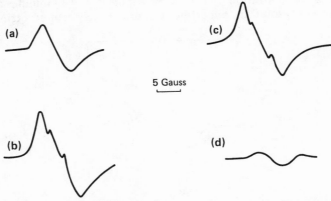

Fig. 4–5. ESR spectrum of: (a) 1% (w/v) fulvic acid solution adjusted to pH 7; (b) same solution 10 min after the addition of H_2O_2; (c) 120 min after the addition of the H_2O_2; (d) 22 h after the addition of the H_2O_2 (Senesi et al., 1977a). 1 gauss = 10^{-4} tesla. Reprinted with permission from *Soil Biology and Biochemistry*, Vol. 9, p. 371–372. Copyright © 1977. Pergamon Press, PLC, Oxford.

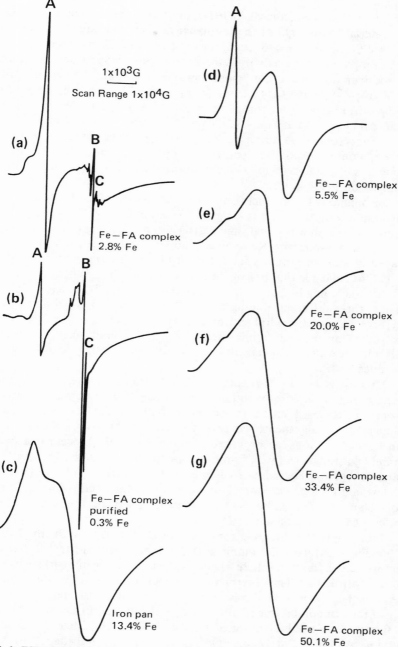

Fig. 4–6. ESR spectra of: (a) and (b) naturally-occurring iron–fulvic acid complexes extracted from an Inceptisol; (c) organic iron-pan extracted from a Spodosol Bh horizon; (d), (e), (f), and (g) laboratory-prepared iron–fulvic acid complexes containing increasing amounts of iron (Senesi et al., 1977b). Reprinted with permission from *Geochimica et Cosmochimica Acta*, Vol. 41, p. 973. Copyright © 1977. Pergamon Press, Inc., Elmsford, NY.

tional, very sharp resonance (labelled C) appears in the spectrum of the purified material to the right of the resonance arising from organic C (labelled B; $g = 2.0039$). Figure 4-6, curve (c) presents the ESR spectrum of an organic iron-pan taken directly from the Bh horizon of a Spodosol. In the iron-pan spectrum the two major resonances (labelled A and C) partially overlap. Figure 4-6, curve (d) to curve (g) present ESR spectra of laboratory-prepared iron–fulvic acid complexes containing increasing amounts of the metal. An inspection of Fig. 4-6, curve (a) to curve (d) shows a stronger intensity for resonance A and a weaker signal for resonance C in the natural material compared to the synthetic complex. As more iron is added, resonances A and C overlap, which means that the two signals are swamped by high iron loading.

Resonance A is a distinct feature of ESR spectra of naturally-occurring iron–humic complexes and is assigned (Senesi et al., 1977b) to an Fe(III)–organic complex in which high-spin Fe(III) ions occupy sites of approximately orthorhombic symmetry. Resonance C appears to be due to Fe(III) ions in octahedral sites with small axial distortions from cubic symmetry. Senesi et al. (1977b) conclude that in iron–humic complexes, iron is bound to humic material (including humic acid, fulvic acid and humin) by at least two, and possibly three different binding mechanisms: (i) Fe(III) is strongly bound and protected by tetrahedral and/or octahedral coordination, and (ii) Fe(III) is adsorbed on external humic surfaces, weakly bound octahedrally. The Fe(III) in the latter form can easily be complexed by other ligands and can be reduced.

Lakatos et al. (1977) reports that all 3d-transition metals form inner sphere chelate complexes with humic acids. Mn(II) ions are coordinated to raw peat or to peat humic acids octahedrally, whereas Cu(II) ions occur in square planar arrangements with two carboxylate and two aliphatic N ligands. Vanadyl (VO^{2+}) ions, according to Lakatos et al. (1977), occur in a square pyramidal arrangement with four oxygen-containing ligands. These authors also note that in acidic solutions diamagnetic Mn(VII), Cr(VI), Mo(VI), and V(V) oxoanions are *reduced* by humic acid to paramagnetic Mn(II), Cr(III), Mo(V), and V(IV) ions but that diamagnetic Cu(I), on the other hand, is *oxidized* to paramagnetic Cu(II).

McBride (1978) disagrees with Lakatos et al. (1977) on the mechanism of bonding of Mn(II) with humic acid. He concludes that $Mn(II)(H_2O)_6$ retains its inner hydration sphere when adsorbed at the organic surface by electrostatic attraction. This interpretation means that an outer sphere rather than an inner sphere complex is formed. The apparent difference in the nature of the Mn bond is due to differences in the hydration states of the samples; one sample had been air-dried whereas the other consisted of an aqueous solution. As far as Cu(II) and VO^{2+} are concerned, McBride (1978) states that although chelation cannot be ruled out as a mechanism in the complexation of these metal ions by humic materials, there is no direct evidence from ESR spectra that more than one single bond is formed between the surface and the metal ion. In a later study on vanadyl ion–fulvic acid interactions,

McBride (1980) suggests that VO^{2+} may be bound to two or more ligands from different molecules.

According to Cheshire et al. (1977), copper in soil humic acid is present partly as a copper–porphyrin type complex but in fulvic acid it occurs in some other complexed form. VO^{2+} occurs in complexed forms in fulvic acid that are more covalent than VO^{2+}-humic acid complexes. In agreement with McBride (1978), Cheshire et al. (1977) conclude that Mn(II) complexes of humic and fulvic acids are highly ionic.

Boyd et al. (1981) studied the mechanism of Cu(II) binding by sewage sludge humic acid (see chapter 9 in this book, Boyd & Sommers). They conclude that Cu(II) forms two equatorial bonds with oxygen donor atoms originating from functional groups of the humic acid. The ESR data also indicate that the two Cu(II)–humic acid oxygen bonds occupy *cis*-positions in the square plane of Cu(II), a result that is consistent with the formation of a Cu chelate.

Thus, a significant amount of novel information on the symmetry and coordination of metals in metal–humic complexes has been obtained with the aid of ESR spectroscopy. By contrast, the method has been used with less success for generating new knowledge on the chemical structure of humic substances.

4-3 PYROLYSIS–MASS SPECTROMETRY (Py-MS)

Pyrolysis techniques in combination with gas chromatography (Py-GC), mass spectrometry (Py-MS) or gas chromatography–mass spectrometry (Py-GC-MS) have been used for the characterization of synthetic polymers, biopolymers, polysaccharides, whole bacteria, and soil organic matter (Meuzelaar et al., 1977). The development of curie-point pyrolysis by Bühler and Simon (1970) had a profound effect on pyrolysis techniques. By utilizing heating rates in excess of $1\,°C\ ms^{-1}$, using microgram samples, and allowing rapid escape of the pyrolysis products from the reaction zone, secondary reactions can be minimized. Under such conditions, pyrolysis products may provide important clues to the chemical structure of the original materials. An important contribution to Py-MS was made by Schulten et al. (1973) who pyrolyzed samples directly in the vacuum of a mass spectrometer and used soft ionization such as field ionization (FI) and field desorption (FD) mass spectrometry for identification purposes.

In a recent review, Schulten (1987) states that the rapid, reproducible, chemical characterization of humic substances by pyrolysis–soft ionization mass spectrometry has the following advantages over pyrolysis–electron ionization mass spectrometry; (i) molecular cleavage during the fast degradation in the ion source of the mass spectrometer results in the formation of high-mass chemical subunits; (ii) short reaction times and small sample weights favor the generation of larger, thermal fragments (possibly chemical building blocks) that can be identified and related to the chemical structure of the starting material; and (iii) the primary thermal fragmentation of the pyrolyzates can be monitored, with the ability to significantly avoid consecutive mass spectrometric fragmentation. Schulten (1987), who has pioneered this approach, recommends heating rates of 0.2 to 10 °C s^{-1} and a temperature range of 50 to 800 °C. Soft ionization techniques such as field ionization, field desorption, chemical ionization, and fast atom bombardment are employed in the positive and negative modes. For the detection and identification of pyrolysis products a time/temperature controlled mass spectrometer is used.

Figure 4–7 shows the Py-FI mass spectrum of a humic acid extracted from the surface horizon of a Udic Boroll in West Germany (Schulten, 1987). The spectrum displays high relative abundances of signals up to m/z values of 117; only weaker signals are detected in the higher mass range (up to m/z = 386). Signals at m/z values of 60, 82, 96, 110, and 126 arise from polysaccharides, whereas the signal at m/z 92 is due to toluene. Nitrogen-containing ions are prominent at m/z values of 79 (pyridine), 92 (methyl pyridine) and 117 (indole). The presence of lignin in the humic acid is indicated by signals at m/z values of 212 (lignin monomer), 302 (lignin dimer) and 344 (lignin dimer). An m/z value of 310 could be due to methoxyphenol-alkylphenol and a value of 386 could result from alkylphenol-alkylphenol-alkyphenol. The tentative identifications of the mass numbers are based on

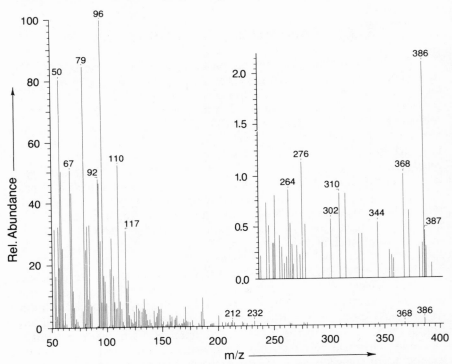

Fig. 4–7. Py-FI mass spectrum of a humic acid extracted from the surface horizon of a Udic Boroll (Schulten, 1987). Reprinted with permission from *Journal of Analytical and Applied Pyrolysis*, Vol. 12, p. 157. Copyright © 1987. Elsevier Science Publishers, Amsterdam.

the work of Schulten et al. (1987). An inspection of Fig. 4–7 indicates that carbohydrate and N-containing components are considerably more volatile than those derived from lignin and plant phenolics. From the data in Fig. 4–7 it appears that a significant portion of the total N in this humic acid is probably heterocyclic.

An impressive application of Py-FIMS was reported by Hempfling et al. (1987) with the collaboration of Schulten. These workers were interested in following litter decomposition and humification in an acidic forest soil (an Eutric Haplorthod) in West Germany. The Py-FIMS spectrum of an L horizon sample under beech (*Fagus sylvatica*) is shown in Fig. 4–8. Thermal degradation products derived mainly from monosaccharides and polysaccharides produce signals at m/z values of 72, 74, 84, 85, 96, 98, 110, 112, 114, 126, 144, and 162, with m/z values of 126, 144, and 162 coming from aldohexose subunits, and m/z values of 114 and 132 originating from aldopentose subunits. Furthermore, coniferyl alcohol and some of its derivatives also give signals at m/z values of 124, 152, 164, 180, and 182, as do fragments of syringyl units with propenoid side chains at m/z values of 168 and 194, and sinapyl alcohol at $m/z = 210$. Dimeric guaiacyl and syringyl fragments of lignin can be seen at m/z values of 260, 272, 284, 298, 312, 326, 342, 358, 372, and 376 and at m/z values between 302 and 386. Dimers of sinapyl alcohol appear at m/z values of 402 and 418. Hempfling et al. (1987) assign signals at m/z values of 136, 166, 178, 192, 208, and 212 to lignin monomers or their fragments, and assign signals at m/z values of 300, 314, 316, 328, 330, 332, 340, 356, 360, 362, 374, 386, 388, and 416 to lignin dimers. The presence of n-fatty acids is indicated by molecular ions at m/z values of 256, 396, and 424; tetrahydroxyflavones, such as kaempferol, occur at $m/z = 286$, sterols give m/z values of 412 and 414, and α-tocopherol gives an m/z value of 430. Lignin trimers are probably also present but so far have not been identified. The homologous series at m/z values of 592, 620, 648, and 676 may be due to wax esters of increasing chain length.

Other applications of Py-MS are concerned with (i) soil organic matter variations as affected by drainage and mull humification (Bracewell & Robertson, 1984), (ii) effects of drainage and illuviation in B horizons (Bracewell & Robertson, 1987a), (iii) tranformations of organic matter in surface organic horizons (Bracewell & Robertson, 1987b), (iv) humification in peat and peaty soils (Bracewell et al., 1980), (v) litter decomposition and humification in forest soils (Hempfling et al., 1987), (vi) decomposition in forest humus layers (Kögel et al., 1987), and (vii) horizon differentiation in organic soils (Schulten, 1987).

4–4 SUPERCRITICAL GAS EXTRACTION (SCGE)

Supercritical gas extraction has been used for the efficient and mild extraction of specific components of coal (Bartle et al., 1975), cellulose (Baumeister & Edel, 1980), foods and tobacco (Hubert & Vitzthum, 1978) and soils (Spiteller, 1982, 1985; Spiteller & Ashauer, 1982). Gases in the super-

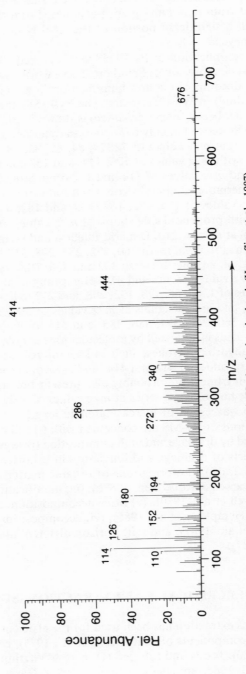

Fig. 4-8. Py-FI mass spectrum of the L horizon of an Eutic Haplorthod under beech (Hempfling et al., 1987).

critical state have physical characteristics which are intermediate between those of the corresponding gases and liquids. The extraction efficiency of supercritical gases arises from the following favorable properties (Bronstrup, 1982): (i) high densities, which are close to those of the corresponding liquids; (ii) higher diffusion constants than the corresponding liquids; and (iii) lower viscosities than the corresponding liquids. It is these properties that give supercritical gases an exceptional ability for penetrating complex matrices to solubilize specific components that are otherwise either insoluble or only slightly soluble in the solvents used.

The method which is described here is the SCGE of soils and humic materials with n-pentane under the following experimental conditions: temperature, 250 °C; pressure, 11.0 MPa; duration, 2 h. A diagram of the custom-built SCG extractor is shown in Fig. 4–9, and a flowsheet of the procedure used for separating the extracts is shown in Fig. 4–10. Supercritical gas extraction was performed on three soils [Bainsville (Aquoll), Armadale (Aquod), and Indian Head (Boroll)] and two humic materials (Bainsville humic acid and Armadale fulvic acid) (Schnitzer et al., 1986).

Yields of n-pentane SCGE extracts from the three soils range from 1.2 to 2.0 g kg^{-1} (Table 4–3), but yields of extracts from humic materials are (per unit weight) about 10 times higher, varying from 15.8 to 19.8 g kg^{-1}. These results indicate that the extracted substances accumulate in the humic materials.

Weight percentages of the initially extracted organic matter range from 1.8 to 3.8 and are higher for the three soils than for the two humic materials. The efficiency of SCGE can be best evaluated by comparing the data in the last vertical column in Table 4–3 with data in the literature determined by conventional extraction with organic solvents. According to Stevenson (1982), ether extracts from most surface soils 1.0% (w/w) or less of the initial organic matter. Similarly, Ogner and Schnitzer (1970) report that the exhaustive, successive extraction of the fulvic acid listed in Table 4–3 with n-hexane, benzene, and ethylacetate removes a total of 0.5% (w/w) of the

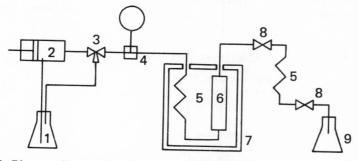

Fig. 4–9. Diagram of supercritical gas extractor. (1) 500-mL Erlenmeyer flask containing solvent; (2) Milton-Roy Mini-Pump, flow rate 60 mL/h; maximum pressure; 36.0 MPa; (3) 3-way valve; (4) pressure gauge, 43.0 MPa; (5) stainless steel tube (1.6 by 0.7 mm); (6) HPLC tube, stainless steel, 50 mL; (7) gas chromatographic oven with temperature control, 25–500 °C; (8) two-way valve; (9) 250-mL Erlenmeyer flask to collect effluent solution (Schnitzer et al., 1986).

original organic matter. Under somewhat milder conditions, Spiteller (1985) applied SCGE with CO_2 to a number of soil samples. The extracts were whitish, consisted mainly of fatty materials, and accounted for 1 to 2% (w/w) of the initial organic matter. Thus, compared with these data, SCGE with *n*-pentane is relatively efficient.

Each extract was separated by analytical gas chromatography and each peak identified by mass spectrometry. The data in Table 4–4 show that most of the extracted alkanes are *n*-alkanes. Only the Armadale soil contains a relatively high proportion of cyclic alkanes. Yields of alkanoic acids (mg kg^{-1}) identified in the SCGE extracts are summarized in Table 4–5. The fol-

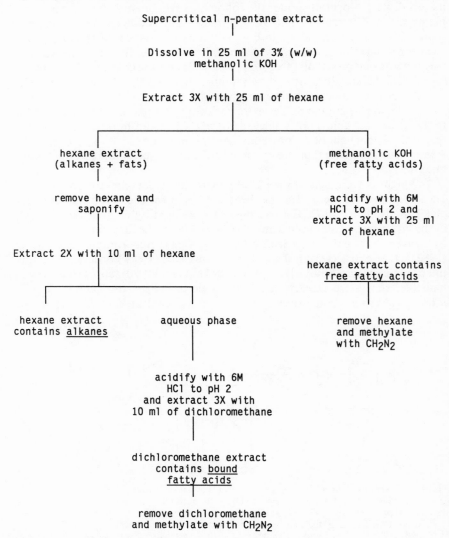

Fig. 4–10. Scheme for the separation of supercritical gas extracts.

Table 4-3. Supercritical gas extraction with n-pentane of soils and humic materials (Schnitzer et al., 1986).

Material	Weight of starting material		Yield of extract	
			Fraction of starting material	Organic matter in starting material
	g	mg	g kg^{-1}	%
Bainsville soil	40.0	78.8	2.0	3.0
Armadale soil	40.0	50.2	1.3	3.8
Indian Head soil	30.0	36.4	1.2	2.5
Bainsville humic acid	1.0	19.8	19.8	2.3
Armadale fulvic acid	2.0	31.5	15.8	1.8

Table 4-4. Yields of alkanes extracted by supercritical gas extraction with n-pentane (Schnitzer et al., 1986).

Material	Type of alkane		Total alkanes	Organic matter in starting material
	Normal	Cyclic		
		mg kg^{-1}		%
Bainsville soil	8.0	0.6	8.6	0.01
Armadale soil	7.7	1.2	8.9	0.03
Indian Head soil	4.1	0.2	4.3	0.01
Bainsville humic acid	198.0	3.4	201.4	0.02
Armadale fulvic acid	32.5	0.0	32.5	0.004

lowing types of alkanoic acids were identified: n-fatty acids, unsaturated fatty acids, branched fatty acids, hydroxy-fatty acids, and α, ω-alkyldicarboxylic acids. Saturated n-fatty acids constitute between 70 to 75% (w/w) of the alkanoic acids identified in the n-pentane SCGE extracts. Unsaturated fatty acids account for <3% (w/w). On the other hand, branched fatty acids make up 12 to 27% (w/w) of the alkanoic acids identified. Concentrations of hydroxyacids and dicarboxylic acids are low.

The different types of compounds which were identified in the SCG extracts are summarized in Fig. 4-11. The data in Fig. 4-11 show the presence in the extract of alkanes and alkanoic acids with up to 32 CH_2 groups, where n stands for the number of CH_2 groups in these compounds.

Figure 4-12 provides information on the distribution of the hydrocarbons identified. The n-C_{26} and n-C_{24} alkanes are the most abundant straight-chain hydrocarbons, while the even-to-odd carbon preference index (EOPI) of all n-alkanes is 1.35. These data indicate a microbiological origin (Jones, 1969) for the n-alkanes in the soils and humic materials. As far as n-fatty acids are concerned, the distribution plots for free n-fatty acids [Fig. 4-13, curve (a)] exhibit pronounced peaks of C_{16} and C_{24} while bound n-fatty acids [Fig. 4-13, curve (b)] show a dominant peak at C_{16}. Since the overall EOPI for all n-fatty acids isolated is 4.15, the origin of the n-fatty acids in the soils and humic substances also appears to be predominantly microbial.

Table 4–5. Yields of alkanoic acids extracted by supercritical gas extraction with n-pentane (Schnitzer et al., 1986).

Material	n-fatty acids Free	n-fatty acids Bound	Unsaturated fatty acids	Branched fatty acids	Hydroxy-fatty acids	$HO_2C-(CH_2)_n-CO_2H$	Total	Organic matter in starting material
			mg kg^{-1}					%
Bainsville soil	114.9	34.0	13.3	39.5	1.8	5.4	208.9	0.3
Armadale soil	54.0	32.1	3.6	14.6	4.6	15.4	124.3	0.4
Indian Head soil	51.9	11.9	1.0	18.9	0	0.8	84.5	0.2
Bainsville humic acid	699.5	417.0	4.2	421.0	61.5	3.4	1756.6	0.2
Armadale fulvic acid	385.7	172.0	2.0	133.9	32.0	14.5	740.1	<0.1

Type of compound	Range of n
$CH_3-(CH_2)_n-CH_3$	12-32
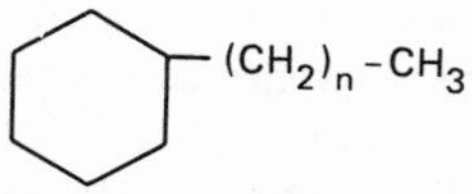 (cyclohexyl)$-(CH_2)_n-CH_3$	19-31
$CH_3-(CH_2)_n-CO_2H$	7-29
(isoalkyl)$()_n^-CO_2H$	12-19
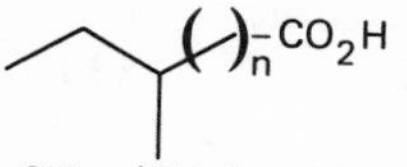 (anteiso)$()_n^-CO_2H$	13-18
$CH_2-(CH_2)_n-CO_2H$ $|$ OH	12-16
$C_{\overline{18}}$	18
$HO_2C-(CH_2)_n-CO_2H$	15-25

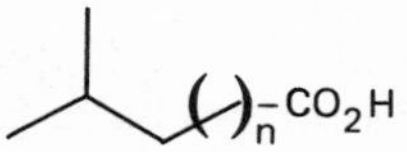

Fig. 4-11. Types of compounds identified in *n*-pentane extracts of three soils and two humic materials.

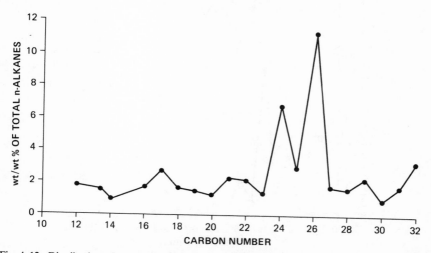

Fig. 4-12. Distribution of *n*-alkanes in the supercritical gas extract of the Armadale soil (Schnitzer et al., 1986).

Effects of solvents of increasing polarity on the SCGE of a soil were recently investigated by Schnitzer and Preston (1987). The solvents used included ethanol and mixtures of ethanol with water. The extracts were analyzed by chemical and ^{13}C-NMR methods. Proportions of alkanes, alkanoic acids, and carbohydrates in the extracts decrease as the polarity of the solvent increases. Conversely, proportions of aromatics increase with increasing polarity of the solvent. ^{13}C-NMR spectra of the SCG extracts are exceptionally well-defined with sharp and distinct peaks revealing fine chemical structure (Fig. 4–14). Major aliphatic signals are present in all spectra at 14.0, 17.5, 19.6, 23.4, 24.9, 31.8, 34.0, 37.3, and 42.5 ppm. These carbons occur in CH_3, CH_2, and CH groups of straight-chain, branched, and cyclic alkanes and alkanoic acids. The resonances at 56.2 ppm are most likely due to C in OCH_3 and those at 58.5 and 60.0 ppm to C in amino acids, peptides and/or proteins. Parts of the strong resonances at 68.7, 69.5, 72.2, and 73.6 ppm may be ascribed to C in carbohydrates. Well-defined aromatic signals can be seen in all spectra at 118.7, 125.1, 129.3, 130.7, 134.6, 135.5, and 140.8 ppm. Resonances due to phenolic C appear at 151.0 and 168.7 ppm, and those due to C in CO_2H groups at 180.3, 182.0, and 182.8 ppm. The ^{13}C-NMR spectrum of a classical humic acid (extracted with $0.5M$ NaOH) is shown in Fig. 4–14, curve (e). This spectrum is very similar to the

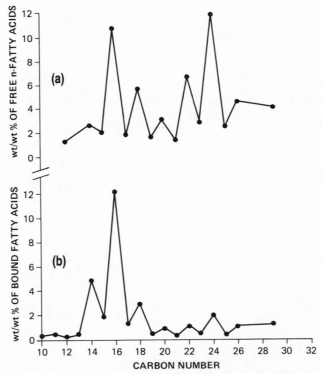

Fig. 4–13. Distribution of free fatty acids (curve a) and of bound fatty acids (curve b) in the supercritical gas extract of the Armadale soil (Schnitzer et al., 1986).

other spectra in Fig. 4–14 except that the signals are not as sharp because of overlapping of individual resonances. The ^{13}C-NMR data are summarized in Table 4–6. Of special interest is the increasing aromaticity of the

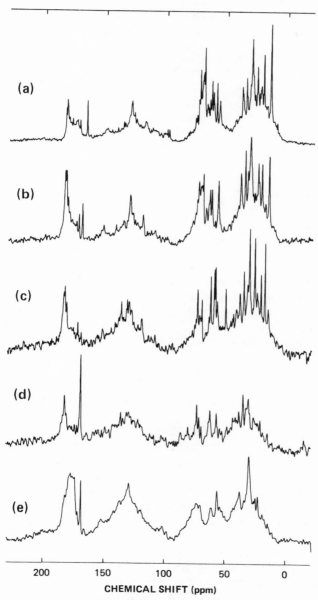

Fig. 4–14. Carbon-13 NMR spectra (in NaOD) of supercritical gas extracts of the Bainsville soil with: (a) EtOH (100% v/v); (b) EtOH-H$_2$O (75:25% v/v; (c) EtOH-H$_2$O (50–50% v/v; (d) EtOH-H$_2$O (25:75% v/v); (e) ^{13}C-NMR spectrum of the Bainsville humic acid extracted with 0.5 M NaOH (classical procedure) (Schnitzer & Preston, 1987).

Table 4-6. Carbon-13 NMR characteristics of SCG extracts obtained with EtOH and EtOH-H$_2$O mixtures from the Bainsville soil and of a classical humic acid extracted from the same soil (Schnitzer and Preston, 1987).

	Carbon distribution				
	Solvent or solvent mixture				
Chemical shift	EtOH (100 v/v %)	EtOH–H$_2$O (75:25 v/v %)	EtOH–H$_2$O (50:50 v/v %)	EtOH–H$_2$O (25:75 v/v %)	Classical humic acid
ppm	wt/wt %				
0–40	37.0	36.9	30.0	23.8	26.4
40–62	16.3	16.3	12.5	15.6	20.5
62–105	16.7	14.6	12.8	12.5	12.8
105–150	16.7	17.0	27.0	26.3	27.3
150–170	5.2	3.8	2.1	10.0	4.1
170–190	8.2	11.7	15.8	12.1	9.1
Aliphatic C (%)	70.0	67.7	55.1	51.8	59.5
Aromatic C (%)	21.9	20.7	29.1	36.2	31.4
Ratio $\dfrac{\text{aliphatic C}}{\text{aromatic C}}$	3.2	3.3	1.9	1.4	1.9
Aromaticity (%)	23.8	23.4	34.5	41.2	34.5

supercritical gas extracts with increasing polarity of the solvent mixture. There is also a parallel increase in phenolic OH + CO$_2$H group (= total acidity, 150–190 ppm) with increase in solvent polarity. Thus, new, important information on the chemical composition and structure of humic materials can be obtained by combining SCGE with ^{13}C-NMR and mass spectrometric methods.

4-5 SUMMARY AND CONCLUSIONS

Solution and *solid-state* ^{13}C-NMR spectroscopy provides information on the average structural composition of humic substances but does not tell us how the components interact with each other and what structural arrangement they produce. ^{13}C-NMR spectra of soil humic materials have shown wide ranges in aromaticities. For example, the aromaticity of a humic acid extracted from an Aquoll is 32.8% (w/w) (Preston & Schnitzer, 1984), while that for a humic acid extracted from a Paleosol is 71.6% (w/w) (Calderoni & Schnitzer, 1984). Well-developed humic acids extracted from Natriborolls have aromaticities of about 50% (w/w) (Schnitzer & Preston, 1986). Likewise, the aliphaticity of humic acids also varies widely, depending on the environmental conditions, and is greater than had been assumed prior to the widespread availability of ^{13}C-NMR spectrometers. Consequently, more attention needs to be given to identifying these aliphatic structures, which in some humic acids are very significant components, but whose chemistry remains largely unknown.

So far, the main applications of ESR spectroscopy to humic materials have been in providing information on the symmetry and coordination of paramagnetic metals in metal–humic acid and metal–fulvic acid complexes. This information is of considerable practical value to soil scientists because it makes it possible to evaluate and possibly predict the availability of these metals to plant roots and microbes. Up to this time ESR spectroscopy has shed little light on the chemical structure of humic materials other than demonstrating that humic substances contain free radicals and that substituted semiquinones are likely contributors to the free radicals in these materials. The failure of ESR spectroscopy to provide significant new structural information on humic substances may be a result of the structural complexity and heterogeneity of humic substances.

The combination of pyrolysis with soft ionization mass spectrometry is a promising technique for identifying major humic components, especially relatively high molecular weight esters and polyesters of long-chain alcohols and fatty acids and other components of waxes, which are sometimes associated with humic substances. With the aid of this technique it is also possible to detect in some humic acids lignin-derived components of considerable molecular complexity.

SCGE is an important method for the relatively mild and efficient extraction of major humic components. Depending on the polarity of the solvent or solvent mixture, it is possible to extract either aliphatic or mainly aromatic components and then characterize the extracts by chemical and spectroscopic methods. Among the latter, ^{13}C-NMR spectroscopy, field ionization mass spectrometry, and field desorption mass spectrometry are especially suitable procedures.

During the next decade we can look forward to the generation of more detailed and specific information on the major components of humic substances. This information will allow us to better understand and possibly use these materials in agriculture. Ultimately, soil scientists and geochemists will have to address the following questions:

1. How do the major humic components interact with each other? Do they associate via low-energy bonds or are stronger bonds, such as covalent bonds, formed between the different components?
2. Do the major components form a stable structural network on which the more transient carbohydrates and proteinaceous materials (and possibly also the more stable alkanes, fatty acids, and esters) are adsorbed?
3. Is this stable structural network mainly aromatic? If not, what are the contributions of aliphatic structures to the network?

It will take a lot of ingenuity, hard work, and the application of the most promising methods available to arrive at satisfactory answers to these questions.

REFERENCES

Atherton, N.M. 1973. Electron spin resonance. John Wiley & Sons, New York.

Bartle, K.D., T.G. Martin, and D.F. Williams. 1975. Chemical nature of a supercritical gas extract of coal at 350 °C. Fuel 54:226–235.

Baumeister, M., and E. Edel. 1980. Athanol-Wasser-Aufschluss. Papier 34:V9–V18.

Blois, M.S., H.W. Brown, and J.E. Maling. 1961. Precision g-value measurements of free radicals in biological systems. p. 117–131. In M.S. Blois et al. (ed.) Free radicals in biological systems. Academic Press, New York.

Boyd, S.A., L.E. Sommers, D.W. Nelson, and D.X. West. 1981. The mechanism of copper (II) binding by humic acid: an electron spin resonance study of a copper (II)-humic acid complex and some adducts with nitrogen donors. Soil Sci. Soc. Am. J. 45:745–749.

Bracewell, J.M., and G.W. Robertson. 1984. Characteristics of soil organic matter in temperate soils by Curie-point pyrolysis-mass spectrometry. I. Organic matter variations with drainage and mull humification in A horizons. J. Soil Sci. 35:549–558.

Bracewell, J.M., and G.W. Robertson. 1987a. Characteristics of soil organic matter in temperate soils by Curie-point pyrolysis-mass spectrometry. II. Effect of drainage and illuviation in B horizons. J. Soil Sci. 38:191–198.

Bracewell, J.M., and G.W. Robertson. 1987b. Characteristics of soil organic matter in temperate soils by Curie-point pyrolysis-mass spectrometry. III. Transformations occurring in surface organic horizons. Geoderma 40:333–344.

Bracewell, J.M., G.W. Robertson, and B.L. Williams. 1980. Pyrolysis-mass spectrometry studies of humification in a peat and a peaty podzol. J. Anal. Appl. Pyrolysis 2:53–62.

Bronstrup, B. 1982. Organosolv-Abbau lignocellulosischer Biomasse in einer Hochdruck-Hochtemperatur-Stromungsapparatur. Ph.D. thesis. Univ. of Oldenburg, West Germany.

Bühler, C., and W. Simon. 1970. Curie-point pyrolysis gas chromatography. J. Chromatogr. Sci. 8:323–329.

Calderoni, G., and M. Schnitzer. 1984. Effects of age on the chemical structure of paleosol humic acids and fulvic acids. Geochim. Cosmochim. Acta 48:2045–2051.

Cheshire, M.V., M.L. Berrow, B.A. Goodman, and C.M. Mundie. 1977. Metal distribution and nature of some Cu, Mn and V complexes in humic and fulvic acid fractions of soil organic matter. Geochim. Cosmochim. Acta 41:1131–1138.

Cheshire, M.V., B.A. Goodman, D.B. McPhail, and G.P. Sparling. 1985. Electron paramagnetic resonance characteristics of the humic acids from a podzol. Org. Geochim. 8:427–440.

Fitzpatrick, J.D., and C. Steelink. 1969. Benzosemiquinone radicals in alkaline solutions of hardwood lignins. Tetrahedron Lett. 57:5041–5044.

Hatcher, P.G., M. Schnitzer, L.W. Dennis, and G.E. Maciel. 1981. Aromaticity of humic substances in soils. Soil Sci. Soc. Am. J. 45:1089–1094.

Hempfling, R., F. Ziegler, W. Zech, and H.R. Schulten. 1987. Litter decomposition and humification in acidic forest soils studied by chemical degradation, IR and NMR spectroscopy and field ionization mass spectrometry. Z. Pflanzenernaehr. Bodenk. 150:179–186.

Hubert, P., and O.G. Vitzthum. 1978. Fluid extraction von Hopfen, Gewurzen und Tabak mit uberkritischen Gasen. Angew. Chem. 90:756–762.

Jones, J.G. 1969. Studies on lipids of soil microorganisms with particular reference to hydrocarbons. J. Gen. Microbiol. 59:145–152.

Kögel, I., P.G. Hempfling, P.G. Hatcher, and H.R. Schulten. 1987. Decomposition in forest humus layers studied by CPMAS ^{13}C NMR, pyrolysis field ionization mass spectrometry and CuO oxidation. Sci. Total Environ. 62:111–120.

Lakatos, B., T. Tibai, and J. Meisel. 1977. EPR spectra of humic acids and their metal complexes. Geoderma 19:319–338.

McBride, M.B. 1978. Transition metal binding in humic acid: an ESR study. Soil Sci. 126:200–209.

McBride, M.B. 1980. Influence of pH and metal ion content on vanadyl ion-fulvic acid interactions. Can. J. Soil Sci. 60:145–149.

Meuzelaar, H.L.C., K. Haider, B.R. Nagar, and J.P. Martin. 1977. Comparative studies of pyrolysis-mass spectra of melanins, model phenolic polymers and humic acids. Geoderma 17:239–252.

Oades, J.M., A.M. Vassallo, A.G. Waters, and M.A. Wilson. 1987. Characterization of organic matter in particle size and density fractions from a red-brown earth by solid state ^{13}C N.M.R. Aust. J. Soil Res. 25:71–82.

Ogner, G., and M. Schnitzer. 1970. The occurrence of alkanes in fulvic acid, a soil humic fraction. Geochim. Cosmochim. Acta 34:921-928.

Preston, G.M., and M. Schnitzer. 1984. Effects of chemical modifications and extractions on the carbon-13 NMR spectra of humic materials. Soil Sci. Soc. Am. J. 48:305-311.

Preston, C.M., and M. Schnitzer. 1987. ^{13}C-NMR of humic substances: pH and solvent effects. J. Soil Sci. 38:667-678.

Rex, R.W. 1960. Electron paramagnetic resonance studies of stable free radicals in lignins and humic acids. Nature 188:1185-1186.

Schnitzer, M. 1978. Humic substances: Chemistry and reactions. p. 1-64. *In* M. Schnitzer and S.U. Khan (ed.) Soil organic matter. Elsevier Publ. Co., Amsterdam.

Schnitzer, M., C.A. Hindle, and M. Meglic. 1986. Supercritical gas extraction of alkanes and alkanoic acids from soils and humic materials. Soil Sci. Soc. Am. J. 50:913-919.

Schnitzer, M., and C.M. Preston. 1986. Analysis of humic acids by solution and solid-state carbon-13 nuclear magnetic resonance. Soil Sci. Soc. Am. J. 50:326-331.

Schnitzer, M., and C.M. Preston. 1987. Supercritical extraction of a soil with solvents of increasing polarities. Soil Sci. Soc. Am. J. 51:639-646.

Schulten, H.R. 1987. Pyrolysis and soft ionization mass spectrometry of aquatic/terrestrial humic substances and soils. J. Anal. Appl. Pyrolysis 12:49-186.

Schulten, H.R., G. Abbt-Braun, and F.H. Frimmel. 1987. Time-resolved pyrolysis field ionization mass spectrometry of humic material isolated from freshwater. Environ. Sci. Technol. 21:349-357.

Schulten, H.R., H.D. Beckey, A.J. Boerboom, and H.L.C. Meuzelaar. 1973. Pyrolysis field desorption mass spectrometry of deoxyribonucleic acid. Anal. Chem. 45:2358-2362.

Senesi, N., Y. Chen, and M. Schnitzer. 1977a. Hyperfine splitting in electron spin resonance spectra of fulvic acid. Soil Biol. Biochem. 9:371-372.

Senesi, N., S.M. Griffith, M. Schnitzer, and M.G. Townsend. 1977b. Binding of Fe^{3+} by humic materials. Geochim. Cosmochim. Acta 41:969-976.

Senesi, N., and M. Schnitzer. 1977. Effects of pH, reaction time, chemical reduction and irradiation on ESR spectra of fulvic acid. Soil Sci. 123:224-234.

Spiteller, M. 1982. Ein neues Verfahren zur Extraktion von organischen Stoffen aus Boden mit uberkritischen Gasen. 1. Mitteilung. Z. Pflanzenernaehr. Bodenk. 145:483-492.

Spiteller, M. 1985. Extraction of soil organic matter by supercritical fluids. Org. Geochem. 8:111-113.

Spiteller, M., and A. Ashauer. 1982. Ein neues Verfahren zur Extraktion von organischen Stoffen aus Boden mit uberkritischen Gasen. 2. Mitteilung. Z. Pflanzenernaehr. Bodenk. 145:567-575.

Steelink, C. 1964. Free radical studies of lignin, lignin degradation products and soil humic acids. Geochim. Cosmochim. Acta 28:1615-1622.

Steelink, C., and G. Tollin. 1962. Stable free radicals in soil humic acid. Biochem. Biophys. Acta 59:25-34.

Stevenson, F.J. 1982. Humus chemistry. John Wiley & Sons, New York.

Sweet, M.S., and E.M. Perdue. 1982. Concentration and speciation of dissolved sugars in river waters. Environ. Sci. Technol. 16:692-698.

Wilson, M.A. 1987. NMR techniques and applications in geochemistry and soil chemistry. Pergamon Press, Oxford.

Chapter 5

Nitrogen in Humic Substances as Related to Soil Fertility[1]

F. J. STEVENSON AND **XIN-TAO HE**, *University of Illinois, Urbana, Illinois*

ABSTRACT

An account is given of the nature and origin of nitrogen (N) in humic substances as a basis for understanding how N behaves in the soil–plant system. A significant amount of the soil N ($>50\%$) occurs as a structural component of humic substances, a portion of which is biologically stable and not readily available to plants. The N of humic substances may occur in the following types of structures: (i) as an amino acid attached to aromatic rings, (ii) as a bridge constituent linking quinone groups together, (iii) as part of a heterocyclic ring, (iv) as an open chain (–NH–, =N-) group, and (v) as peptides and proteins held through H-bonding. From 20 to 35% of the N applied to soils as fertilizers is retained in the soil in organic forms at the end of the growing season. This residual fertilizer N becomes increasingly unavailable for plant uptake during subsequent seasons, ultimately attaining equilibrium with the native humus N. The process whereby N becomes stabilized is discussed from the standpoint of the chemical nature of the stabilized N and the long-term N balance of the soil.

---

[1]Contribution from the Department of Agronomy, University of Illinois and the Illinois Agricultural Experiment Station, Urbana, IL 61801 (project 15-339). Supported in part by TVA grant 50003A.

The N content of humic substances ranges from < 1% for fulvic acids (FA) to over 5% for humic acids (HA). According to Haworth (1971), some of this N occurs as peptides or proteins attached to the central core through H-bonding. A portion of the N may also occur as amino acids covalently bonded to quinone or phenol rings through the amino group. A considerable portion of the N (up to 60%) cannot be accounted for in known forms (e.g., amino acids, amino sugars) and is assumed to exist as an integral component of the humic macromolecule, such as in heterocyclic rings or as a bridge constituent.

The significance of humic substances to the N fertility of the soil arises from the fact that much of the organic N in soil resists attack by microorganisms and is thereby relatively unavailable to plants. Also, as a consequence of mineralization–immobilization turnover by microorganisms, a portion of the fertilizer N applied to soil becomes stabilized by incorporation into humic substances. Results of field trials using the stable isotope ^{15}N have shown that about one-third of the fertilizer N remains behind in organic forms after the first growing season, only a small fraction (< 15%) of which becomes available to plants during the second growing season. Stabilization through incorporation into the structures of humic substances appears likely. Thus, a knowledge of the chemical nature of the N in humic substances will not only provide a better understanding of the factors affecting the availability of the indigenous soil N but may lead to the development of improved management practices for the efficient and environmentally acceptable use of fertilizer N.

In this chapter, consideration is given to the chemical nature of the N in humic substances, with emphasis being given to the availability of soil and fertilizer N to higher plants. Recent reviews of the N of humic substances as related to soil fertility include those of Schnitzer (1985) and Stevenson (1982, 1985a, b).

5–1 CHEMICAL NATURE OF THE NITROGEN IN HUMIC SUBSTANCES

One of the major approaches for characterizing the N in humic substances is by acid or base hydrolysis. In a typical procedure, the sample is heated with $3M$ or $6M$ HCl for periods from 12 to 24 h, after which the N is separated into the following fractions:

Acid insoluble-N	Nitrogen remaining in residue following hydrolysis; usually obtained by difference (total N − hydrolyzable N).
NH_3-N	Ammonia recovered from the hydrolysate by steam distillation with MgO.
Amino acid-N	Usually determined by the ninhydrin-CO_2 or ninhydrin-NH_3 methods; recent workers have favored the latter.

| Amino sugar-N | Steam distillation with phosphate–borate buffer at pH 11.2 and correction for NH_3-N; colorimetric methods are also available. |
| Hydrolyzable unknown-N (HUN fraction) | Hydrolyzable N not accounted for as NH_3, amino acids, or amino sugars; part of this N occurs as a non-α-amino N in arginine, tryptophan, lysine, and proline. |

Typical results for the distribution of N in hydrolysates of humic and fulvic acids are given in Table 5-1. For the preparations analyzed, from 20 to 45% of the N occurred as amino acids; from 2 to 8% occurred as amino sugars. In comparison with fulvic acids, a much lower percentage of N in the humic acids occurred as NH_3-N. The fulvic acids contained little, if any, acid insoluble-N, whereas a much higher percentage of the N occurred in hydrolyzable unknown compounds (HUN fraction). Schnitzer et al. (1983) subjected the acid hydrolysates of soil humic acid and fulvic acid to gel filtration and obtained several fractions in which >90% of the N occurred in unknown forms. A significant amount of the unknown N may occur as purine and pyrimidine bases (Cortez & Schnitzer, 1979a, b).

5-1.1 Amino Acids and Amino Sugars

The main identifiable compounds in hydrolyates of humic and fulvic acids are amino acids and amino sugars. Other nitrogenous biochemicals are also present, but specialized techniques are required for their separation and identification. The amino acid content of humic acids may decrease with an increase in "degree of humification" (Tsutsuki & Kuwatsuka, 1978, 1979).

Data given in Table 5-1 also show that humic acids extracted from soil with $0.5M$ NaOH have higher N contents than those extracted with $0.1M$ $Na_4P_2O_7$. Furthermore, a higher percentage of the N in the alkali-extracted humic acids was accounted for as amino acids.

Part of the amino acid-N of humic acids may exist as peptides or proteins linked to the central core by H-bonding (Haworth, 1971). Peptide-like substances have been determined in humic acids by chromatographic assay of proteinaceous constituents liberated by cold hydrolysis with concentrated HCl (Cheshire et al., 1967; Piper & Posner, 1968). It has also been demonstrated that amino acids are released through the action of proteolytic enzymes (see review of Ladd & Jackson, 1982). Infrared spectra of some, but not all, humic acids show absorption bands typical of the peptide linkage (Otsuki & Hanya, 1967; Stevenson & Goh, 1971). Evidence for N in loosely bound forms has come from the observation that the N content of humic acids can be lowered considerably by passage through a cation exchange resin in the H-form (Simonart et al., 1967; Sowden & Schnitzer, 1967). Protein-rich fractions have been obtained from humic acids by extraction with phenol (Biederbeck & Paul, 1973; McGill et al., 1975; McGill and Paul, 1976).

Table 5-1. Distribution of the forms of N in humic and fulvic acids.

Extractant	N	Acid insoluble	NH$_3$	HUN	Amino acids	Amino sugars	Reference
				—%—			
			Humic acids				
0.1M Na$_4$P$_2$O$_7$ (6)†	1.79–2.63	41.3–59.0	8.8–12.8	5.2–10.6	19.5–34.5	2.6–5.0	Bremner (1955)
0.5M NaOH (5)	2.31–3.74	32.6–43.7	8.4–13.7	4.7– 8.6	32.2–44.7	3.4–8.1	Bremner (1955)
0.5M NaOH (4)	2.11–2.69	35.9–50.8	8.2–14.0	16.2–21.8	22.1–26.5	1.8–3.9	Rosell et al. (1978)
			Fulvic acids‡				
0.5M NaOH (3)	3.37–3.89	--	15.1–19.3	41.3–54.9	26.4–34.2	3.6–5.2	Khan & Sowden (1972)

† Numbers in parentheses indicate number of samples analyzed.
‡ Results are for dialyzed preparation.

5–1.2 Structural Nitrogen of Humic Substances

A major proportion of the N of humic acids, ranging from 30 to 60%, is recovered as acid insoluble-N. This N may occur as a structural component of the molecule, such as a bridge constituent linking quinone groups together (I, II).

I   II

Additional quantities of amino acids can be recovered from humic acids by subjecting the residues from the initial acid hydrolysate to a second hydrolysis with $2.5M$ NaOH (Piper & Posner, 1968, 1972; Griffith et al., 1976). Sodium amalgam reduction of the alkali-treated residue has been shown to lead to a further release of amino acids (Piper & Posner, 1968). Acid hydrolysis would be expected to remove amino acids bound by peptide bonds (III) as well as those linked to quinone rings (IV). On the other hand, amino acids bonded directly to phenolic rings (V) may not be released without subsequent alkaline hydrolysis. The effect of alkaline hydrolysis was believed to be due to oxidation of the amino phenol to the quinonimine form, with subsequent hydrolysis of the imine and release of the amino acid.

III   IV   V

An interesting result was obtained by Aldag (1977), who observed an increase in amino acid N at the expense of the HUN fraction when an acid hydrolysate of humic acid ($6M$ HCl for 24 h) was further heated with $6M$ HCl containing 3% H_2O_2 (Table 5–2). The increased release of amino acids was attributed to the presence of phenolic–amino acid addition products in the hydrolysate (e.g., see structure V). Hydrolysis of the initial acid-insoluble residue with 3% H_2O_2–$6M$ HCl led to a further release of amino acids as well as other N forms. The increase in NH_3 released by hydrolysis in the presence of 3% H_2O_2 can be accounted for, in part, by oxidation of organic N compounds, such as amino sugars.

In a study of the incorporation of ^{14}C-labeled glycylglycine into humic acids, Perry and Adams (1972) found that the N-terminal glycine residue was more resistant to hydrolysis than was the C-terminal residue. In a subsequent

Table 5-2. Distribution of the forms of N in humic acid as influenced by different hydrolysis treatments (adapted from Aldag, 1977).

Hydrolysis condition	Form of N				
	Acid insoluble	NH_3	HUN	Amino acid	Amino sugar
	% of total N				
1. With 6M HCl, 24 h	38.5	13.8	11.0	34.2	2.4
2. Hydrolysate from (1) further hydrolyzed with 3% H_2O_2-6M HCl	38.5	18.2	6.8	35.2	1.2
3. Residue from (1) rehydrolyzed with 3% H_2O_2-6M HCl	27.6	21.8	11.6	37.8	1.2

study, Adams and Perry (1973) found that maximum incorporation of amino acids into humic acids occurred at a pH that coincided with the apparent dissociation constants of their α-amino groups. Incorporation took place into forms that were both acid hydrolyzable and nonhydrolyzable, the latter accounted for 10 to 20% of the total. Stepanov (1969) had earlier demonstrated chemical fixation of amino acid N by humic acids. Confirmation for the chemical binding of peptides by humic acids has been obtained by differential ultraviolet spectroscopy (Müller-Wegener, 1984).

It seems likely, therefore, that part of the acid insoluble-N in humic substances may occur in the form of N-phenyl amino acids resulting from the bonding between amino groups and aromatic rings (see also Witthauer & Klocking, 1971; and references contained therein). For peptides, the amino acid directly attached to the aromatic ring may be less available to plants and microorganisms than amino acids of the peripheral peptide chain, as depicted in Fig. 5-1.

A high percentage of the N of fulvic acids (40–55%) occurs in the HUN fraction. It is possible that some of this N exists in linkages represented by structures (I) to (V); some may occur as non-α-amino acid N of such amino acids as arginine, tryptophan, lysine, and proline. Schnitzer and Hindle (1980) found that from 17 to 59% of the unknown N of some soil humic and fulvic acids could be converted to NH_3 and N-gases by mild chemical oxidation with peracetic acid.

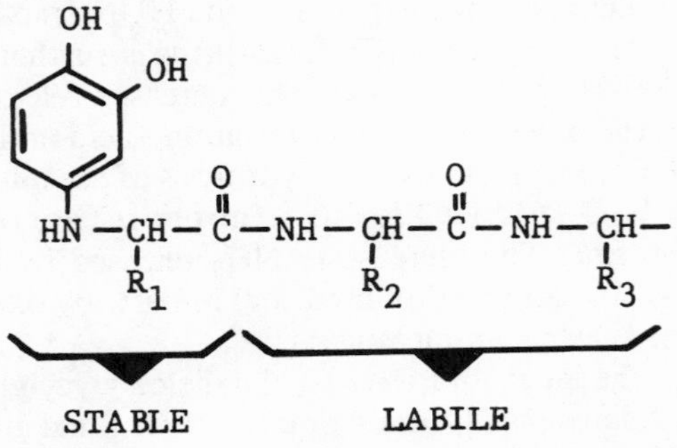

Fig. 5-1. Biological availability of amino acid-N in peptide–phenol complexes.

The fulvic acid fraction of soil organic matter [the alkali-soluble, acid-soluble component; see chapter 2 (Malcolm)] contains a broad spectrum of organic N compounds, including amino acids and amino sugars. Some of this N may not be a structural component of fulvic acid per se, but may exist as an impurity. Thus far, few attempts have been made to establish the origin of the various organic N compounds in hydrolysates of crude fulvic acid extracts. Differences in N distribution patterns observed for the fulvic acid fraction from different sources may partially reflect variations in the amount of true fulvic acids that are present relative to nonhumic substances. In some fulvic acid preparations, most of the N appears to be present as amino acids (Otsuka, 1975; Sequi et al., 1975). In others, rather large amounts of NH_3-N have been observed (Stevenson, 1960; Batsula & Krupskiy, 1974).

The origin of the NH_3 in acid hydrolysates of humic and fulvic acids is unknown. Kickuth and Scheffer (1976) concluded that the pseudo-amide N (hydrolyzable NH_3 not accounted for as asparagine and glutamine) was derived from the $=NH$ of quinonimines. The reaction of amino acids with humic substances (Loginow, 1967) and phenols (Haider et al., 1965; Ladd & Butler, 1966) has been shown to lead to incorporation of N into structures that yield NH_3 when the products are subjected to acid hydrolysis. An alternative source of the NH_3 is from melanoidins formed by reaction of amino acids with reducing sugars, as suggested by Stevenson (1960).

In summary, it can be said that the N of humic substances exists in several forms, some being labile and readily utilized by microorganisms (e.g., H-bonded peptides and proteins) and some being recalcitrant, or nearly so. Significant amounts of the N associated with humic and fulvic acids cannot be accounted for in known compounds. This N may occur in the following types of linkages.

1. As a free amino (-NH_2) group.
2. As an open chain (-NH-, =N-) group.
3. As part of a heterocyclic ring, such as an -NH- of indole and pyrrole or the -N= of pyridine.
4. As a bridge constituent linking quinone rings together.
5. As an amino acid attached to aromatic rings in such a manner that the amino acid is not released by acid hydrolysis.

As one might expect, humic and fulvic acids contain the same forms of N that are obtained when soils are subjected to acid hydrolysis (Bremner, 1955; Khan & Sowden, 1972; Batsula & Krupskiy, 1974; Aldag, 1977; Rosell et al., 1978; Tsutsuki & Kuwatsuka, 1978). However, N distribution patterns vary somewhat. In comparison with unfractionated soil, a lower percentage of the N in humic acids occurs as NH_3-N and in the HUN fraction; a higher percentage occurs as amino acid N and as acid insoluble-N. Results obtained by Rosell et al. (1978) for the distribution of N in some Argentine soils and their humic acids are shown in Fig. 5-2.

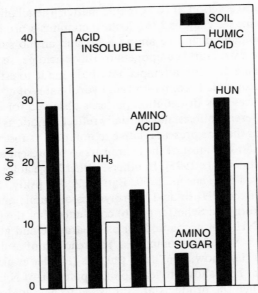

Fig. 5–2. Comparison of the distribution of the forms of N by acid hydrolysis of soils and their humic acids. From Stevenson (1982) as adapted from data of Rosell et al. (1978).

5–1.3   Effects of Soil Management Practices

The assumption is sometimes made that the structural N of humic substances (i.e., see structures I and II) represents the more stable component of soil organic N. On this basis, one would expect major changes to occur in the distribution of the forms of N when soils are subjected to intensive cultivation, namely, large losses of N from amino acids and other more readily available forms at the expense of the acid-insoluble fraction. In practice, this has not been shown to be the case (see review by Stevenson, 1982). Surprisingly, neither long-term cropping nor the addition of organic amendments to the soil greatly affects the relative distribution of the forms of N. All forms, including the acid-insoluble fraction, appear to be biodegradable (Ivarson & Schnitzer, 1979; Malik & Haider, 1982). The observation that management practices have little effect on the relative distribution of the various forms of N emphasizes that chemical fractionation of soil N by acid hydrolysis procedures is of little practical value as a means of testing soils for available N or for predicting crop yields (Cornfield, 1957; Keeney & Bremner, 1964, 1966; Porter et al., 1964; Cornforth, 1968; Kadirgamathaiyah & MacKenzie, 1970; Moore & Russell, 1970; Giddens et al., 1971; Osborne, 1977).

Results of these and other studies show rather convincingly that humic substances are not completely inert but are constantly changing as a consequence of the activities of microorganisms. Under steady-state conditions, mineralization of native humus is compensated for by synthesis of new humus. During humification, the N of amino acids and other amino compounds is incorporated into the structures of humic and fulvic acids, in which form it is not readily available to plants (see next section of this chapter).

5-2 STABILIZATION OF APPLIED FERTILIZER NITROGEN

As noted earlier, a significant fraction of the fertilizer N applied to the soil, of the order of 20 to 35%, remains behind in organic forms after the first growing season. No more than 15% of this residual N becomes available to plants during the second growing season, and availability decreases even further for succeeding crops (see reviews of Stevenson, 1985a, b). A similar effect has been noted for the N of crop residues.

A key factor affecting the efficiency with which fertilizer N is consumed by crops is the progressive stabilization of N by conversion to more resistant humus forms. The net result is a decrease in the "labile" pool of potentially available N in the soil.

5-2.1 The Humification Process

Although a number of explanations can be given for the low availability of residual fertilizer N in soils, there is little doubt but that stability is due in part to incorporation of N into the complex structures of humic substances, for which two mechanisms can be given.

1. The turnover of N through mineralization–immobilization leads to incorporation of N into the melanins of fungi. Many species of fungi belonging to the *Imperfecti* group are known to synthesize dark-colored, humic-like substances from simple carbon substrates. These macromolecules have many of the properties of soil humic substances, including N content, resistance to microbial decomposition, molecular weight, and the nature of the compounds released by chemical degradation procedures. Russell et al. (1983) noted close similarities in the infrared spectra of humin, humic acid, and fungal pigments, from which they concluded that well-humified soil organic matter was to a large extent microbial in origin.
2. The N of amino acids, peptides, and proteins is stabilized through reactions involving quinones and other carbonyl-containing substances. As noted below, the humification of organic remains in soil leads to incorporation of N into the complex structures of humic and fulvic acids.

A popular view at the present time is that humic substances in soil are formed by a multiple stage process that includes: (i) decomposition of plant polymers into simple monomers, (ii) metabolism of the monomers by microorganisms with an accompanying increase in the soil biomass, (iii) repeated recycling of the biomass C (and N) with synthesis of new cells, and (iv) concurrent polymerization of reactive monomers into high molecular weight sub-

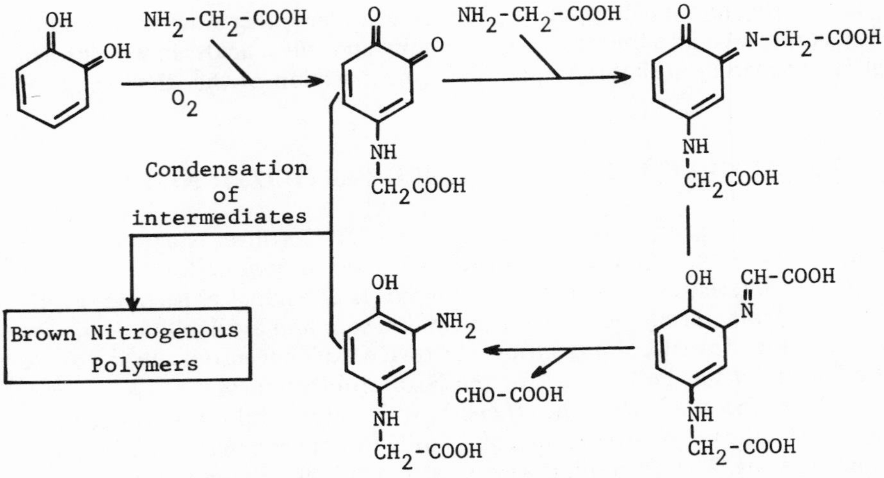

OH
OH NH_2-CH_2-COOH   O / O NH_2-CH_2-COOH   O / N-CH_2-COOH

O_2

Condensation
of
intermediates

NH
CH_2COOH

NH
CH_2COOH

Brown Nitrogenous
Polymers

OH
NH_2

CHO-COOH

NH
CH_2-COOH

OH   CH-COOH
‖
N

NH
CH_2-COOH

Fig. 5-3. Formation of brown nitrogenous polymers as depicted by the reaction between glycine and 1,2-benzoquinone.

stances. Polyphenols derived from lignin, together with those synthesized by microorganisms, can be converted to quinones, such as through the action of polyphenol oxidase enzymes. The quinones then polymerize in the presence or absence of amino compounds (amino acids, NH_3, etc.) to form brown-colored macromolecules. The general sequence is shown in Fig. 5-3. The net effect of this series of reactions is that the N of amino acids is incorporated into complex structures, such as in heterocyclic rings and as a bridge between quinone and aromatic rings (see structures I and II).

5-2.2 Composition and Fate of Residual Fertilizer Nitrogen

Several approaches using the stable isotope ^{15}N have been used to characterize the immobilized N in soil, including fractionations based on acid hydrolysis and partitioning into humic and fulvic acids by classical alkali extraction. A noteworthy feature of the acid hydrolysis studies is that much of the N cannot be accounted for in known forms but is recovered in the acid-insoluble, HUN, and NH_3 fractions (see section 5-2.3 of this chapter). With time, more and more of the residual N is converted to these unknown forms, and, within a matter of years, the composition of the fertilizer N is indistinguishable from that of the native humus N (Allen et al., 1973; Smith et al., 1978).

Results of the field study of Allen et al. (1973) are typical in that an average of one-third of the applied ^{15}N was accounted for in the surface soil after the first growing season, the remainder having been taken up by plants or lost through leaching and denitrification. The soil was Brenton silt loam (Aquic Argiudoll). Isotope-ratio analysis revealed that most of the residual N had been incorporated into organic forms. Comparison of the distribution pattern for the fertilizer-derived ^{15}N with that of the native humus N

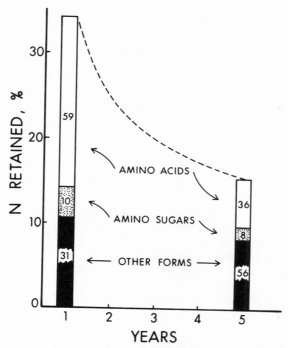

Fig. 5-4. Composition and fate of residual ^{15}N in Brenton silt loam (Aquic Arguidoll) as deduced from data of Allen et al. (1973). The composition of the residual ^{15}N at the 5-yr sampling period was very similar to that of the native humus N. For a discussion of the nature of the unknown forms of N, representing 31 and 56% of the residual ^{15}N for the first and fifth years, respectively, see text.

showed that a considerably higher proportion of the fertilizer N occurred in the form of amino acids (59.0 vs. 36.0%) and amino sugars (9.9 vs. 8.0%); lower proportions occurred as hydrolyzable NH_3 (10.6 vs. 18.1%), as acid insoluble-N (10.3 vs. 21.7%), and in the HUN fraction (10.2 vs. 16.2%). When the plots were resampled 4 yr later, the fertilizer N remaining in the soil, representing one-sixth of that initially applied, had a composition very similar to that of the native humus N. As shown in Fig. 5-4, the mean residence time, or average life, for the N retained after the first season was about 5 yr. The N retained after this period was postulated to have mean residence time of 25 yr; thereafter, the mean residence time would be the same as for the native humus, or an estimated 200 to 800 yr. From this calculation, it can be seen that a small fraction of the fertilizer N applied during any given growing season will remain in the soil for a long time, perhaps centuries.

Results similar to those noted above were obtained by Smith et al. (1978), who found that fertilizer N was initially incorporated into such compounds as amino acids and subsequently into more stable forms. As compared to the native soil N, more of the residual fertilizer N occurred as an amino acid-containing fraction, with small amounts being accounted for as hydrolyzable NH_3 and insoluble N. Equilibrium with the native soil N had not been achieved in 3 yr.

Residual fertilizer N has been partitioned into the classical humus fractions (e.g., humic acid, fulvic acid, and humin) by alkali extraction (Wojcik-Wojtkowiak, 1978; Gamzikow, 1981; Rudelov & Pyranishnikov, 1982). The relative amounts of N recovered in humic and fulvic acids were found by Wojcik-Wojtkowiak (1978) to depend on several factors, including form of applied N. McGill et al. (1975) isolated humic and fulvic acids from ^{15}N-labeled soils and observed a higher degree of labeling in the fulvic acid fraction.

Finally, it should be noted that results obtained with ^{15}N-labeled substrates complemented in a most satisfactory way those for [14]C. From 30 to 37% retention of applied C has been observed at the end of the first growing season, depending on soil and climatic conditions (Jenkinson, 1965; Führ & Sauerbeck, 1968; Oberländer & Roth, 1968; Wagner, 1968; Shields & Paul, 1973). In the experiments conducted by Jenkinson (1965), approximately 33% of the applied C remained behind in the soil after the first year. This residual C had a mean residence time of about 4 yr; thereafter, the mean residence time approached that for the native humus. Much of the N required to maintain a constant C/N ratio will come from applied fertilizer N; part may come from the soil N pool.

5-2.3 Nature and Composition of Newly Immobilized Nitrogen

Additional information regarding the biological conversion of N to stable humus forms has been provided by studies in which the organic N has been labeled with ^{15}N by short-term incubation of the soil with inorganic ^{15}N and a suitable C substrate. Results of these studies have confirmed that stabilization of N occurs very rapidly, and that a significant portion of the immobilized N occurs in forms not readily available to plants or microorganisms.

In the case of the study of Kai et al. (1973), the soil was incubated for periods of up to 20 wk with [15]NO_3 and three C substrates (glucose, straw, and cellulose). For all C substrates, an initial net immobilization of ^{15}N was followed by a period of net mineralization. At the point of maximum incorporation of N into the biomass, there was a distinct difference in the percentage distribution of the forms of organic N between the newly immobilized ^{15}N and the native humus N, as determined by acid hydrolysis. Higher percentages of the immobilized N occurred as amino acids and in the HUN fraction; a lower percentage was found as acid-insoluble N. With time, the percentage of the organic N as acid-insoluble N and NH_3-N increased, while the percentage as amino acids and in the HUN fraction decreased. Similar trends have been observed by Ahmad et al. (1973), and by Kelley and Stevenson (1985), among others.

Results of a study in which the organic matter of a typical Illinois Mollisol (Flanagan silt loam, Aquic Argiudoll) was labeled with ^{15}N by short-term incubation of soil with glucose as the C substrate are shown in Fig. 5-5 (Stevenson & Kelley, 1985). Utilization of the glucose was accompanied

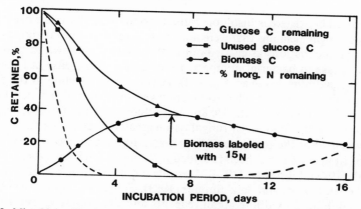

Fig. 5–5. Microbial utilization of glucose with incorporation of inorganic ^{15}N into organic forms. The soil was Flanagan silt loam, a typical Illinois Mollisol. Soil at the point of maximum incorporation of ^{15}N into the biomass was partitioned into the classical humic fractions (see text). Adapted from Stevenson and Kelley (1985).

by rapid immobilization of applied ^{15}N and conversion to organic forms. Soil at the point of maximum conversion of glucose C into the biomass was extracted sequentially with $0.15M$ $Na_4P_2O_7$ and $0.1M$ KOH, following which the N was partitioned into the classical humic fractions (He et al., 1988a).

As can be seen from Fig. 5–6, higher percentages of the applied ^{15}N, as compared to the native soil N, were accounted-for in the two extracts (22 vs. 16% by $0.15M$ $Na_4P_2O_7$ and 27 vs. 22% by $0.1M$ KOH, respectively). About 49% of the applied ^{15}N was removed by the sequential extraction, as compared to 38% for the native soil N. Although somewhat less than for the native soil N (59%), the percentage accounted for as humin was still surprisingly high (46%). One explanation for this result is that much of the newly

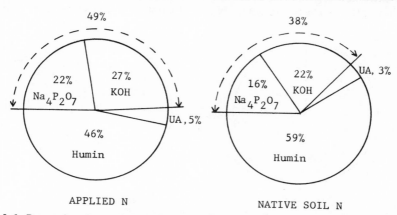

Fig. 5–6. Recoveries of recently immobilized and native soil N by sequential extraction with $0.15M$ $Na_4P_2O_7$ and $0.1M$ KOH (see Fig. 5–5). UA refers to the N unaccounted-for (He et al., 1988a). Reprinted with permission from Soil Biology and Biochemistry, Vol. 20, p. 75–81. Copyright © 1988. Pergamon Press, PLC, Oxford.

immobilized N existed as insoluble cellular components of microorganisms, including fungal melanins that were not extractable under the conditions used. Short-term incubations as carried out in this study are known to lead to an initial increase in bacterial numbers, followed by an increase in fungal organisms (maximum between 4 to 8 d). A subsequent decline in the fungal population would be expected to be followed by a second increase in bacterial numbers (see He et al., 1988b, and references contained therein). On a weight basis, the amount of fungal tissue greatly exceeds that for bacteria.

Evidence for the presence of humic acid-like substances in alkaline extracts of fungal tissue has been provided by Kang and Felbeck (1965), Schnitzer et al. (1973), Filip et al. (1974), Meuzelaar et al. (1977), Tan et al. (1978), Russell et al. (1983), and Saiz-Jiminez (1983). Fungal melanins are not completely solubilized by dilute alkali reagents and the insoluble residues have properties similar to those of the humin fraction of soil organic matter (Russell et al., 1983). Ortiz de Serra et al. (1973) found that a higher proportion of the N in fungal humic acids occurred in the form of amino acids as compared to soil humic acids.

The solubilized organic matter (see Fig. 5-6) was further partitioned into humic and fulvic acids (extracts combined). As shown in Fig. 5-7, the percentage of the extracted ^{15}N accounted for in the humic acid fraction (24%) was somewhat more than for the native soil N (16%). The results definitely show that: (i) the immobilized ^{15}N was partitioned into all humus fractions, and (ii) a significant fraction of the newly immobilized N was indistinguishable from the native humus N. Results similar to those noted above have been obtained in fractionation studies carried out by Chichester (1970), Anderson et al. (1974), and McGill et al. (1975).

Chichester (1970) fractionated ^{15}N-labeled soil based on soil particle size and then extracted each particle fraction with $0.5M$ Na$_4$P$_2$O$_7$ at 100°C. The extracted organic N was further fractionated by acid hydrolysis. Distribution patterns for the fertilizer N and the native soil N were very similar,

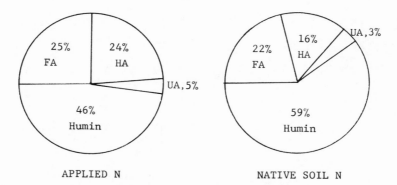

Fig. 5-7. Comparison between the distribution patterns of recently applied and native soil N in the HA, FA, and humin fractions of soil organic matter. The HAs and FAs were obtained by sequential extraction with $0.15M$ Na$_4$P$_2$O$_7$ and $0.1M$ KOH (extracts combined). UA refers to the N unaccounted-for (adapted from He et al., 1988a).

a result that was attributed to conversion of the fertilizer-derived N into complexes similar in composition and properties to the pre-existing soil N.

The fractionation scheme used by McGill et al. (1975) is shown in Fig. 5–8. At the time when the microbial population reached maximum (incubation period of 5 d), the added ^{15}N extracted by $Na_4P_2O_7$ represented about 20% of the total added ^{15}N and the ratio of the added ^{15}N in fulvic acid relative to that in humic acid was >2:1. Up to 10% of the total ^{15}N was accounted for in the humic acid fraction, about 50% of which was removed by phenol extraction and assumed to consist of proteinaceous material held to the humic acid through H-bonding. He et al. (1988b) found that the aqueous phenol was more selective in removing recently immobilized ^{15}N from soil than reagents used for extraction of soil organic matter (i.e., $0.15M$ $Na_4P_2O_7$ and $0.1M$ NaOH).

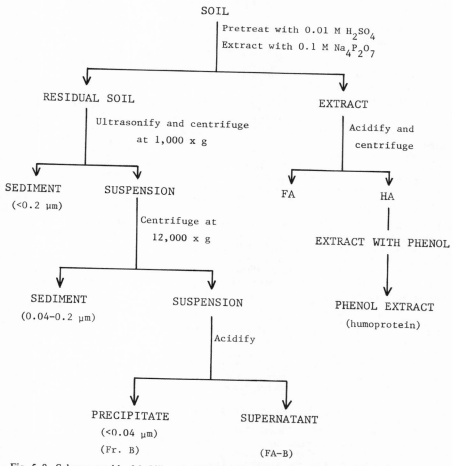

Fig. 5–8. Scheme used by McGill et al. (1975) to fractionate immobilized ^{15}N into soluble organic and clay-bound forms.

5-3 SUMMARY

Despite the progress that has been made in recent years, much more needs to be done before our knowledge of soil N is complete. As much as one-half of the N in humic substances has not been accounted for in known forms. The nature of this unknown N will remain obscure until more is known about the chemical structures of humic and fulvic acids. Other major problems remaining to be solved include: (i) mechanisms whereby N is stabilized by incorporation into humic substances, (ii) resistance of soil N complexes to attack by soil organisms, (iii) availability of complex forms of N to higher plants, and (iv) long-term fate of residual fertilizer N in soil.

REFERENCES

Adams, W.A., and D.R. Perry. 1973. The effect of pH on the incorporation of amino acids into humic acid extracted from soil. J. Soil Sci. 24:18–25.

Ahmad, Z., Y. Yahiro, H. Kai, and T. Harada. 1973. Factors affecting immobilization and release of nitrogen in soil and chemical characteristics of the nitrogen newly immobilized. IV. Soil Sci. Plant Nutr. (Tokyo) 19:287–298.

Aldag, R.W. 1977. Relations between pseudo-amide nitrogen and humic acid nitrogen released under different hydrolysis conditions. p. 293–299. In Soil Org. Matter Stud., Proc. Symp. 1976 (Braunschweig, 6–10 Sept.). Int. Atomic Energy Agency, Vienna, Austria.

Allen, A.L., F.J. Stevenson, and L.T. Kurtz. 1973. Chemical distribution of fertilizer nitrogen in soil as revealed by nitrogen-15 studies. J. Environ. Qual. 2:120–124.

Anderson, D.W., E.A. Paul, and R.J. St. Arnaud. 1974. Extraction and characterization of humus with reference to clay-associated humus. Can. J. Soil Sci. 54:317–323.

Batsula, A.A., and N.K. Krupskiy. 1974. The forms of nitrogen in humus substances of some virgin and developed soil of the left bank of the Ukraine. Sov. Soil Sci. 6:456–462.

Biederbeck, V.O., and E.A. Paul. 1973. Fractionation of soil humate with phenolic solvents and purification of the nitrogen rich portion with polyvinylpyrrolidone. Soil Sci. 115:357–366.

Bremner, J.M. 1955. Studies on soil humic acids: I. J. Agric. Sci. 46:247–256.

Cheshire, M.V., P.A. Cranwell, C.P. Falshaw, A.J. Floyd, and R.D. Haworth. 1967. Humic acid: II. Structure of humic acids. Tetrahedron 23:1669–1682.

Chichester, F.W. 1970. Transformations of fertilizer nitrogen in soils. II. Total and ^{15}N-labelled nitrogen of soil organo-mineral sedimentation fractions. Plant Soil 33:437–456.

Cornfield, A.H. 1957. Effect of 8-years fertilizer treatment on the protein-nitrogen content of four cropped soils. J. Sci. Food Agric. 8:509–511.

Cornforth, I.S. 1968. The potential availability of organic nitrogen fractions in some West Indian soils. Exp. Agric. 4:193–201.

Cortez, J., and M. Schnitzer. 1979a. Purines and pyrimidines in soils and humic substances. Soil Sci. Soc. Am. J. 43:958–961.

Cortez, J., and M. Schnitzer. 1979b. Nucleic acid bases in soils and their association with organic and inorganic soil components. Can. J. Soil Sci. 59:277–286.

Filip, Z., K. Haider, H. Beutelspacher, and J.P. Martin. 1974. Comparisons of IR spectra from melanins of microscopic soil fungi, humic acids and model phenol polymers. Geoderma 11:37–52.

Führ, F., and D. Sauerbeck. 1968. Decomposition of wheat straw in the field as influenced by cropping and rotation. p. 241–250. In Isotopes and radiation in soil organic matter studies. Tech. Meet. FAO/IAEA, Vienna, Austria.

Gamzikov, G.P. 1981. Fertilizer nitrogen transformation: An interaction with West-Siberian soils. p. 194–197. In P. Dutil and F. Jacquin (ed.) Colloque: humus-azote. Station de Science du Sol, I.N.R.A., Charlons-Sar-Marne, France.

Giddens, J., R.D. Hauck, W.E. Adams, and R.N. Dawson. 1971. Forms of nitrogen and nitrogen availability in fescuegrass sod. Agron. J. 63:458–460.

Griffith, S.M., F.J. Sowden, and M. Schnitzer. 1976. The alkaline hydrolysis of acid-resistant soil and humic acid residues. Soil Biol. Biochem. 8:529-531.

Haider, K., L.R. Frederick, and W. Flaig. 1965. Reaction between amino acid compounds and phenols during oxidation. Plant Soil 22:49-64.

Haworth, R.D. 1971. The chemical nature of humic acid. Soil Sci. 111:71-79.

He, X-T., F.J. Stevenson, R.L. Mulvaney, and K.R. Kelley. 1988a. Incorporation of newly immobilized ^{15}N into stable organic forms in soil. Soil Biol. Biochem. 20:75-81.

He, X-T., F.J. Stevenson, R.L. Mulvaney, and K.R. Kelley. 1988b. Extraction of newly immobilized ^{15}N from an Illinois Mollisol using aqueous phenol. Soil Biol. Biochem. 20:857-862.

Ivarson, K.C., and M. Schnitzer. 1979. The biodegradability of the "unknown" soil-nitrogen. Can. J. Soil Sci. 59:59-67.

Jenkinson, D.S. 1965. Studies on the decomposition of plant material in soil: I. J. Soil Sci. 16:104-115.

Kadirgamathaiyah, S., and A.F. MacKenzie. 1970. A study of soil nitrogen organic fractions and correlation with yield response of sudan-sorghum hybrid grass on Quebec soils. Plant Soil 33:120-128.

Kai, H., Z. Ahmad, and T. Harada. 1973. Factors affecting immobilization and release of nitrogen in soil and chemical characteristics of the nitrogen newly immobilized. III. Soil Sci. Plant Nutr. (Tokyo) 19:275-286.

Kang, K.S., and G.T. Felbeck, Jr. 1965. A comparison of the alkaline extract of tissues of *Aspergillus niger* with humic acids from three soils. Soil Sci. 99:175-181.

Keeney, D.R., and J.M. Bremner. 1964. Effect of cultivation on the nitrogen distribution in soils. Soil Sci. Soc. Am. Proc. 28:653-656.

Keeney, D.R., and J.M. Bremner. 1966. Characterization of mineralizable nitrogen in soils. Soil Sci. Soc. Am. Proc. 30:714-719.

Kelley, K.R., and F.J. Stevenson. 1985. Characterization and extractability of immobilized ^{15}N from the soil microbial biomass. Soil Biol. Biochem. 17:517-523.

Kahn, S.U., and F.J. Sowden. 1972. Distribution of nitrogen in fulvic acid fraction extracted from the black Solonetzic and black Chernozemic soils of Alberta. Can.J. Soil Sci. 52:116-118.

Kickuth, R., and F. Scheffer. 1976. Constitution and role as plant nutrients of pseudo-amide nitrogen in humic acids. (In German). Agrochimica 20:373-386.

Ladd, J.N., and J.H.A. Butler. 1966. Comparison of some properties of soil humic acids and synthetic phenolic polymers incorporating amino derivatives. Aust. J. Soil Res. 4:41-54.

Ladd, J.N., and R.B. Jackson. 1982. Biochemistry of ammonification. p. 173-228. In F.J. Stevenson (ed.) Nitrogen in agricultural soils. Agron. Monogr. 22. ASA, Madison, WI.

Loginow, W. 1967. Effect of humic acids on the deamination of amino acids. (In Polish). Pamiet. Pulawski 29:3-43.

Malik, K.A., and K. Haider. 1982. Decomposition of ^{14}C-labelled melanoid fungal residues in a marginally sodic soil. Soil Biol. Biochem. 14:457-460.

McGill, W.B., and E.A. Paul. 1976. Fractionation of soil and ^{15}N nitrogen to separate the organic and clay interactions of immobilized N. Can. J. Soil Sci. 56:203-212.

McGill, W.B., J.A. Shields, and E.A. Paul. 1975. Relation between carbon and nitrogen turnover in soil. Organic fractions of microbial origin. Soil Biol. Biochem. 7:57-63.

Meuzelaar, H.L.C., K. Haider, B.R. Nagar, and J.P. Martin. 1977. Comparative studies of pyrolysis-mass spectra of melanins, model phenolic polymers, and humic acids. Geoderma 17:239-252.

Moore, A.W., and J.S. Russell. 1970. Changes in chemical fractions of nitrogen during incubation of soils with histories of large organic matter increase under pasture. Aust. J. Soil Res. 8:21-30.

Müller-Wegener, V. 1984. Wechselwirkungen von Huminstoffen mit Peptiden. Z. Pflanzenernaehr. Bodenkd. 147:110-124.

Oberländer, H.E., and K. Roth. 1968. Transformations of ^{14}C-labeled plant material in soils under field conditions. p. 351-361. In Isotopes and radiation in soil organic matter studies. Tech. Meet. FAO/IAEA, Vienna, Austria.

Ortiz de Serra, M.I., F.J. Sowden, and M. Schnitzer. 1973. Distribution of nitrogen in fungal "humic acids". Can. J. Soil Sci. 53:125-127.

Osborne, G.J. 1977. Chemical fractionation of soil nitrogen in six soils from southern New South Wales. Aust. J. Soil Res. 15:159–165.

Otsuka, H. 1975. Accumulated state of humus in the soil profile, and sugar, uronic acid and amino acid contents and amino acid composition in fulvic acids, part 5. (In Japanese). J. Sci. Soil Manure, Japan 46:180–184. From Soil Sci. Plant Nutr. (Tokyo) 21:420–421.

Otsuki, A., and T. Hanya. 1967. Some precursors of humic acid in recent lake sediments suggested by infrared spectra. Geochim. Cosmochim. Acta 31:1505–1515.

Perry, D.R., and W.A. Adams. 1972. The incorporation of (^{14}C) labelled glycine into extracted humic acid. p. 59–67. In Report Welsh Soils Discussion Group. no. 13. University College, Aberystwyth, U.K.

Piper, T.J., and A.M. Posner. 1968. On the amino acids found in humic acids. Soil Sci. 106:188–192.

Piper, T.J., and A.M. Posner. 1972. Humic acid nitrogen. Plant Soil 36:595–598.

Porter, L.K., B.A. Stewart, and H.J. Hass. 1964. Effects of long-time cropping on hydrolyzable organic nitrogen fractions in some Great Plains soils. Soil Sci. Soc. Am. Proc. 28:368–370.

Rosell, R.A., J.C. Salfeld, and H. Sochtig. 1978. Organic components in Argentine soils: I. Nitrogen distribution in soils and their humic acids. Agrochimica 22:98–105.

Rudelov, Ye V., and D.N. Pyranishnikov. 1982. Dynamics of immobilized nitrogen. Sov. Soil Sci. (Engl. Transl.) 14:40–45.

Russell, J.D., D. Vaughan, D. Jones, and A.R. Fraser. 1983. An IR spectroscopic study of soil humin and its relationship to other soil humic substances in fungal pigments. Geoderma 29:1–12.

Saiz-Jimenez, C. 1983. The chemical nature of melanins from Coprinus spp. Soil Sci. 136:65–74.

Schnitzer, M. 1985. Nature of nitrogen in humic substances. p. 303–325. In G.R. Aiken et al. (ed.) Humic substances in soil, sediment, and water. Wiley-Interscience, New York.

Schnitzer, M., and D.A. Hindle. 1980. Effect of peracetic acid oxidation on N-containing components of humic materials. Can. J. Soil Sci. 60:541–548.

Schnitzer, M., P.R. Marshall, and D.A. Hindle. 1983. The isolation of soil humic acid components rich in "unknown" nitrogen. Can. J. Soil Sci. 63:425–433.

Schnitzer, M., M.I. Ortiz de Serra, and K. Ivarson. 1973. The chemistry of fungal humic acid-like polymers and of soil humic acids. Soil Sci. Soc. Am. Proc. 37:229–236.

Sequi, P., G. Guidi, and G. Petruzzelli. 1975. Distribution of amino acid and carbohydrate components in fulvic acid fractionated on polyamide. Can. J. Soil Sci. 55:439–445.

Shields, J.A., and E.A. Paul. 1973. Decomposition of ^{14}C-labelled plant material under field conditions. Can. J. Soil Sci. 53:297–306.

Simonart, P., L. Batistic, and J. Mayaudon. 1967. Isolation of protein from humic acid extracted from soil. Plant Soil 27:153–161.

Smith, S.J., F.W. Chichester, and D.E. Kissel. 1978. Residual forms of fertilizer nitrogen in field soils. Soil Sci. 125:165–169.

Sowden, F.J., and M. Schnitzer. 1967. Nitrogen distribution in illuvial organic matter. Can. J. Soil Sci. 45:111–116.

Stepanov, V.V. 1969. Reaction of humic acids with some nitrogen-containing compounds. Sov. Soil Sci. (Engl. Transl.) 2:167–173.

Stevenson, F.J. 1960. Chemical nature of the nitrogen in the fulvic fraction of soil organic matter. Soil Sci. Soc. Am. Proc. 24:472–477.

Stevenson, F.J. 1982. Organic forms of soil nitrogen. p. 67–122. In F.J. Stevenson (ed.) Nitrogen in agricultural soils. Agron. Monogr. 22. ASA, Madison, WI.

Stevenson, F.J. 1985a. Nitrogen transformations in soil: A perspective. p. 7–26. In K.A. Malik et al. (ed.) Nitrogen and the environment. Nuclear Inst. for Agriculture and Biology, Faisalabad, Pakistan.

Stevenson, F.J. 1985b. Cycles of soil: C, N, P, S, micronutrients. Wiley-Interscience, New York.

Stevenson, F.J., and K.M. Goh. 1971. Infrared spectra of humic acids and related substances. Geochim. Cosmochim. Acta 35:471–483.

Stevenson, F.J., and K.R. Kelley. 1985. Stabilization, chemical characteristics and availability of immobilized nitrogen in soils. p. 239–259. In K.A. Malik et al. (ed.) Nitrogen and the environment. Nuclear Inst. for Agriculture and Biology, Faisalabad, Pakistan.

Tan, K.H., P. Sihanonth, and R.L. Todd. 1978. Formation of humic acid-like compounds by the entomycorrhizal fungus, Pisolithus tinctorius. Soil Sci. Soc. Am. J. 42:906–908.

Tsutsuki, K., and S. Kuwatsuka. 1978. Chemical studies of soil humic acids: III. Soil Sci. Plant Nutr. (Tokyo) 24:29–38.

Tsutsuki, K., and S. Kuwatsuka. 1979. Chemical studies of soil humic acids: IV. Soil Sci. Plant Nutr. (Tokyo) 25:29–38.

Wagner, G.W. 1968. Significance of microbial tissue to soil organic matter. p. 197–205. *In* Isotopes and radiation in soil organic matter studies. Tech. Meet. FAO/IAEA, Vienna, Austria.

Witthauer, J., and R. Klocking. 1971. Bundungasrten des Stickstoffs in Huminsauren. Arch. Acker-Pflanzenbau Bodenkd. 15:577–588, 663–670.

Wojcik-Wojtkowiak, D. 1978. Nitrogen transformations in soil during humification of straw labeled with ^{15}N. Plant Soil 49:49–55.

Chapter 6

Roles of Organic Matter, Minerals, and Moisture in Sorption of Nonionic Compounds and Pesticides by Soil

CARY T. CHIOU, *U.S. Geological Survey, Denver, Colorado*

ABSTRACT

The sorption of nonionic organic compounds and pesticides on soil is here related to solute and solvent properties and to the following soil characteristics: organic matter content, mineral matter, and moisture content. A wide range of sorption behavior can be accounted for by considering the soil to be a dual sorbent, in which the mineral fraction of the soil functions as a conventional solid adsorbent and the organic matter functions as a partition medium. In aqueous systems, adsorption on mineral matter is suppressed by water, and the uptake by soil consists primarily of solute partitioning into the organic matter; this model agrees with the observed dependence of uptake by soil on soil organic matter content, the linearity of the isotherms, the small heat of soil sorption, and the absence of competition between solutes. By contrast, sorption from nonpolar organic solvents on dry soils is attributed to adsorption on soil minerals. Here the specific interaction of the adsorbate polar groups overcomes the weaker adsorptive competition by the (nonpolar) solvents, while the partition effect into soil organic matter is minimized by the relatively high solvency of the medium. Soil uptake from nonpolar solvents is depressed by soil moisture and approaches zero when the soil becomes fully saturated with water. The markedly higher sorption of organic vapors by dry and subsaturated soils, relative to that by wet soils, is ascribed to mineral adsorption, which predominates over the simultaneous uptake by partitioning into the organic matter, in which the mineral adsorptivity is influenced by the clay type and content. An increase of ambient humidity or soil moisture sharply depresses the vapor uptake because of adsorptive displacement by water on mineral surfaces. A small residual vapor uptake on water-saturated soil is attributed to the partitioning into the soil organic phase, which is similar to sorption by soil from aqueous systems. The effect of humidity on the activity and toxicity of nonionic organic pesticides in soil can be accounted for by the effect of humidity on sorption of the pesticide. The transport and fate of similar organic pollutants in the environment should likewise be influenced by ambient humidity.

The mechanism of sorption of nonionic organic compounds and pesticides by soils has been a subject of profound interest for many decades. Research in this field was especially active between the 1950s and 1970s because of increasing use of organic pesticides, which called for a need to assess the effectiveness and persistence of the chemicals in association with the soil. Since the 1970s, a growing concern over the likelihood and extent of environmental contamination by organic products from agricultural and industrial applications has further stimulated research in this subject. The development of this field of research has now come to the point where the individual roles of soil organic matter and minerals in sorption of nonionic organic compounds and pesticides by soil can be placed in a much better perspective. This enables us to reexamine the old and new data for consistency and to assess the behavior of organic pollutants and pesticides introduced into the environment.

Relative to nonionic organic compounds, the mechanism of sorption of ionic organic compounds by soil has been better characterized. For instance, cationic compounds (such as paraquat, diquat, and certain s-triazines in cationic forms) are known to sorb by ion exchange with cations in soils and clays (Weber et al., 1965; Tahoun & Mortland, 1966; Weber, 1966; Knight & Tomlinson, 1967; Weed, 1968); organic anions usually exhibit relatively insignificant sorption on soil by ion exchange or by other mechanisms (Frissel & Bolt, 1962; Harris & Warren, 1964; Weber et al., 1965). Since there is relatively little ambiguity concerning the mechanism of soil uptake of ionic species, the present discussion will focus on the sorptive interactions of nonionic compounds with soils.

In the following text covering the mechanistic roles of soil organic matter and minerals in the uptake of organic compounds and pesticides by soil, some terminology is used frequently in referring to the mechanism involved. The term *sorption* is commonly used to denote uptake of a solute or a vapor by soil without reference to a specific mechanism. The term *adsorption* onto a solid refers to a condensation of vapor or solute (both being referred to as *adsorbate*) on the surface or interior pores of a solid (*adsorbent*). While no explicit distinction is made between different types of adsorption in this chapter, it is worth pointing out that physical adsorption is governed by London–van der Waals forces (the same forces that are responsible for adhesion of a molecule to a bulk solid) between the adsorbate and adsorbent. When the forces leading to adsorption are related to chemical bonding forces, the adsorption is referred to as *chemisorption*. The distinction between these two types of adsorption is not always a sharp one; adsorption of polar molecules onto polar solids may fall under either classification depending on the adsorption energy. In contrast to adsorption, the terms *partition* or *partitioning* are used to describe a model in which the sorbed material dissolves in an organic phase by forces common to solution (e.g., by van der Waals forces). This is analogous to the extraction of a solute from water into an organic phase. When the organic phase is a solid (e.g., soil organic matter), partitioning is distinguished from adsorption by the homogeneous distribution of the sorbed material throughout the entire volume of the solid phase, analogous to, for example, the partitioning of benzene into polystyrene. In adsorption, the adsorbate occupies only the surface of the solid and does not permeate its entire volume. The equilibrium partition coefficient of a compound between an organic phase and water is a function of its solubilities in the two solvent phases, and the magnitudes of the individual solubilities are strongly influenced by interactions of the functional groups between the solute and individual solvents.

Finally, as an aid to the discussion of the sorptive behavior of the compounds with soil, the common names, chemical names, and structures of those pesticides identified in this chapter are shown in Table 6–1.

OBJECTIVE

This article presents an overview of the sorptive interactions of nonionic compounds and pesticides with soils and their relations to the activity of the compounds. Detailed consideration will be given to the relative roles of soil organic matter and minerals and the effect of water and other solvents to account for differences in soil behavior under various system conditions. Throughout this discussion, the term *organic matter* will be used to refer to the bulk of the organic content in soil. While differences in composition between humic and nonhumic components in organic matter can influence the overall behavior of the soil organic matter, separate treatments of their contributions to soil sorption are not possible at this time because practically all sorption studies have been carried out in the past with intact soils.

6–1 HISTORICAL BACKGROUND

The complex and heterogeneous nature of soil or sediment compositions has undoubtedly been a major retarding factor in the development of this field of science for some time, especially with regard to the individual functions of soil organic matter and minerals. The common concept adopted by earlier researchers in the study of pesticide–soil interactions was to regard the soil either as a single adsorbent or as a mixed adsorbent of some kind, analogous to some well-defined adsorbents. While such a view accounts for sorption data that were observed under certain conditions, it runs into difficulties in attempting to explain the sorptive behavior of the soil in aqueous systems, i.e., for the water-saturated soil. The adsorptive character of the soil was recognized indisputably, for instance, in the earlier study of vapor uptake of methyl bromide (Chrisholm & Koblitsky, 1943), chloropicrin (Stark, 1948) and ethylene dibromide (Hanson & Nex, 1953; Wade, 1954) by dry and hydrated soils and clays, in which the uptake of the organic vapors was suppressed by the moisture content in soils and clays. Similarly, the uptake of parathion and lindane from hexane solution by soils was found to be suppressed by soil moisture (Yaron & Saltzman, 1972; Chiou et al., 1985).

These observations indicate that organic compounds (pesticides) and water compete for adsorption on soil (mineral) surfaces. In addition, the uptake of organic compounds from the vapor phase and from hexane solution on dry soils and clays gives nonlinear isotherms characteristic of the adsorptive nature of the soil and clay.

On the other hand, the uptake of nonionic organic compounds and pesticides by soils in aqueous systems displays a set of unique features. Most notably, the extent of soil uptake for given organic compounds and pesticides shows a strong dependence on the organic matter content of the soil (Sherburne & Freed, 1954; Bailey & White, 1964; Goring, 1967; Lambert, 1968; Hamaker & Thompson, 1972; Saltzman et al., 1972; Savage & Wauchope, 1974; Choi & Chen, 1976; Filonow et al., 1976; Browman & Chesters, 1977; Felsot & Dahm, 1979; Chiou et al., 1979, 1983; Karickhoff et al., 1979; Chiou, 1981; Karickhoff, 1981; Means et al., 1982; Nkedi-Kizza et al., 1983). The uptake of organic vapors by wet soils shows similar effects (Wade, 1954;

Table 6–1. Common names, chemical names, and structures of the organic compounds and pesticides involved in sorption by soil (sediment) mentioned in the text.

Common name	Chemical name	Structure
β-BHC	1,2,3,4,5,6-Hexachloro-cyclohexane,β isomer	
Carbofuran	2,3-Dihydro-2,2-dimethyl-7-benzofuranyl methylcarbamate	

(continued on next page)

Table 6-1. Continued.

Common name	Chemical name	Structure
Chloropicrin	Trichloronitromethane	
Chlorpyrifos	O,O-Diethyl-O-(3,5,6-tri-chloro-2-pyridyl) phosphorothioate	
2,4-D	2,4-Dichlorophenoxy-acetic acid	
DDE	2,2-bis (p-Chlorophenyl)-1,1-dichloroethene	
DDT	2,2-bis (p-Chlorophenyl)-1,1,1-trichloroethane	
Diazinon	O,O-Diethyl O-(2-iso-propyl-6-methyl-4-pyrimidinyl) phos-phorothioate	
Dichlofenthion	O,O-Diethyl O-2,4-di-chlorophenyl phos-phorothioate	
Dieldrin	1,2,3,4,10,10-Hexachloro-6,7-eposy-1,4,4a,5,6,7,8,8a-octahydro-exo-1,4-endo-5,8-dimethano-naphthalene	

(continued on next page)

Table 6–1. Continued.

Common name	Chemical name	Structure
Diquat	1,1'-Ethylene-2,2'-bi-pyridylium ion	
Diuron	3-(3,4-Dichlorophenyl)-1,1-dimethylurea	
EDB	1,2-Dibromoethane	BrH_2C-CH_2Br
Fensulfothion	O,O-Diethyl O-[4-(methyl-sulfinyl) phenyl] phosphorothioate	
Fonofos	O-Ethyl S-phenyl ethyl-phosphonodithioate	
Heptachlor	1,4,5,6,7,10,10-Hepta-chloro-4,7,8,9-tetra-hydro-4,7-endo-methyleneindene	
Lindane	1,2,3,4,5,6-Hexachlorocy-clohexane,γ isomer	
Monuron	3-(p-Chlorophenyl)-1,1-dimethyl urea	
Paraquat	1,1'-Dimethyl-4,4'-bi-pyridinium ion	

(continued on next page)

Table 6–1. Continued.

Common name	Chemical name	Structure
Parathion	*O,O*-Diethyl *O-p*-nitro-phenyl phosphoro-thioate	
PCB	Polychlorinatedbiphenyl	
Phenanthrene	Phenanthrene	
Phorate	*O,O*-Diethyl *S*-(ethyl-thiomethyl) phos-phorodithioate	
Phorate sulfone	*O,O*-Diethyl *S*-[(ethyl-thiomethyl) sulfone] phosphorodithioate	
Phorate sulfoxide	*O,O*-Diethyl *S*-[(ethyl-thiomethyl) sulfoxide] phosphorodithioate	
Pyrene	Pyrene	
Terbufos	*O,O*-diethyl *S*-[(1,1-di-methylethyl) thio-methyl] phosphoro-dithioate	
Terbufos sulfone	*O,O*-Diethyl *S*-[(1,1-di-methylethyl)thio-methyl sulfone] phos-phorodithioate	
Terbufos sulfoxide	*O,O*-Diethyl *S*-[(1,1-di-methylethyl) thio-methyl sulfoxide] phosphorodithioate	

Leistra, 1970). The importance of soil organic content in sorption by wet soils is demonstrated by the relative invariance of the sorption data of given organic compounds among soils, or size fractions of soil, when normalized on the basis of the organic matter content in the soil sample (Goring, 1967; Lambert, 1968; Hamaker & Thompson, 1972; Kenaga & Goring, 1980; Chiou, 1981; Karickhoff, 1981; Schwarzenbach & Westall, 1981). The equilibrium sorption isotherms of compounds on hydrated soils are all essentially linear (Wade, 1954; Swoboda & Thomas, 1968; Leistra, 1970; Saltzman et al., 1972; Yaron & Saltzman, 1972; Chiou et al., 1979, 1983, 1985; Karickhoff et al., 1979), and are not strongly temperature dependent, giving only relatively small exothermic heats (Mills & Biggar, 1969; Spencer & Cliath, 1970; Hamaker & Thompson, 1972; Yaron & Saltzman, 1972; Pierce et al., 1974; Chiou et al., 1979, 1985). Moreover, soil uptake of binary solutes from water shows no apparent competition between the solutes (Chiou et al., 1983, 1985) in contrast to the strong competitive effects that are found in sorption by dry soil from the vapor phase and from organic solvents (Chisholm & Koblitsky, 1943; Stark, 1948; Hanson & Nex, 1954; Wade, 1954; Spencer et al., 1969; Spencer & Cliath, 1970; Yaron & Saltzman, 1972; Chiou et al., 1985; Chiou & Shoup, 1985).

In spite of the fact that the observed relationship between soil uptake and organic matter content had greatly simplified predictions of the uptake of nonionic organic compounds from water by soils, there was no general agreement regarding the sorptive mechanism associated with soil organic matter. Many researchers assumed that organic matter functions as a high-surface-area adsorbent (Bailey & White, 1964; Haque, 1975; Browman & Chesters, 1977) capable of adsorbing nonionic organic compounds by hydrophobic interactions (Weber & Weed, 1974; Carringer et al., 1975; Browman & Chesters, 1977; Hassett et al., 1981; Stevenson, 1982; Mingelgrin & Gerstl, 1983). The hydrophobic adsorption concept, however, is not in keeping with general adsorption criteria and with soil sorption data to be discussed later in this chapter.

In considering the strong uptake of monuron on muck soil in aqueous systems as reported by Sherburne and Freed (1954), Hartley (1960) speculated that a solvent action of the "oily constituent" of the soil organic matter might be responsible for the enhanced soil uptake (while making no effort to distinguish this effect from surface adsorption). This view was further emphasized by Goring (1967) in discussing the distribution of organic toxicants between soil water and soil solids. While it is quite reasonable that the oily constituent in organic matter would exhibit such a solvent effect, the role of soil organic matter as a whole (or that of the nonoily components) remained to be specified. Lambert (1967, 1968) later assumed the role of soil organic matter to be analogous to that of a solvent medium in partition chromatography, and suggested a correspondence between sorption coefficient and solvent–water distribution coefficient. Swoboda and Thomas (1968) noted that the soil–water equilibrium isotherm of parathion was linear over a wide range of equilibrium concentrations in water and suggested that part of the parathion might be "adsorbed" as a liquid dissolved in the organic fraction of the soil similar to the partitioning process in liquid–liquid extraction. Further evidence confirming the partitioning of the pesticide between organic matter and water was not available at that time, however, and consequently the validity and significance in soil uptake by partitioning into soil organic matter in aqueous and nonaqueous systems was not duly recognized. Subsequently, numerous studies illustrated the existence of linear relations between the sorption coefficients of organic compounds and pesticides in aqueous systems, on the basis of the soil organic matter content (K_{om}) or the soil organic C content (K_{oc}), and the corresponding octanol–water partition coefficients (K_{ow}) as a means of estimating sorption coefficients from K_{ow} values (Hamaker & Thompson, 1972; Briggs, 1973, 1981; Chiou et al., 1979, 1983; Karickhoff et al., 1979; Kenaga & Goring, 1980; Means et al., 1980, 1982; Hassett et al., 1981; Karickhoff, 1981; McCall et al., 1983; Dzombak & Luthy, 1984).

Chiou et al. (1979, 1981, 1983, 1984) took account of the sorption data in aqueous systems and presented evidence for the presumed partitioning (solubilization) of nonionic organic compounds into soil organic matter as the major process of sorption from water. In addition to the recognized dependence of soil sorption on organic matter content, they noted that equilibrium isotherms in water are essentially linear up to high relative concentrations (ratios of equilibrium concentrations to solute solubilities); moreover, the equilibrium heats of sorption for solutes are less exothermic than heats of solute condensation from water (i.e., the reverse heats of solution), and the system shows a lack of solute competition. The inability of the soil mineral fraction to adsorb nonionic organic compounds from water is attributed to the strong dipole interaction of minerals with water, as pointed out by Goring (1967), which excludes the organic compounds from this portion of the soil. In keeping with the hypothesis of solute partitioning into the organic matter as the dominant mechanism of soil uptake, solutes with lower water solubilities as liquids or supercooled liquids show higher equilibrium sorption coefficients (K_{om}) but lower limiting sorption capacities. By application of the Flory–Huggins model to account for solute solubility in (amorphous) soil organic phase, Chiou et al. (1983) established a partition equation to account for the magnitudes of the observed sorption coefficients. This analysis led to the recognition that the primary factor affecting the sorption coefficients of slightly water-soluble organic compounds is the solubility of the compounds (as liquids or supercooled liquids) in water. The frequently observed empirical correlation between sorption coefficient (K_{om}) and octanol–water partition coefficient (K_{ow}) was recognized to be merely a consequence of the fact that water solubility is the major determinant of both K_{om} and K_{ow} values (Chiou et al., 1982, 1983).

The described sorptive characteristics of nonionic organic solutes on soil from aqueous and nonaqueous systsems can be reconciled by the postulate that the soil behaves as a dual sorbent, rather than a single or mixed adsorbent. The mineral matter in soil behaves as a conventional solid adsorbent and the organic matter as a partition medium. The linear isotherms and other characteristics as observed in aqueous systems are attributed to solute partitioning into the organic matter, with a concomitant suppression of adsorption by water on mineral matter.

The dual functionality of the soil has been further illustrated in a comparison of the sorptive characteristics of parathion and lindane from water and from organic–solvent solution (Chiou et al., 1985). According to the model, the high solubility of organic solutes in organic solvents minimizes the partition effect with soil organic matter, and therefore the net uptake of the solute should depend mainly on the solute's ability to compete with the solvent for adsorption on soil minerals. The results from Chiou et al. (1985) indicate that the sorption capacity of parathion from hexane on a dehydrated (Woodburn) soil (fine-silty, mixed, mesic Aquultic Argixeroll) is more than two orders of magnitude greater than that from aqueous solution on the same soil at equal relative concentrations in the two solvents. While the equilibrium isotherms for parathion uptake by soil in aqueous systems are linear, the isotherms from hexane solution are distinctly nonlinear, and the uptake is sharply reduced by the addition of water as recognized earlier by Yaron and Saltzman (1972). The parathion sorption from hexane on dry soil is accompanied by high molar exothermic heats (which vary with parathion loading) in contrast to relatively low and virtually constant molar heats of sorption from aqueous solution. Adsorptive competition is found between parathion and lindane in hexane systems, in contrast to the absence of such an effect in aqueous systems.

Lindane likewise shows a markedly higher uptake by soil from hexane solution than from water and displays a nonlinear isotherm. However, the sorption capacity in hexane solution on dry soil is much greater for parathion than for lindane, which is attributed to strong polar interactions of the polar groups in parathion with (polar) mineral surfaces. By contrast, these two compounds are shown to give relatively comparable uptakes in aqueous systems because of their comparable water solubilities, in keeping with the presumed solute partitioning into soil organic matter as the dominant sorption process. The contrasting effects of soil minerals and organic matter in sorption from water and from hexane are further demonstrated by a sharp reduction of the uptake of parathion from hexane on a dry peat soil (510 g organic matter kg^{-1} soil and low in mineral content) in comparison to that on the relatively mineral-rich Woodburn soil (19 g organic matter kg^{-1} soil and high in mineral content). This behavior is fundamentally different from that in aqueous systems.

With the foregoing brief account of the roles of soil organic matter and minerals and the effect of humidity, it is now fitting to examine in more detail soil sorption data of nonionic organic compounds in aqueous and nonaqueous systems for consistency. For better appreciation of the sorptive behavior of the soil in these systems, we shall first consider some important features associated with partition and adsorption equilibria. In a later section, we will then deal with the sorption of organic compounds and pesticides by soil from the vapor phase in relation to ambient humidity, a subject of practical importance in the understanding of the activity of organic contaminants and pesticides in the environment.

6-2 THEORETICAL CONSIDERATIONS OF ADSORPTION AND PARTITION INTERACTIONS

There are a number of useful textbooks and articles dealing with adsorption theory and with mathematical models for interpreting interactions of organic compounds with various types of adsorbents in vapor and solution systems (Brunauer, 1945; Kipling, 1965; Adamson, 1967; Manes, 1980; Gregg & Sing, 1982; Everett, 1983). Since it is beyond the scope of this chapter to discuss various adsorption models, the reader is advised to consult the cited literature. Rather, we shall give consideration here to some distinct characteristics of adsorption and partition equilibria to facilitate recognition of their mechanistic differences. In view of the fact that our common notion of partition equilibria comes largely from experience associated with the solvent–solvent extraction of solutes, it is desirable herein to elucidate the basic parameters that influence the partition constant in solvent–solvent systems, as well as other thermodynamic properties related to the solute partitioning. To keep the treatment of the subject in a proper manner, we shall first consider the partition equilibrium of organic solutes in ordinary solvent–solvent systems, in which the solute partition constants can be reasonably interpreted by conventional solution theory. Afterwards, we shall extend our treatment to systems involving a macromolecular organic phase as a partition medium by taking into account the effect of the size disparity between solute and macromolecular phase on solute solubility and partition constant. The latter treatment gives an immediate application to the understanding of solute partitioning into soil organic matter. We shall choose water as a common solvent in both systems. Many symbols are used in the theoretical development that follows and these symbols and their definitions are compiled in Table 6–2.

The partition coefficient of a slightly water-soluble organic compound in a solvent–water mixture, in which the solvent has a small solubility in water, can be derived by reference to the conventional solution theory for a two-phase mixture (Chiou et al., 1982), giving the result

$$\log K = -\log S_w - \log \overline{V_0^*} - \log \gamma_0^* - \log (\gamma_w/\gamma_w^*) \qquad [1]$$

where K is the solute partition coefficient (i.e., the ratio of the solute concentration in the organic phase to that in water); S_w is the molar water solubility of the solute as liquid or supercooled liquid at the system temperature (mol/L); $\overline{V_0^*}$ is the molar volume of water-saturated organic phase (L/mol); γ_0^* is the solute activity coefficient (Raoult's law convention) in water-saturated solvent phase; γ_w is the solute activity coefficient in water; and γ_w^* is the solute activity coefficient in solvent-saturated water. For a solute that is sparingly soluble in water, all the activity-coefficient terms are practically constant. Likewise, the molar volume term ($\overline{V_0^*}$) is constant, which is determined by the molar volume of the pure solvent corrected for water miscibility. The $\log (\gamma_w/\gamma_w^*)$ term corrects for the enhancement of solute solubility in water due to saturation of the solvent, which is relatively insignificant

Table 6-2. Glossary of symbols used in this chapter.

Symbol	Definition
C_e	Solute concentration in water or an organic solvent in equilibrium with soil or sediment.
C_e^*	Solute concentration in water in equilibrium with soil or sediment, taking into account the effect of dissolved organic matter in water on the apparent solute concentration.
γ_w	Solute activity coefficient in water (Raoult's law convention).
γ_w^*	Solute activity coefficient in water containing a dissolved organic solvent or dissolved soil organic matter, which modifies the solubility behavior of the solute.
γ_o^*	Solute activity coefficient in the water-saturated organic phase.
$\Delta \overline{G}^\circ$	Molar free energy change in the sorption of a solute at the standard state. See the text for the definition of the "standard state".
$\Delta \overline{H}^\circ$	Molar enthalpic change in the sorption of a solute at the standard state.
$\Delta \overline{H}$	Molar heat of sorption of a solute from a solvent onto a sorbent (soil or sediment).
$\Delta \overline{H}_d$	Molar heat of desorption of a solute from a sorbent into a solvent ($\Delta \overline{H}_d = -\Delta \overline{H}$).
$\Delta \overline{H}_f$	Molar heat of fusion of a solid compound.
$\Delta \overline{H}_o$	Molar heat of solution of a solute in an organic phase.
$\Delta \overline{H}_v$	Molar heat of vaporization of a compound.
$\Delta \overline{H}_w$	Molar heat of solution of a solute in water.
K	Solute partition coefficient between an organic phase and water.
K°	Solute partition coefficient as defined by the hypothetical "ideal line." See the definition following Eq. [1] and Eq. [3] in the text.
K_{ow}	Solute partition coefficient in octanol–water systems.
K_{om}	Solute partition coefficient between bulk organic matter in soil (sediment) and water.
K_{om}^*	K_{om}, taking into account the contribution of dissolved organic matter to the concentration (solubility) of the solute in water.
K_{oc}	K_{om} expressed in terms of the organic carbon content in the soil (or sediment) organic matter.
K_{oc}^*	K_{oc}, taking into account the contribution of dissolved organic carbon to the concentration (solubility) of the solute in water.
K_{dom}	Solute partition coefficient between dissolved organic matter and water.
P	Equilibrium partial pressure of a compound in sorption experiments.
P°	Saturation vapor pressure of a compound.
Q	Uptake of a compound (pesticide) by a unit mass of soil or sediment.
R	Gas constant.
ρ	Density of the soil organic matter.
S_w	Solubility of a solute in water.
S_{om}	Solubility of a solute in soil (or sediment) organic matter.
$\Delta \overline{S}^\circ$	Molar entropic change in the sorption of a solute at the standard state.

(continued on next page)

Table 6–2. Continued.

Symbol	Definition
T	System temperature (Kelvin).
$\overline{V}$	Molar volume of a solute.
$\overline{V}_0^*$	Molar volume of the water-saturated organic phase.
X	Concentration of dissolved organic matter in water.
χ	Flory–Huggins interaction parameter for a solute in a polymer (macromolecular) phase.
χ_H	Enthalpic contribution to χ.
χ_S	Entropic contribution to χ.

for solutes that have high water solubility in systems in which the solvent has a low solubility in water. The use of supercooled liquid solubility rather than solid solubility for solid solutes in Eq. [1] is to correct for the melting-point effect that affects only the solute solubility, but not the partition coefficient because this effect cancels in the two solvent phases (Mackay, 1977; Chiou et al., 1982).

Several important points associated with the partition equilibrium of relatively water-insoluble organic solutes in solvent–water mixtures are worth noting. First, the magnitudes of K for organic solutes are determined to a large extent by the reciprocals of S_w values, rather than by the solubilities in the organic solvent (i.e., the magnitudes of γ_0^*). The effect of γ_0^* on K is generally quite small relative to that of S_w for organic solutes because of their high degrees of miscibility with the organic solvent, as is typically found for nonionic organic compounds in octanol–water systems. As a result, one finds a linear relationship between log K_{ow} (octanol–water) and log S_w for a wide range of chemical classes (Chiou et al., 1977, 1982; Mackay et al., 1980; Miller et al., 1985). A list of the log K_{ow} and log S_w values for some aromatic solutes is given in Table 6–3. In Fig. 6–1, a plot of the observed log K_{ow} values against the corresponding log S_w for some organic solutes is presented in comparison with the "ideal line," which is defined as log K_{ow}° $= -\log S_w - \log \overline{V}_0^*$, with log $\overline{V}_0^* = -0.92$ for water-saturated octanol. The difference between log K_{ow}° and log K_{ow} gives a measure of log γ_0^* + log (γ_w/γ_w^*), or essentially of log γ_0^* for most solutes since log (γ_w/γ_w^*) is significant only for those solutes that are extremely insoluble in water (such as DDT and hexachlorobenzene).

Second, the partition coefficient (K) of a slightly water-soluble solute in a solvent–water mixture is largely independent of the solute concentration below the limit of the solute solubility. That this is true can be attributed to the fact that the activity coefficient of the solute in aqueous solution (which is inversely related to S_w) remains practically constant if the solubility is small and that the γ_0^* value is generally small and not sensitive to the solute concentration in the organic phase [because of high compatibility between solute and solvent (Chiou, 1981)]. It may thus be expected that a linear relation between solute concentration in the organic phase and that in the water phase should exist over a wide range of solute concentration relative to the

Table 6-3. Octanol–water partition coefficients and liquid and supercooled liquid solubilities of aromatic compounds (reproduced from data of Chiou et al., 1982).†

Compound	$\log S_w$, mol/L‡	$\log K_{ow}$
Aniline	−0.405	0.90
o-Toluidine	−0.817	1.29
m-Toluidine	−0.853	1.40
N-Methylaniline	−1.28	1.66
N,N-Dimethylaniline	−2.04	2.31
o-Chloroaniline	−1.53	1.90
m-Chloroaniline	−1.37	1.88
Benzene	−1.64	2.13
Toluene	−2.25	2.69
Ethylbenzene	−2.84	3.15
Propylbenzene	−3.30	3.68
Isopropylbenzene	−3.38	3.66
1,3,5-Trimethylbenzene	−3.09	3.42
t-Butylbenzene	−3.60	4.11
Fluorobenzene	−1.80	2.27
Chlorobenzene	−2.36	2.84
Bromobenzene	−2.55	2.99
Iodobenzene	−2.78	3.25
o-Xylene	−2.72	2.77
m-Xylene	−2.73	3.20
p-Xylene	−2.73	3.15
Diphenylmethane	−4.07	4.14
o-Dichlorobenzene	−2.98	3.38
m-Dichlorobenzene	−3.04	3.38
p-Dichlorobenzene	(−3.03)	3.39
1,2,4-Trichlorobenzene	−3.57	4.02
Biphenyl	(−3.88)	4.09
2-Chlorobiphenyl	(−4.57)	4.54
3-Chlorobiphenyl	−5.16	4.95
Naphthalene	(−3.08)	3.36
2-Methylnaphthalene	(−3.69)	4.11
Phenanthrene	(−4.48)	4.57
Anthracene	(−4.63)	4.54
Pyrene	(−5.24)	5.18
Hexachlorobenzene	(−5.57)	5.50
p,p'-DDT	(−6.74)	6.36

† Reprinted with permission from *Environmental Science and Technology*, Vol. 16, p. 8–9, Copyright © 1982. American Chemical Society, Washington, DC.
‡ Solubilities are the 20 to 25 °C values. Numbers enclosed with parentheses are supercooled liquid solubilities, calculated based on their solid solubilities with correction of the melting point (Chiou et al., 1982).

solute's solubility. This feature provides a useful basis for distinguishing a partition equilibrium from an adsorption equilibrium.

The third distinct feature in partition equilibria is the relative independence of the individual solute partition constants in binary or multiple solute systems, as is generally experienced in the solvent–solvent extraction of multiple solute components. The partition constant of a solute would be significantly affected by the presence of other solutes only when the solvency of the solvent media is significantly affected by very high concentrations of the other solutes. Such cases are, however, relatively rare. In the normal concentration range it has been shown, for example, that the individual solute

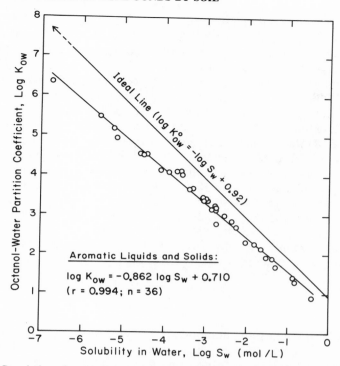

Fig. 6-1. Correlation of octanol–water partition coefficient and water solubility of aromatic liquids and solids from Table 6-3. Reproduced from Chiou et al. (1982), with permission from *Environmental Science and Technology*, Vol. 16, p. 8–9. Copyright © 1982. American Chemical Society, Washington, DC.

partition coefficients in octanol–water systems determined with a combination of solutes are essentially the same as when determined singly (Schellenberg et al., 1984). In adsorption, competition between solutes for adsorbent surfaces or specific sites of the adsorbent leads to reduction of the adsorption capacities of the individual solutes in a multisolute system relative to their capacities as single solutes (Manes, 1980; Everett, 1983). This reduction is generally very marked for the solutes in a multisolute system at the concentration range in which the individual single-component adsorption capacities of the solutes are relatively comparable.

Another unique characteristic that is associated with the partitioning of organic compounds into an organic phase is the observation that the limiting sorption capacities for such systems tend to be smaller with compounds exhibiting higher partition coefficients (or isotherm slopes). For nonionic solutes the high partition constants (and low water solubilities) tend to correlate strongly with molecular weight; therefore, solutes with high molecular weights tend to have lower limiting sorption capacities (although the effect is considerably less than directly proportional). The inverse relationship between limiting capacity (at water saturation) and molecular weight is opposite to what one would expect for adsorption on a solid, where one would expect either a constant limiting adsorption capacity (for pore filling, as on activated carbon) or an increase in adsorption capacity (for open surfaces). By contrast, the reduction of solubility in the organic phase with increasing molecular weight of the solute (which magnifies any incompatibility between solute and organic solvent) is just what one would expect for a partition process. This point is illustrated by a comparison of the solubilities and octanol–water partition coefficients of DDT and benzene. The K_{ow} value of DDT is about 4 orders of magnitude greater than that of benzene, whereas the S_w of DDT as a supercooled liquid is about 5 orders of magnitude lower than that of benzene. While DDT has a solubility in water-saturated octanol of about 32 g/L (or 0.090 mol/L) as a solid at 24 °C (the estimated supercooled solubility is about 10 times greater), benzene is completely miscible with the solvent (Chiou et al., 1982).

Lastly, it is useful to consider the equilibrium heat associated with the partitioning of a solute. The molar heat of transfer $(\Delta\overline{H})$ of a solute from water to an organic phase by partitioning is simply equal to the difference of the heats of solution of the solute in the two solvent phases (Chiou et al., 1979; Chiou, 1981). That is

$$\Delta\overline{H} = \Delta\overline{H}_o - \Delta\overline{H}_w \qquad [2]$$

where $\Delta\overline{H}_o$ is the heat of solution in the organic phase and $\Delta\overline{H}_w$ is the heat of solution in water, both of which are usually positive. Low solubilities of organic compounds in water (i.e., poor compatibilities with water) are commonly associated with large $\Delta\overline{H}_w$ values. The corresponding $\Delta\overline{H}_o$ terms should be much smaller for organic compounds because of their enhanced compatibilities with organic solvents. Equation [2] applies for both solid and liquid compounds. For solids, $\Delta\overline{H}_o$ and $\Delta\overline{H}_w$ include heats of fusion $(\Delta\overline{H}_f)$ as a part of heats of solution, which are constant for given solid substances. Therefore, $\Delta\overline{H}$ reflects only the difference of the heats of incompatibility of the dissolved liquid or supercooled liquid solute in the solvent media. According to Eq. [2], the heat of partition of an organic solute would be relatively small and less exothermic than its heat of condensation from water $(-\Delta\overline{H}_w)$. This expectation is in keeping with the fact that the partition coefficients of organic compounds in octanol–water (and other solvent–water) systems show relatively small inverse dependence on temperature (Chiou, 1981). By contrast, adsorption of a solute from dilute solution on an adsorbent exhibits a higher exothermic heat that results both from condensation of the solute on the surface and from (exothermic) interaction of the solute with the surface. Therefore, $\Delta\overline{H}$ for adsorption must be more exothermic than the corresponding heat of solute condensation from the solution (i.e., the reverse heat of solution); otherwise, the adsorption would be extremely weak. The difference in equilibrium heat between partition and adsorption is expected to be especially marked for solid compounds that have large $\Delta\overline{H}_f$ values, because $\Delta\overline{H}_f$ adds to the heat of solution, which becomes part of the heat of adsorption but has no effect on partitioning (as $\Delta\overline{H}_f$ cancels in partition equilibria). In a later section, we shall investigate the heat effect associated with the sorption of DDT (which has a large $\Delta \overline{H}_f$) from water by soil or sediment as a basis for distinguishing partitioning from adsorption.

Up to this point, we have considered many characteristics pertinent to the partition equilibria of nonionic organic compounds in a solvent–water system. While these features are common, we should note, however, that Eq. [1] gives a correct expression of the magnitude of the partition constant (K) only for systems in which the size disparity between solute and solvent is relatively small. Equation [1] loses its rigor when the size disparity between solute and solvent is sufficiently large. This may be seen from its improper implication that the partition constant of a solute should greatly decrease with a large increase of the solvent molar volume (or molecular weight), which is not consistent with known high athermal solubilities of small organic molecules in (amorphous) polymeric substances (Flory, 1942, 1953; Huggins, 1942). Consequently, Eq. [1] must be corrected by taking account of the effect of size disparity on the solubility of a solute in a solvent for solute partitioning into a macromolecular organic phase (such as soil organic matter). The modified partition equation has been developed by Chiou et al. (1983) by applying the Flory–Huggins theory for the solute solubility in an amorphous polymer. On the assumption that the molar volume of the solute is negligibly small compared to that of the polymer, one gets

$$\log K = -\log S_w \overline{V} - \log \rho - (1 + \chi)/2.303 - \log (\gamma_w/\gamma_w^*) \qquad [3]$$

in which $\overline{V}$ is the molar volume of the solute (L mol^{-1}), ρ is the density of the organic phase introduced to express the concentration of the solute in the organic phase on a mass-to-mass basis since the volume of amorphous material is often more difficult to determine, and χ is the Flory–Huggins interaction parameter (dimensionless), that is, the sum of excess enthalpic (χ_H) and excess entropic (χ_S) contributions to the incompatibility of the solute in a polymer. The value of χ_S depends presumably on the characteristics of the polymer network that affect the accessibility and flexibility of the polymer segment, and is usually determined empirically. Scott (1949) suggested that $\chi_S \simeq 0.25$ for high polymers, and this value was adopted by Chiou et al. (1983) for soil organic matter. The $\chi_H/2.303$ term in Eq. [3], equivalent to the $\log \gamma_o^*$ term in Eq. [1], expresses the extent of solute incompatibility with the macromolecular phase. The term $\log (\gamma_w/\gamma_w^*)$, as defined earlier, expresses the magnitude of solubility enhancement of the solute in water resulting from solubilization of a fraction of the macromolecular organic substance. As before, an "ideal line" can be established by letting $\log K^\circ = -\log S_w \overline{V} - \log \rho - (1 + \chi_S)/2.303$ and the difference between $\log K^\circ$ and $\log K$ gives a measure of the effect of $[\chi_H/2.303 + \log (\gamma_w/\gamma_w^*)]$, relative to $-\log S_w \overline{V}$, on $\log K$. In situations where the molar-volume ratio between solute and solvent ($\overline{V}/\overline{V}_o^*$) is not negligible, a more general partition equation should be used instead (Chiou, 1985).

Equation [3] has been applied in analyzing sorption (partition) equilibria of nonionic organic compounds between soil organic matter and water (Chiou et al., 1983). A more detailed discussion of the partition equilibrium with soil organic matter will be given later in this chapter. It is proper to point out, however, that analysis of the partition data by application of Eq. [3] leads to essentially the same conclusion as recognized in ordinary solvent–water systems that S_w is the major factor affecting the partition coefficients (K) of the compounds with the soil organic phase.

With the above consideration of the characteristic difference between adsorption and partition, we shall now examine the sorptive behavior of soils in aqueous and nonaqueous systems in relation to the roles of soil organic matter and mineral constituents.

6–3 SORPTION BY SOIL IN AQUEOUS SYSTEMS

6–3.1 Soil-water Equilibrium Characteristics

In water solution, the equilibrium isotherms for the sorption of non-ionic organic compounds on soils or sediments are usually virtually linear, as has been demonstrated in a number of studies (see, e.g., Lambert, 1968; Swoboda & Thomas, 1968; Mills & Biggar, 1969; Hamaker & Thompson, 1972; Yaron & Saltzman, 1972; Pierce et al., 1974; Chiou et al., 1979, 1983; Karickhoff et al., 1979; Means et al., 1980, 1982; Mingelgrin & Gerstl, 1983; Chiou & Shoup, 1985). Similar isotherm linearity has been found in soil uptake of pesticides from the vapor phase on moist (water-saturated) soils (Leis-

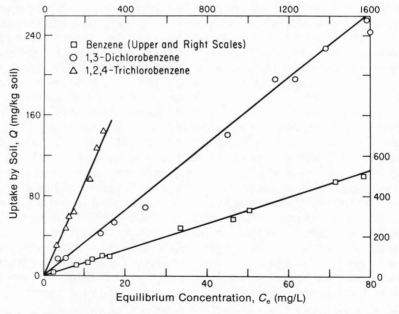

Fig. 6–2. Soil-water equilibrium isotherms of benzene, 1,3-dichlorobenzene, and 1,2,4-trichlorobenzene on Woodburn soil (1.9% organic matter) at 20 °C. Reproduced from Chiou et al. (1983) with permission from *Environmental Science and Technology*, Vol. 17, p. 229. Copyright © 1983. American Chemical Society, Washington, DC.

tra, 1970; Spencer & Cliath, 1970). In some studies where slight isotherm curvatures were shown (see, e.g., Saltzman et al., 1972; Mingelgrin & Gerstl, 1983), the extent of the reported isotherm nonlinearity (either concave upward or downward) appears to be within the normal range of data scatter that cannot be distinguished from a linear plot. This is especially true with soil samples that contain relatively low organic matter contents, because in such cases determinations of small changes of aqueous concentrations before and after sample equilibration (as commonly used for determining uptake by soil) are subject to greater experimental errors. The fact that the curvatures in some allegedly nonlinear isotherms (Mingelgrin & Gerstl, 1983) are quite small and that they are also not consistently concave upward or concave downward further supports the above interpretation. While in previous studies the isotherms presented were often confined to a range considered representative of practical situations, such linear isotherms also extend to high relative concentrations for sparingly water-soluble nonionic organic compounds and pesticides. Figure 6–2 shows typical linear isotherms for the sorption of benzene, 1,3-dichlorobenzene, and 1,2,4-trichlorobenzene from water on a Woodburn soil at 20°C. The soil contains 1.9% organic matter. The linearity of the benzene isotherm extends to about $C_e/S_w = 0.90$, where C_e and S_w are the equilibrium solute concentration and solubility in water, respectively. Similar linear isotherms for many other compounds on a Willamette silt loam (fine-silty, mixed, mesic pachic Ultic Argixeroll) (16 g organic matter kg^{-1} soil) are shown in Fig. 6–3, in which, for example, 1,2-dichlorobenzene shows linear sorption with C_e/S_w up to 0.95.

The high linearity range of the isotherm, together with the recognized dependence of sorption on soil organic matter in aqueous systems, gives strong evidence that soil sorption of nonionic organic compounds from water is effected mainly by partitioning in (rather than by adsorption on) the soil or-

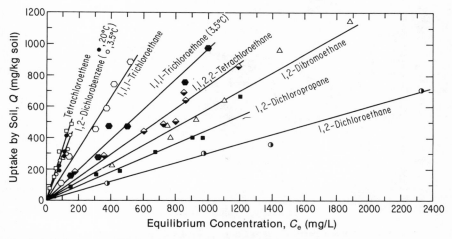

Fig. 6–3. Soil-water equilibrium isotherms of selected organic liquids on Willamette silt loam (1.6% organic matter) at 20°C. Reproduced from Chiou et al. (1979) with permission from *Science*. Copyright © 1979 by the AAAS.

ganic matter. Additional data will be presented later to further substantiate this view. Although the partition concept was suggested by Swoboda and Thomas (1968) based on the linear parathion sorption by soil from water, it has, however, not gained wide acceptance presumably because of a lack of other supporting data. As a matter of fact, there appears to be some misconception about the implication of linear isotherms in the literature. The linearity of sorption isotherms, as frequently observed for sparingly water-soluble organic compounds and pesticides on soil, was often thought to be a manifestation of the solutes' low concentrations in water that restrict soil "adsorption" to a linear range (Mingelgrin & Gerstl, 1983; MacIntyre & Smith, 1984). Although adsorption isotherms of single solutes (and vapors) are necessarily linear at low relative concentrations (C_e/S_w), commonly referred to as the "Henry's-law region," the observed isotherm linearity extending over almost the entire range of relative concentration (as with soil uptake from water) should not be confused with the linear portion of a nonlinear adsorption isotherm. To make this point evident, one may, for example, compare the adsorption isotherms of 1,2-dichlorobenzene, 1,1,1-trichloroethane, 1,2-dichloropropane, and 1,2-dibromoethane on an activated carbon (Fig. 6–4) with their sorption isotherms on a Willamette silt loam (Fig. 6–3). One notes that the isotherms of the compounds on activated carbon are linear only at very low equilibrium concentrations relative to their solubilities (which are 148, 1360, 3570, 3520 mg L^{-1}, respectively), whereas the sorption isotherms on soil show no obvious indication of curvature even at concentrations approaching saturation.

An important corollary of linear isotherms on soils in aqueous systems is the implication that the molar heat of sorption of the compound is constant and independent of loading. This effect can be readily illustrated by

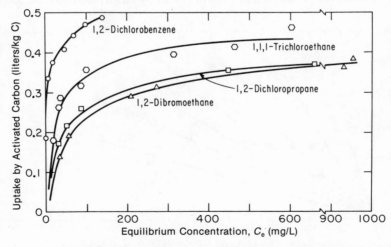

Fig. 6–4. Adsorption isotherms of selected organic liquids from water on Pittsburgh CAL (12 × 40) activated carbon at 20°C. The adsorption capacity is expressed in adsorbate volume per kilogram of carbon based on the bulk density of the adsorbate. Reproduced from Chiou (1981) with permission from Academic Press, Orlando, FL.

thermodynamic calculations of the equilibrium heat of sorption of a compound from its linear isotherms at two temperatures, as have been shown in many studies (Mills & Biggar, 1969; Spencer & Cliath, 1970; Yaron & Saltzman, 1972; Chiou et al., 1979). Figure 6-5 gives a schematic plot of the linear isotherms of a compound at temperatures T_1 and T_2 (K), with $T_2 > T_1$. Let Q denote the mass of the compound taken up by a unit mass of soil (or, more correctly, by a unit mass of soil organic matter), C_e the aqueous concentration of the compound in equilibrium with given Q, and S_w the solubility of the compound. Each linear isotherm is assumed to extend over a wide range of equilibrium concentrations relative to solubility at the system temperature. The isotherms are drawn such that the soil uptake at T_1 is greater than that at T_2, as is usually observed for most organic compounds and pesticides; however, a reverse temperature dependence can take place if the compound shows abnormal (i.e., exothermic) heat of solution in water, as shown for 1,1,1-trichloroethane (Chiou et al., 1979).

The molar isosteric heat of soil sorption at a given uptake capacity can be obtained by use of the classical Clausius–Clapeyron equation. At capacity Q_A, for example, the equation gives

$$\Delta\overline{H}\,(Q_A) = -R \ln\,[C_e(A,T_2)/C_2(A,T_1)]/(1/T_1 - 1/T_2) \qquad [4]$$

where R is the gas constant, and $C_e(A,T_2)$ and $C_e(A,T_1)$ are the equilibrium concentrations corresponding to Q_A at temperatures T_2 and T_1, respectively. The molar heat of sorption at capacity Q_B, i.e., $\Delta\overline{H}(Q_B)$, can be similarly determined by substituting $C_e(B,T_2)$ for $C_e(A,T_2)$ and $C_e(B,T_1)$ for $C_e(A,T_1)$ in Eq. [4]. Because the isotherms are linear, one finds that $C_e(A,T_2)/C_e(A,T_1) = C_e(B,T_2)/C_e(B,T_1)$, or that $\Delta\overline{H}(Q_A) = \Delta\overline{H}(Q_B)$. By repeating the same calculations at any loadings, one thus concludes that the molar heat of sorption is constant and independent of the loading capacity. Because of the linearity of the isotherms, the concentration ratio in Eq. [4]

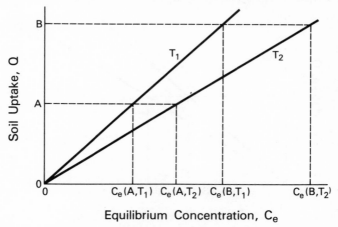

Fig. 6-5. Schematic plot of linear soil uptake from water (Q) vs. equilibrium concentration (C_e) at two arbitrary temperatures T_1 and T_2 with $T_2 > T_1$.

is also equal to the ratio of the sorption coefficient (the slope of the isotherm) at T_1 to that at T_2.

For most organic compounds and pesticides, the calculated molar heats of sorption by soil in aqueous systems are generally less exothermic than $-\Delta\overline{H}_w$ and are relatively independent of sorption capacities (Mills & Biggar, 1969; Yaron & Saltzman, 1972; Pierce et al., 1974; Chiou et al., 1979; Chiou et al., 1985), despite their high magnitudes of sorption. The same is true in the sorption of organic compounds from the vapor phase by moist (water-saturated) soils, such as found for ethylene dibromide (Wade, 1954) and lindane (Spencer & Cliath, 1970), in which heats of sorption are less exothermic than heats of vapor condensation ($-\Delta\overline{H}_v$). These observations are inherently consistent with the idea that the uptake of nonionic organic compounds by water-saturated soils results primarily from partitioning into the soil organic matter.

The relation between $\Delta\overline{H}$ and $\Delta\overline{H}_w$ for soil sorption in aqueous systems can be readily explained by the temperature dependence of the normalized isotherms, as shown in Fig. 6-6, in which the relative solute concentration (C_e/S_w) is used for the abscissa. By such a normalized plot, one usually finds a reverse temperature dependence for organic solutes in soil sorption (Mills & Biggar, 1969; Yaron & Saltzman, 1972); i.e., at given loading on soil

$$C_e(T_1)/S_w(T_1) > C_e(T_2)/S_w(T_2) \qquad [5]$$

or

$$S_w(T_2)/S_w(T_1) > C_e(T_2)/C_e(T_1) \qquad [6]$$

which can be expressed as

$$d \ln S_w/dT > d \ln C_e/dT. \qquad [7]$$

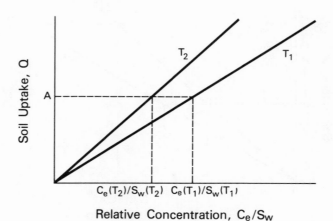

Fig. 6-6. Schematic plot of linear soil uptake from water (Q) vs. relative concentration (C_e/S_w) at two arbitrary temperatures T_1 and T_2 with $T_2 > T_1$.

Since one finds that

$$d \ln S_w/dT = \Delta \overline{H}_w/RT^2 \qquad [8]$$

and that

$$d \ln C_e/dT = \Delta \overline{H}_d/RT^2 = -\Delta \overline{H}/RT^2 \qquad [9]$$

with $\Delta \overline{H}_d$ denoting the heat of desorption (i.e., $\Delta \overline{H}_d = -\Delta \overline{H}$), one therefore gets

$$\Delta \overline{H} > -\Delta \overline{H}_w \qquad [10]$$

which means that the heat liberated in soil sorption of the solute is smaller than its reverse heat of solution in water. This agrees with the findings with DDT (Pierce et al., 1974), lindane and β-BHC (Mills & Biggar, 1969), parathion (Yaron & Saltzman, 1972), and 1,1,1-trichloroethane (Chiou et al., 1979).

Equation [10] is consistent with the expectation for a partition process shown in Eq. [2]. Since the heats of solution of organic compounds in an organic phase ($\Delta \overline{H}_o$) are generally small due to their improved compatibilities, $\Delta \overline{H}$ will be small for compounds with low $\Delta \overline{H}_w$ values and may even become positive (endothermic) for compounds with abnormal (negative) $\Delta \overline{H}_w$ values. Chiou et al. (1979) show, for example, that the $\Delta \overline{H}$ for 1,2-dichlorobenzene sorption by soil from water is nearly zero because of its low $\Delta \overline{H}_w$ and the $\Delta \overline{H}$ for 1,1,1-trichloroethane is positive because of its negative $\Delta \overline{H}_w$ in the temperature range of 3.5 to 20 °C. One may thus conclude from these observations that for systems in which the $\Delta \overline{H}$ values are negative, such exothermic heats derive primarily from condensation of the solutes from water ($-\Delta \overline{H}_w$) and that, with $\Delta \overline{H}_o$ being normally positive, the interactions between soil organic matter and solute ($\Delta \overline{H}_o$) are normally endothermic (as for the heat of solution) rather than exothermic in nature; one gets small and nearly constant exothermic heats for nonionic organic solutes as a result of the difference between $\Delta \overline{H}_o$ and $\Delta \overline{H}_w$.

The sorption data of p,p'-DDT in aqueous systems are especially worth noting in that DDT has a relatively high heat of fusion, about 25 kJ mol^{-1} (Plato & Glasgow, 1969), and is highly incompatible with water (as reflected in its low water solubility), which should make $\Delta \overline{H}_w$ much greater than 25 kJ mol^{-1}. The sorption coefficient of DDT with soil or sediment expressed on the basis of the organic matter content of soil or sediment (K_{om}) is approximately 1.5×10^5 (Shin et al., 1970; Pierce et al., 1974), while the heat of sorption ($\Delta \overline{H}$) is about -8.4 to -16.8 kJ mol^{-1} (Pierce et al., 1974). Based on these values, one can calculate the standard entropy change for the transfer of DDT from water to the soil organic matter as

$$\Delta \overline{G}^\circ = -RT \ln K_{om} \qquad [11]$$

and

$$\Delta \overline{S}^\circ = (\Delta \overline{H}^\circ - \Delta \overline{G}^\circ)/T \qquad [12]$$

where $\Delta \overline{G}^\circ$ is the (molar) standard free energy change for the transfer of one mole (or a unit mass) of the solute from water at unit concentration to the organic matter phase at unit concentration, and $\Delta \overline{H}^\circ$ and $\Delta \overline{S}^\circ$ are the corresponding standard enthalpic and entropic changes. Now, if one takes $K_{om} \simeq 1.5 \times 10^5$ and $\Delta \overline{H}^\circ = \Delta \overline{H} \simeq -12.6$ kJ mol^{-1} for DDT at $T = 298$ K, one gets $\Delta \overline{S}^\circ \simeq 58$ J K^{-1} mol^{-1}. Although the calculated $\Delta \overline{S}^\circ$ for DDT is subject to some uncertainty because of the inaccuracy of the $\Delta \overline{H}$ used, the magnitudes of the data would nonetheless show a gain in entropy for the transfer of DDT at the stated standard state. Such a gain in entropy deviates from the expectation for a decrease in entropy for adsorption of a trace component from solution. We shall consider later in this chapter the heat effect associated with soil sorption of organic compounds in nonaqueous systems.

Accountability of the partition model for soil organic matter in aqueous systems can be further verified in terms of the estimated magnitude of the solute solubility in the organic matter. Since the isotherm is practically linear, the solubility of a solute in the organic matter can be determined by

$$S_{om} = K_{om} \cdot S_w \qquad [13]$$

where S_{om} is the solubility in organic matter and S_w the solubility in water. For solid DDT with $S_w = 5.5$ μg L^{-1} and $K_{om} \simeq 1.5 \times 10^5$ at 25 °C, one therefore gets $S_{om} \simeq 830$ mg kg^{-1}, or 0.83 g kg^{-1}. By comparison, the solubility of DDT in pure octanol is about 42 g L^{-1} (Chiou et al., 1982), which is some 50 times its solubility in soil organic matter. The relatively low estimated solubility of DDT in soil organic matter is to be expected for a solid compound in a high-molecular-weight amorphous solid material. It is evident from these data that the very high sorption coefficient of DDT is primarily due to its extremely low solubility in water, which results in the high partition coefficient. In other words, if one finds S_{om} to be greater than, or nearly the same as, the solute solubility in a good solvent (such as octanol), the validity of the assumed partitioning with soil organic matter should then be seriously questioned.

Assuming partitioning, one expects organic compounds with higher solubility in water to give lower K_{om} but higher S_{om} values because these compounds are usually also more compatible with organic solvents. As an example, benzene having $S_w = 1780$ mg L^{-1} gives a $K_{om} \simeq 18$ (Chiou et al., 1983), and its S_{om} value calculated from Eq. [13] is about 32 g kg^{-1} which is about 40 times greater than for DDT (due partly to the fact that benzene is a liquid) and is reconcilable with the observation that benzene is completely miscible with octanol and most organic solvents. Thus, although the S_{om} values for given solutes would vary somewhat among various soil or sediment organic matter samples due to their compositional differences, the magnitudes of the S_{om} values illustrated are largely in the range to be expected for the solubilities of organic compounds in high-molecular-weight amorphous material. In their attempts to rationalize differences in soil uptake of organic compounds from water, Mingelgrin and Gerstl (1983) suggest that the less polar the compound the more it will tend to "adsorb" on a "hydrophobic" surface (organic matter) from a polar solvent (water), while removing solvent molecules from that surface. This adsorption view is in obvious contradiction to the results shown above, namely that the limiting capacity of DDT in soil organic matter is far less than that of benzene, or those of relatively polar phenols based on the reported sorption data of Boyd (1982).

Consideration of the sorption characteristics of individual solutes from a binary or a multicomponent system provides another means of distinguishing partitioning from adsorption. Studies of this nature in soil sorption of organic compounds and pesticides are relatively few in the literature. In their study of the simultaneous sorption of pyrene and phenanthrene from aqueous solution by sediments, Karickhoff et al. (1979) contended that there was no discernible sorptive interaction between the two compounds (although the authors did not make an explicit comparison of the single-solute and binary-solute isotherms). Boyd (1982) compared the Freundlich K constants of several phenols derived from single-component and multicomponent sorption experiments, and found that in binary and ternary systems the individual K values for some phenols were reduced by 10 to 30% while little change was noted for others. The observed changes for phenols could result partly from difficulties in accurately determining relatively small changes of the aqueous concentrations before and after equilibration ($\leq 20\%$). Chiou et al. (1983, 1985) reported the absence of sorptive competition between m-dichlorobenzene and 1,2,4-trichlorobenzene and between parathion and lindane in soil sorption of the binary solutes from water. The isotherm data for parathion and lindane are illustrated in Fig. 6-7. Overall, these data do not manifest the strong competitive effect between solutes that is found in adsorption of multiple solutes from solution (Manes, 1980; Everett, 1983).

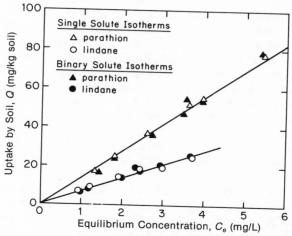

Fig. 6–7. Equilibrium sorption isotherms of parathion and lindane as single and binary solutes from water on Woodburn soil at 20 °C. Reproduced from Chiou et al. (1985) with permission from *Organic Geochemistry*, Vol. 8, p. 11–13. Copyright © 1985. Pergamon Press, Inc., Elmsford, NY.

As mentioned earlier, the fact that the soil does not display significant adsorption of nonionic organic compounds in aqueous systems is presumably the effect of the strong adsorptive competition of water for soil minerals, which results in displacement of the organic compounds. On the other hand, low solubilities (i.e., poor compatibilities) of the organic compounds in water enhance solute partitioning (solubilization) in the soil organic matter phase, which in effect accounts for the large difference between the sorption coefficients of DDT and benzene as discussed previously. The ineffectiveness of minerals in adsorbing organic compounds from water has been documented, for instance, by the very weak uptake of phenol on goethite (Yost & Anderson, 1984), 2,4-D on montmorillonite (Haque & Sexton, 1968), lindane and dieldrin on sand (Boucher & Lee, 1972), lindane on Ca-bentonite (Chiou et al., 1984), parathion on clay (Saltzman et al., 1972), and DDT on montmorillonite (Pierce et al., 1974). Obviously, because a trace amount of organic matter in unpurified clay minerals would have a significant impact on uptake by minerals from aqueous solution, the neglect of a small organic content in clays in such studies could lead to improper interpretations of the results. The significant uptake of 2,4-D by unpurified illite clay as reported by Haque and Sexton (1968) might be an artifact of the organic impurities in the clay. For instance, the sorption of 2,4-D on a fulvic acid–clay complex reported by Khan (1974) was greater than that on montmorillonite as reported by Haque and Sexton (1968).

We now consider in more detail the effect of water solubility (S_w) on the sorption (partition) coefficient with organic matter (K_{om}) in relation to model Eq. [3] and the relationship between K_{om} and the solvent (octanol)–water partition coefficient. Table 6-4 shows a list of the log K_{om} values of 12 aromatic compounds with Woodburn soil (19 g organic matter kg^{-1} soil), the respective log S_w values, molar volumes ($\overline{V}$), and log K_{ow} values of the compounds, as reported by Chiou et al. (1983). A plot of log K_{om} vs. log $S_w\overline{V}$ is shown in Fig. 6-8, in comparison with the hypothetical "ideal line" obtained by assuming $\rho = 1.2$ and $\chi_S = 0.25$ for organic matter and log $(\gamma_w/\gamma_w^*) = 0$. With given log $S_w\overline{V}$ of a compound, the difference between log K_{om}° from the ideal line and experimental log K_{om} defines the magnitude of $\chi_H/2.303 + \log (\gamma_w/\gamma_w^*)$. The magnitude of log (γ_w/γ_w^*) is relatively small for compounds with log $K_{om} \leq 3$, assuming that the amount of organic matter released from soil into water after soil-water equilibration is < 100 mg L^{-1} in water (Chiou et al., 1984; Gschwend & Wu, 1985) (the actual amount varies with the soil sample and experimental conditions). As an approximation, we assume log $(K_{om}^\circ/K_{om}) \simeq \chi_H/2.303$ in the present analysis.

Comparison of the magnitude of $-\log (S_w\overline{V})$ with that of log (K_{om}°/K_{om}) indicates that $-\log (S_w\overline{V})$ is the major determinant of log K_{om} for the organic compounds, leading to a highly linear correlation of log K_{om} with log $(S_w\overline{V})$ as shown in Fig. 6-8. For the 12 compounds, the regression equation gives

$$\log K_{om} = -0.813 \log (S_w\overline{V}) - 0.993 \qquad [14]$$

Table 6-4. Water solubilities (S_w), molar volumes $(\overline{V})$, octanol–water partition coefficients (K_{ow}) and soil organic matter–water distribution coefficients (K_{om}) of selected organic solutes (reproduced from data of Chiou et al., 1983).[†]

Compound	log S_w[‡]	$\overline{V}$	log $S_w\overline{V}$	log K_{om}	log K_{ow}
	mol L^{-1}	L mol^{-1}			
Benzene	−1.64	0.0894	−2.69	1.26	2.13
Anisole	−1.85	0.109	−2.82	1.30	2.11
Chlorobenzene	−2.36	0.102	−3.35	1.68	2.84
Ethylbenzene	−2.84	0.123	−3.75	1.98	3.15
1,2-Dichlorobenzene	−2.98	0.113	−3.98	2.27	3.38
1,3-Dichlorobenzene	−3.04	0.114	−3.98	2.23	3.38
1,4-Dichlorobenzene	(−3.03)	0.118	−3.96	2.20	3.39
1,2,4-Trichlorobenzene	−3.57	0.125	−4.47	2.70	4.02
2-PCB	(−4.57)	0.174	−5.33	3.23	4.51
2,2'-PCB	(−5.08)	0.189	−5.57	3.68	4.80
2,4'-PCB	(−5.28)	0.189	−5.97	3.89	5.10
2,4,4'-PCB	(−5.98)	0.204	−6.67	4.38	5.62

† Reprinted with permission from *Environmental Science and Technology*, Vol. 17, p. 229. Copyright © 1983. American Chemical Society, Washington, DC.

‡ The listed solubilities are the 20 to 25 °C values. The numbers in parentheses are the supercooled liquid solute solubilities estimated according to the method described in Chiou et al. (1982).

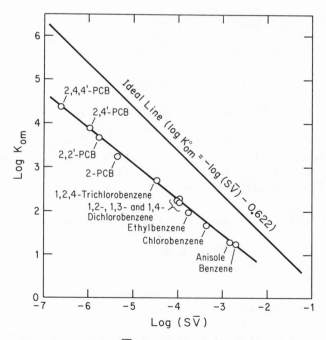

Fig. 6-8. Plot of log K_{om} vs. log $S_w\overline{V}$ for selected organic compounds from Table 6-4. Reproduced from Chiou et al. (1983) with permission from *Environmental Science and Technology*, Vol. 17, p. 229. Copyright © 1983. American Chemical Society, Washington, DC.

with $r^2 = 0.995$, where for the solid solutes the S_w values (mol L^{-1}) used are the estimated supercooled liquid solubilities for reasons stated in the earlier section. Since the variation of $\overline{V}$ among solutes is quite small compared to that of S_w, a linear relation should also exist between log K_{om} and log S_w, as has been widely recognized (Chiou et al., 1979, 1983; Karickhoff et al., 1979; Kenaga & Goring, 1980; Means et al., 1980; Briggs, 1981; Hassett et al., 1981; Karickhoff, 1981; Boyd, 1982). Using the data in Table 6-4, the correlation equation is

$$\log K_{om} = -0.729 \log S_w + 0.001 \qquad [15]$$

with $r^2 = 0.996$. While the slope and intercept values in Eq. [14] and Eq. [15] could vary somewhat with the type of organic matter in soil or sediment, the number of data points and the range of log K_{om} covered in a correlation study would also affect the slope and intercept values. Consequently, extrapolation of log K_{om} based on a correlation equation from a given set of data to regions far beyond the experimental data could lead to erroneous results, and should therefore be avoided.

That water insolubility is the major determinant of K_{om} is an expected result in the partition equilibria of organic compounds between an organic phase and water, despite the fact that soil organic matter is not nearly as good a partition medium as ordinary organic solvents such as octanol (and therefore log K_{om} should be much smaller than log K_{ow}). For this reason, the log K_{om} value would be linearly related to log K_{ow}, as illustrated for various organic compounds and pesticides with different soils and sediments, as mentioned earlier. The data in Table 6-4 give

$$\log K_{om} = 0.904 \log K_{ow} - 0.779 \qquad [16]$$

with $n = 12$ and $r^2 = 0.989$. For compounds with log $K_{ow} = 2$ to 5, the K_{om} values are approximately one order of magnitude lower than the corresponding K_{ow} values. When the sorption coefficient is expressed in terms of soil organic C (K_{oc}), equivalent correlation equations can be obtained with $K_{oc} = 1.72 K_{om}$, i.e., by assuming that the organic matter contains about 58% C (Hamaker & Thompson, 1972).

6-3.2 Effect of Dissolved Organic Matter on Sorption Coefficient

For many highly water-insoluble compounds, the term log (γ_w/γ_w^*) appears to have a more significant impact on log K_{om} in soil–water systems (Eq. [3]) than, say, on log K_{ow} in octanol–water systems (Eq. [1]). In soil sorption, this term expresses the enhancement of the apparent water solubility of a compound by a fraction of organic matter dispersed into water. The presence of a high-molecular-weight humic material in water even at trace quantities can significantly enhance the apparent water solubility of some otherwise extremely insoluble organic compounds by a partition-like interaction with the microscopic organic environment of the humic material (Chiou et al., 1984, 1986; Gschwend & Wu, 1985). A dissolved low-molecular-weight solvent (e.g., octanol) cannot effectively function as a microscopic partition medium because of its size limitation (therefore, its influence on solute solubility is by a mixed-solvent effect rather than by partitioning). As a result, the apparent concentration (or solubility) of a solute in water containing high-molecular-weight organic matter (C_e^*) may be greater than that in pure water (C_e) as shown by the relation

$$C_e^* = C_e + XK_{dom} C_e \qquad [17]$$

where X is the total mass of dissolved and particulate-bound organic matter per unit volume of water (for simplicity, X is collectively termed as the concentration of dissolved organic matter) and K_{dom} is the partition coefficient of the solute between dissolved organic matter (DOM) and water, which is a function of the type of solute and the composition of DOM involved.

Wershaw et al. (1969) showed that the apparent solubility of DDT in 0.5% soil sodium humate solution is >200 times greater than that in pure water (5.5 μg L^{-1} for the latter), giving log K_{dom} > 4.30. The log K_{dom} values of DDT derived with some soil or sediment humic acids as DOM lie in the range of 4.8 to 5.1 (Carter & Suffet, 1982; Chiou et al., 1986). It thus appears that dissolved soil or sediment humic acids are nearly as effective as the bulk soil or sediment organic matter in concentrating organic compounds on a unit weight basis. Therefore, in systems containing, say, X = 30 mg L^{-1} of soil or sediment humic acid in water and with log K_{dom} = 4.8, the apparent K_{om}^* value of DDT based on uncorrected aqueous concentrations (C_e^*) would be about 3 times lower than the true K_{om}. Gschwend and Wu (1985) investigated the apparent sorption (partition) coefficients of two highly insoluble PCB isomers with sediment samples in relation to the total organic C of the nonsettling sediment particles in aqueous solution. They show, with arguments similar to those contained in Eq. [17], that when precautions are taken to either eliminate or account for the nonsettling microparticles in water the calculated true log K_{oc} (or log K_{om}) values remain constant for both compounds over a wide range of sediment-to-solution ratios (whereas the apparent log K_{oc}^* values vary appreciably).

It would thus seem that the relatively large variation of the log K_{om} values as commonly recognized for highly water-insoluble organic compounds with different soils and sediments (Means et al., 1982; Karickhoff, 1984) could be partly related to their different extents of water solubility enhancement by dissolved and suspended organic matter. It would also appear that the significant enhancement of the apparent K_{om} value of DDT with soils following ether and alcohol extraction of the soils as reported by Shin et al. (1970) might be more reasonably explained by the reduction of the soluble fraction of soil organic matter, rather than by their assumption that the solvents remove lipids from organic matter and hence enhance the accessibility of DDT to the organic matter. Obviously, this solubility-enhancement effect by DOM (and hence the effect on K_{om} or K_{oc}) would diminish with an increase of the solute's water solubility because of the corresponding decrease in K_{dom}, as noted with lindane and 1,2,3-trichlorobenzene relative to the effect with DDT and some PCBs (Chiou et al., 1984, 1986; Caron et al., 1985). A more detailed analysis of the functional relationship of K_{dom} with the water solubility of organic compounds and with the source and composition of the humic material (humic and fulvic acids derived from soil and aquatic origins) has been made by Chiou et al. (1986, 1987).

6-4 SORPTION BY SOIL IN ORGANIC-SOLVENT SYSTEMS

6-4.1 Influence of Solvents on Sorption by Soil Minerals and Organic Matter

There are a few reported studies on the sorption of organic compounds and pesticides by soil from organic-solvent solution that are useful for illustrating the soil sorptive behavior. On the basis of the assumed roles of soil minerals and organic matter as discussed, one would immediately expect that the uptake of nonionic organic compounds and pesticides from organic solvents by partitioning into soil organic matter should be minimal because of the good solvency of the solution phase (Chiou et al., 1985). Thus, the extent of soil uptake would be controlled mainly by the effectiveness of the compounds to compete with the solvent for adsorption on soil minerals. Because of the inherent polarity of inorganic minerals, the adsorptivity of a compound (solute or solvent) is expected to be strongly influenced by polar interactions with the mineral. The sorptive behavior of the soil in organic-solvent systems therefore provides a useful basis for characterizing the mineral adsorptivity of water-unsaturated soils.

Hance (1965) reported an interesting set of experimental data on the sorption of diuron from aqueous and petroleum solutions by an oxidized soil (about 30 g organic matter kg^{-1} soil) and a soil organic material (760 g organic matter kg^{-1} soil). The sorption of diuron by the oxidized soil from the petroleum solution was found to be markedly greater than from aqueous solution, whereas the sorption by the organic material showed the opposite behavior. Based on these findings, Hance concluded that there is a competition between diuron and water for adsorption sites in soil under aqueous slurry conditions and that diuron competes more effectively for soil organic matter than for soil mineral surfaces. The high uptake of diuron from petroleum solution by the oxidized soil is due apparently to the strong (competitive) adsorption of the polar solute (diuron) on soil minerals at the expense of the relatively weak competition of a nonpolar solvent for adsorption on minerals. On the other hand, the high uptake of diuron by the soil organic material in aqueous solution is in keeping with the solute partitioning into the organic matter, with concomitant suppression by water of adsorption on soil minerals.

Mills and Biggar (1969) have also demonstrated characteristic differences in the sorption of lindane by Venado clay (500 g montmorillonite and 60 g organic matter per kg soil) and Staten peat muck (220 g organic matter kg^{-1} soil) from aqueous and hexane solutions. The observed sorption capacities in aqueous systems are largely proportional to the organic matter content in the soil samples in accord with the partition model. In hexane solution, the sorption by oven-dried Venado clay is markedly enhanced over sorption by dry peat muck. Moreover, whereas the equilibrium molar heats of lindane sorption in aqueous systems with both soil samples are less exothermic than the reverse heat of solution of lindane in water ($-\Delta \overline{H}_w$), the observed heats of sorption from hexane solution are much more exothermic than the reverse heat of solution in hexane. Again, the presumed weak adsorption of the relatively nonpolar hexane on minerals allows lindane to compete favorably for adsorption on mineral matter from hexane (while partitioning with organic matter is minimized), and consequently the Venado clay shows a much greater sorption capacity.

Yaron and Saltzman (1972) studied the soil uptake of parathion from a wide range of organic solvents and the effect of soil water content on the parathion uptake from these systems. Parathion shows a high uptake on dry soils from hexane, a lower uptake from benzene, and virtually no uptake from such polar solvents as methanol, ethanol, acetone, chloroform, ethyl acetate, and dioxane. While the uptake from hexane by dry and nearly dry soils is markedly higher than from water, such uptake is strongly suppressed by humidity and approaches zero when the soils become water-saturated (as compared with the finding that parathion exhibits definitive uptake on the same soils from aqueous solution). The saturation water content ranges from about 2% by weight for the sandy Mivtahim soil (6% clay and 3 g organic matter kg^{-1} soil) to 17% for the clay-rich Har Barqan soil (56% clay and 19 g organic matter kg^{-1} soil). These observations led the authors to suggest different mechanisms governing parathion "adsorption" in aqueous soil suspensions and in hydrated soil–organic solvent systems. They assumed that in soil–water systems, the solvent (water) is preferentially adsorbed but small amounts of parathion diffuse through water films and are adsorbed as they approach the colloid surfaces.

In consideration of the parathion sorption data reported by Yaron and Saltzman (1972), there is little doubt that the mechanism of sorption by soil in organic–solvent systems is fundamentally different from that in aqueous systems. The results for the aqueous system are reconcilable with the assumed solute partitioning into the soil organic matter. The high uptake of parathion from hexane by dry soils is attributable to strong adsorption on soil minerals (mainly clay components), on which the specific interactions of parathion's polar groups reduce competition by the relatively nonpolar hexane (while partitioning into the organic matter is minimized by the good solvency of hexane). Therefore, as expected, the parathion uptake from hexane would be depressed by humidity because of the strong adsorptive competition of water for minerals (in this case, water is considered as a competing solute), which leads eventually to nearly a complete suppression of parathion uptake when the soils become fully water saturated. By comparison, one notes that parathion exhibits a definitive uptake on soil from aqueous solution because the poor solvency of water (in addition to its suppression of mineral adsorption) makes partitioning of parathion into the organic matter a favorable process. The failure of the soils to sorb parathion from polar organic solvents is due apparently to the fact that such solvents minimize solute (parathion) adsorption on minerals because of their polarity and reduce solute partitioning into organic matter by their good solvency.

6–4.2 Effects of Temperature, Moisture, and Solute Polarity

Another interesting aspect that was recognized in the study of Yaron and Saltzman (1972) is the increasing parathion uptake from hexane with increasing temperature on partially hydrated soils, as opposed to the decreasing sorption with increasing temperature in aqueous systems. Such differences led Mingelgrin and Gerstl (1983) to postulate that the heat of "adsorption" of a solute from solution can be either exothermic or endothermic and consequently that the associated entropy change for solute adsorption can be either negative or positive.

The observed temperature dependence of parathion sorption from hexane with partially hydrated soils warrants careful consideration. As mentioned, a net adsorption of single vapors or single solutes from a solution (to the extent that the concentration of the solute on the adsorbent surface is enhanced over that in the solution phase) must be accompanied by exothermic heat. While it is possible for the solute to exhibit an endothermic heat of adsorption by competition with the solvent for adsorbent, adsorption of the solute in this case must be very weak because of the strong adsorption of the solvent (which makes the solute adsorption endothermic), and this is supposedly the case for the very weak adsorption of a nonionic organic compound on "pure minerals" in aqueous systems. In a binary-solute system, adsorption of the energetically weaker adsorbate may give rise to an anomalous temperature effect even though the weaker adsorbate nevertheless exhibits an exothermic interaction with adsorbent in such a binary-solute system. In the case of the parathion uptake from hexane solution by partially hydrated soils, parathion (a weaker adsorbate) competes with water (a more powerful competitor) for adsorption on soil minerals. An increase in temperature weakens the energetic interaction of minerals with water to a greater extent than with parathion (as the heat of adsorption per unit area is presumably greater for water), therefore assisting the latter to compete more favorably for adsorption (Chiou et al., 1981). Additionally, the solubility of water in hexane should increase more with temperature than the solubility of parathion in hexane, which would further decrease water adsorption with increasing temperature.

In the partially hydrated soil–hexane systems as reported by Yaron and Saltzman (1972), the amounts of water present only partially saturate the available mineral sites. One may therefore expect water adsorption on soil minerals to decrease with increasing temperature and this decrease would make more sites available for (exothermic) adsorption of parathion. If one does not take into account the temperature dependence of the number of adsorption sites, one may mistakenly conclude that the adsorption of parathion is endothermic in these systems. The temperature dependence of parathion adsorption is more reasonably explained on the basis of the higher exothermic heat of adsorption per unit surface area for water than for parathion. In all likelihood, the adsorption of parathion per unit available surface is also decreasing with increasing temperature, but not as rapidly as the concomitant increase in the number of water-free sites, thus resulting in a spurious temperature dependence of parathion adsorption. When excess water is added to fully saturate mineral sites, the adsorption of parathion from hexane should then be nearly completely suppressed, in which case it would be difficult to detect the temperature dependence for parathion uptake. Interpretations given by Mingelgrin and Gerstl (1983) on the temperature dependence of parathion uptake for partially hydrated soils by assumption of an endothermic heat of adsorption (and consequently an increase in entropy for parathion adsorption) appear to be theoretically untenable.

Chiou et al. (1985) made further studies of the sorption of parathion and lindane from aqueous and hexane solutions to substantiate the roles of soil minerals and organic matter in uptake by soil, using two soils that differ markedly in mineral–organic content. The two soils used were Woodburn soil (19 g organic matter kg^{-1} soil, 68% silt, and 21% clay) and Lake Labish peat soil (510 g organic matter kg^{-1} soil, 36% silt, and 3.5% clay). In aqueous systems, both parathion and lindane show linear isotherms, and there is no apparent sorptive competition between the two solutes, as illustrated for Woodburn soil in Fig. 6–7. In dry soil–hexane systems, the sorption of parathion (as well as lindane) is nonlinear and much greater than the corresponding uptake from aqueous solution. The parathion uptake from hexane on oven-dried and air-dried Woodburn soils (the latter contains about 2.5% water) is shown in Fig. 6–9.

As mentioned, comparison of soil uptake from solution is more appropriately made on the basis of relative concentration (rather than absolute concentration) which corrects for differences in solubility of the solute in different solvents. Since the solubility of parathion in hexane is much higher (5.74 × 10^4 mg L^{-1} at 20 °C and 8.56 × 10^4 mg L^{-1} at 30 °C) than in water (about 12 mg L^{-1} at 20 °C), it is important in comparing isotherms to extend the parathion sorption from hexane to sufficiently high absolute concentrations. The data in Fig. 6–9 show that over a relative concentration of parathion between 0 and 0.01 at 20 °C, the isotherm exhibits pronounced curvature, with capacities more than two orders of magnitude greater than in aqueous systems. Such curvature is not evident in the study of Yaron and Saltzman (1972) because their study was limited to very low equilibrium concentrations, which obviously fell within the limiting Henry's law region. The isosteric heat of parathion uptake calculated from 20° and 30 °C isotherms with dry Woodburn soil is highly exothermic (more than the reverse heat of parathion heat of solution in hexane, $-\Delta \overline{H}_h$) and varies with parathion loading capacity, as shown in Fig. 6–9(B).

The high parathion uptake on dry soil, the nonlinear isotherm, and the large exothermic heat of sorption all support the contention that adsorption

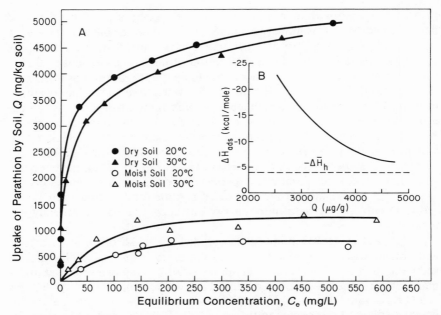

Fig. 6-9. (A) Equilibrium sorption isotherms of parathion from hexane on dry and partially hydrated Woodburn soil at 20 and 30 °C. The partially hydrated (moist) soil contains about 2.5% water. (B) Isosteric heats of parathion sorption on dry Woodburn soil as calculated from the isotherms. Reproduced from Chiou et al. (1985) with permission from *Organic Geochemistry*, Vol. 8, p. 11–13. Copyright © 1985. Pergamon Press, Inc., Elmsford, NY.

on soil minerals is the dominant mechanism in sorption from hexane, as previously discussed. The high net exothermic heat per mole of parathion adsorption (up to 20 kcal mol^{-1}) may be attributed to the simultaneous interactions of many polar groups in parathion with mineral surfaces. Thus, on a per mole basis the heat of parathion adsorption can be greater than for water, whereas the heat of adsorption per unit mineral surface should be greater for water to account for the suppression of parathion adsorption by water. Figure 6–9(A) also shows suppression of parathion adsorption by water with the air-dried Woodburn soil (approximately 2.5% by weight) and an anomalous temperature effect of parathion adsorption by the partially hydrated Woodburn soil at 20 and 30 °C (the water saturation capacity is estimated to be about 5% by weight) for reasons stated earlier. When more water is added to the air-dried soil to reach the saturation point, the parathion uptake from hexane solution is virtually totally suppressed.

The sorption data of lindane from hexane on dry and partially hydrated (Woodburn) soils exhibit essentially the same patterns as with parathion, except that the uptake of lindane is significantly lower at equal relative concentration and is more sharply reduced by the water content (Fig. 6–10); the solubility of lindane in hexane at 20 °C is 1.26×10^4 mg L^{-1} (relative to 7.8 mg L^{-1} in water at 25 °C). With about 2.5% water in soil, the lindane uptake is reduced nearly 25 times relative to the capacity with dry soil, much

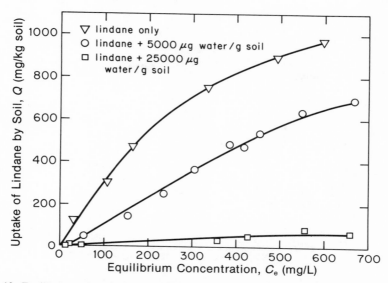

Fig. 6-10. Equilibrium sorption isotherms of lindane from hexane on dry and partially hydrated Woodburn soil at 20°C. Reproduced from Chiou et al. (1985) with permission from *Organic Geochemistry*, Vol. 8, p. 11–13. Copyright © 1985. Pergamon Press, Inc., Elmsford, NY.

more than with parathion, and further addition of water to the saturation point suppresses lindane uptake to a nondetectable level. These differences are consistent with the low polarity of lindane relative to parathion, making lindane a less potent adsorbate, and thus a much weaker competitor than parathion against water for adsorption on mineral surfaces. While the water content in soil suppresses the uptake of parathion and lindane from hexane, a similar competitive effect also occurs between parathion and lindane in their simultaneous sorption from hexane on dry (Woodburn) soil. This effect is illustrated in Fig. 6–11, where the uptake of lindane decreases with an increase of parathion uptake, in keeping with the presumed dominance of mineral adsorption of nonionic organic solutes from nonpolar organic solvents on dry and partially hydrated soil.

To amplify the contribution of mineral adsorption in dehydrated soils, Chiou et al. (1985) made a further comparison of the parathion and lindane uptake from hexane by dry Woodburn soil with that by dry Lake Labish peat soil, with the two soils differing greatly in their organic–mineral contents (Fig. 6–12). The uptake capacities of both compounds on Lake Labish peat soil are markedly lower than on Woodburn soil, due mainly to their differences in mineral (clay) content, as opposed to the findings that soil sorption from water solution is controlled by the organic matter content. The uptake capacities for both parathion and lindane with Woodburn soil are about 20 times greater than with Lake Labish peat soil, which is more than the sixfold difference in clay content of the soils. This discrepancy could be due to differences in clay type of the soils and/or to the possibility that part of the clay surfaces of the organic-rich peat soil is not accessible to adsor-

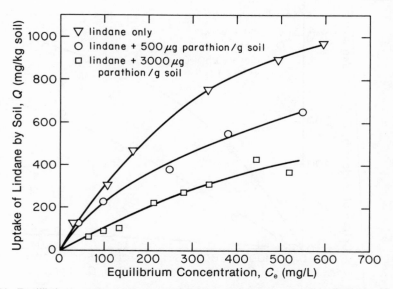

Fig. 6–11. Equilibrium sorption isotherms of lindane from hexane in the absence and presence of parathion as a competing solute on Woodburn soil at 20 °C. Reproduced from Chiou et al. (1985) with permission from *Organic Geochemistry*, Vol. 8, p. 11–13. Copyright © 1985. Pergamon Press, Inc., Elmsford, NY.

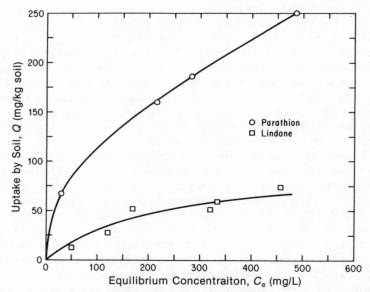

Fig. 6–12. Equilibrium sorption isotherms of parathion and lindane as single solutes from hexane on Lake Labish peat soil at 20 °C. Reproduced from Chiou et al. (1985) with permission from *Organic Geochemistry*, Vol. 8, p. 11–13. Copyright © 1985. Pergamon Press, Inc., Elmsford, NY.

bates because of occupation by the organic matter. Nevertheless, the sorption data with these two soils clearly demonstrate the dominance of mineral adsorption of organic compounds from hexane for dry and partially hydrated soils.

6-5 SORPTION BY SOIL FROM VAPOR PHASE AND EFFECT OF HUMIDITY

Much of the earlier interest in the sorption of organic compounds by water-unsaturated soils appears to result from the agricultural application of fumigants and pesticides for pest control, which produced a need to evaluate the stability and persistence of the compounds in soils under field conditions. Shipinov (1940) found that the isotherms for the vapor uptake of hydrogen cyanide on dry soils were nonlinear (Type II BET isotherm). Stark (1948) reported similar sorption isotherms for the vapor uptake of chloropicrin on dry soils and found a close correlation between the amounts of uptake and soil clay contents. The chloropicrin sorption decreased with increasing soil-moisture content. Hanson and Nex (1953) found that at a moisture content substantially below the wilting point, ethylene dibromide (EDB) was strongly sorbed by the soil, but that sorption decreased sharply to a minimum near the wilting point. Wade (1954) studied the EDB sorption by three soils having significantly different clay–organic C contents at moisture contents corresponding to the usual field range. Over the field moisture range, the EDB sorption by the soils was unaffected by soil-water content and the EDB isotherms were found to be linear instead, with the soil sorption capacities being closely proportional to the respective soil organic C contents (no correlation was found with clay content). Wade also noted much greater EDB sorption on dry soils; moisture sharply suppressed the EDB sorption up to the point corresponding to the saturation water content of the soil and a further increase in moisture content beyond that point produced little additional effect.

Jurinak (1957a, 1957b) and Jurinak and Volman (1957) studied the vapor uptake of EDB and 1,2-dibromo-3-chloropropane on dehydrated clays (which still retained small amounts of water) in relation to clay types and exchanged cations. The extent of sorption was related to the external clay surface areas of specific clays and the sorption data fit the BET adsorption model. The BET monolayer capacity of EDB increased from 17.5 g kg^{-1} for nonexpanding Ca-kaolinite (1.0% water) to 72.1 g kg^{-1} for expanding Ca-montmorillonite (2.3% water). Meanwhile, Call (1957) studied the dependence on relative humidity (R.H) of EDB vapor uptake on several soils and on a Ca-montmorillonite. On dry soils and on soils equilibrated with low ambient humidity (RH < 10%), the EDB isotherms were Type-II shape, in which clay soils displayed greater sorption capacity than soils low in clay content. An increase of RH from 0 to 50% progressively suppressed the EDB sorption (no data were available beyond RH = 50%) with a concomitant change of the isotherm shape toward linearity. The isotherms with Ca-montmorillonite were similar to those with soils, except that when RH increased from 0 to 10% there was a sharp increase in EDB sorption, which was presumably due to expansion of the clay layers with creation of additional surface. Beyond RH = 10%, EDB sorption decreased with increasing RH, and at RH = 90% the sorption capacity was reduced to a small fraction of the dry-clay sorption. That the soils did not show additional EDB sorption on increasing RH from 0 to 10% suggests that the clay minerals in the studied soils were mainly of nonexpanding types. By these observations, Call suggested that sorption of EDB on soils and clays containing low humidities resulted from adsorption on soil (mineral) surfaces, whereas sorption on wet soils and soils containing high humidities were attributed to dissolution in soil water or adsorption onto a water surface.

Leistra (1970) determined the vapor uptake of cis- and trans-1,3-dichloropropene on three types of soils, i.e., humus sand (550 g organic matter kg^{-1} soil), peat sand (180 g organic matter kg^{-1} soil), and peat (950 g organic matter kg^{-1} soil), with moisture contents of 170, 410, and 1200 g kg^{-1} of dry soil, respectively. The isotherms for all three soils were highly linear, with the soil-to-vapor distribution coefficients again being proportional to the respective organic matter contents; the calculated sorption coefficients based on soil organic content were largely independent of the soils for each vapor. The sorption coefficients (K_{om}) of both compounds also exhibited a relatively small temperature dependence, with $\Delta \overline{H}$ being nearly constant and <1 kcal mol^{-1} exothermic, as determined by Hamaker and Thompson (1972). The data suggest that the soils used by Leistra were already water-saturated (note that the amount of water required to saturate the soil in sorption is obviously much smaller than the field water saturation capacity, and appears to be close to the water content at the wilting point of the soil. Thus the saturation water content in the present discussion should not be confused with the field water saturation capacity).

In light of the described vapor sorption data, it is evident that dry and partially hydrated soil minerals (especially, clays) are powerful adsorbents for organic compounds and that the contribution by clay adsorption greatly exceeds the concurrent partitioning into the organic matter on dry mineral soils. Apparently, at saturation-water contents, the adsorptive power of soil minerals for organic compounds is largely lost because of strong competitive adsorption of water (Goring, 1967; Chiou et al., 1981), leaving partitioning with soil organic matter as the dominant mechanism. The fact that the organic–vapor uptake by dry soils is closely related to clay content rather than organic matter content suggests that dry clay is more powerful per unit weight in uptake of organic compounds by adsorption than organic matter in uptake by partitioning, while the reverse becomes true with the hydrated soils. The observed suppression of vapor uptake by moisture is essentially the same phenomenon as is found in the depression of parathion and lindane uptake by water from hexane solution. The only difference is that the organic solvent (hexane) also minimizes partitioning in the organic matter (causing the solute uptake to approach zero at full water saturation). Such partitioning is largely unaffected by moisture in sorption from the vapor phase (although it is possible that the presence of moisture in organic matter could modify its partition interactions with organic compounds).

Results relating the equilibrium dieldrin and lindane vapor concentrations in soil with soil-water contents, as carefully determined by Spencer et al. (1969) and Spencer and Cliath (1970), clearly illustrate the relative roles of soil minerals and organic matter in soil sorption. At soil water contents of $<2.2\%$ on Gila silt loam (Typic Torrifluvent) (6 g organic matter kg^{-1} soil), the equilibrium vapor densities of lindane (at about 50 mg kg^{-1} soil) and dieldrin (at 100 mg kg^{-1} soil) were significantly lower than the corresponding saturation vapor densities of the pure compounds, suggesting that the amounts of the pesticides applied were much below the saturation limits on the soil. An increase in the soil-water content to $>3.9\%$, however, resulted in a sharp increase of the equilibrium vapor concentrations, which became equal in magnitude to the saturation vapor concentrations of the pure compounds and stayed unchanged with increasing water content up to the field capacity (17%). The results with dieldrin at 30 and 40 °C are shown in Fig. 6–13.

The fact that applied dieldrin at 100 mg kg^{-1} soil and lindane at 50 mg kg^{-1} soil show below-saturation vapor concentrations when soil water content is low and display saturation vapor concentrations when soil is wet manifests the distinct roles of soil minerals and organic matter. The low vapor concentrations associated with dry soil may be explained in terms of strong mineral adsorption, which overrides the effect of partitioning with organic matter. Upon wetting, water displaces the pesticide from soil minerals by

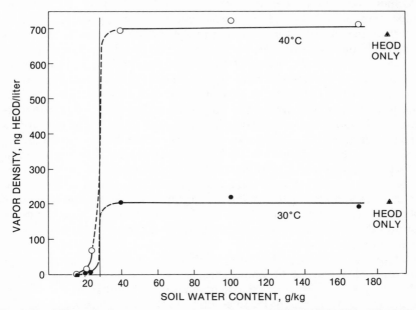

Fig. 6–13.   Effect of soil-water content on vapor density of HEOD (dieldrin) in Gila silt loam at 100 mg HEOD/kg soil. Reproduced from Spencer et al. (1969) with permission from American Society of Agronomy.

adsorptive competition and, as a result, the amount of pesticide in soil becomes more than is required to saturate the organic matter, giving rise to saturation vapor concentrations. One would note that 100 mg of dieldrin kg^{-1} soil in Gila soil (6 g organic matter kg^{-1} soil) corresponds to about 17 g of dieldrin per kilogram of organic matter, and this loading appears to be greater than the solubility of dieldrin in organic matter (S_{om}) as dieldrin has a relatively high melting point (176 °C). Similarly, a loading of 50 mg lindane kg^{-1} Gila soil corresponds to 8.3 g organic matter kg^{-1} soil, which is slightly greater than the calculated S_{om} value by Eq. [13] for lindane at 25 °C using $S_w = 7.8$ mg L^{-1} (Weil et al., 1974) along with an estimated K_{om} of about 900 for lindane with Gila soil. This analysis accounts further for the findings of Spencer and Cliath (1970) that the sorption capacity of the hydrated Gila soil for lindane is identical to that of the same soil containing 3.9% or 10% water.

Spencer and Cliath (1970) also found the sorption isotherms of lindane on Gila soil containing about 3.9% water at 20, 30, and 40 °C to be practically linear at equilibrium vapor concentrations extending to saturation capacities. The heat of vapor sorption calculated from the temperature dependence of the isotherms was less exothermic than the reverse heat of vapor condensation of lindane itself. These observations again reflect the dominance of partitioning in the organic matter in sorption of nonionic organic compounds by a hydrated soil.

To substantiate the effect of humidity on the mechanism and capacity of soil sorption of organic compounds, Chiou and Shoup (1985) determined the vapor sorption of benzene, chlorobenzene, m-dichlorobenzene, p-dichlorobenzene, 1,2,4-trichlorobenzene, and water on dry Woodburn soil, and of benzene, m-dichlorobenzene, and 1,2,4-trichlorobenzene as functions of relative humidity. Isotherms for all compounds on dry soil in a normalized plot of Q vs. P/P^o (where P and P^o are equilibrium and saturation vapor pressures, respectively) were distinctly nonlinear, with water showing the greatest capacity (Fig. 6–14). The observed data closely fit the BET adsorption model. In addition to the isotherm nonlinearity, the sorption capacities of organic vapors on dry soil were found to be about two orders of magnitude greater than for the same compounds on the same soil in aqueous systems as shown in Fig. 6–2 (when the data therein are normalized in the form of Q vs. C_e/S_w). The markedly greater sorption on dry soil is attributed to strong adsorption on soil minerals, which predominates over the simultaneous uptake by partitioning into the organic matter. The difference in sorption capacity between the dry and hydrated Woodburn soil samples illustrates the powerful suppression by water of adsorption of organic compounds on soil minerals.

Chiou and Shoup (1985) showed that the sorption of benzene, m-dichlorobenzene, and 1,2,4-trichlorobenzene by Woodburn soil was progressively depressed by increasing relative humidity. The results for m-dichlorobenzene are shown in Fig. 6–15. In addition to the reduced capacities in the presence of water vapor, the isotherms for organic vapors at RH $\geq 50\%$ also assume a practically linear shape at $P/P^o \leq 0.5$, which may also be at-

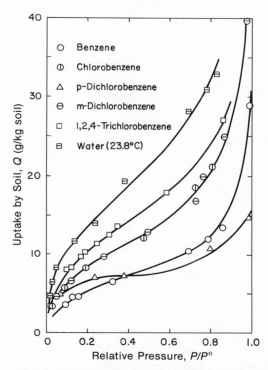

Fig. 6-14. Uptake of organic vapors and water on dry Woodburn soil vs. relative vapor pressure at 20 °C. Reproduced from Chiou and Shoup (1985) with permission from *Environmental Science and Technology*, Vol. 19, p. 1198-1199. Copyright © 1985. American Chemical Society, Washington,DC.

tributed to adsorptive displacement of organic compounds by water at highly active mineral surfaces. As adsorption of the compounds on soil minerals is further decreased by increasing amounts of water, the contribution to soil uptake by partitioning into the organic matter becomes more of a controlling factor, enhancing the range of the isotherm linearity. At about 90% RH, the sorption capacities of the compounds fall into a range close to those found in aqueous systems. These results are essentially in keeping with the assumed roles of soil minerals and organic matter for sorption on the soil in aqueous and organic-solvent systems, and with the reported sorption behavior from the vapor phase of other pesticides on dry and hydrated soils and clays.

The dependence of soil sorption capacity on water content or ambient humidity can be used to account for the variation of the equilibrium vapor concentration of pesticides, which may potentially occur in the field with changes in soil humidity (especially with surface soils). For example, as illustrated in Fig. 6-15, incorporation of a small amount of a compound in a dry or partially hydrated soil gives a very low relative vapor concentration or pressure (i.e., chemical activity). In the case of m-dichlorobenzene on Woodburn soil at, say, 400 mg kg^{-1} soil, the equilibrium partial pressure

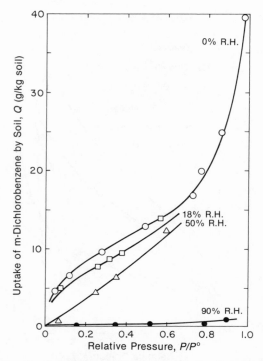

Fig. 6-15. Vapor uptake of *m*-dichlorobenzene on Woodburn soil as a function of relative humidity at 20 °C. Reproduced from Chiou and Shoup (1985) with permission from *Environmental Science and Technology*, Vol. 19, p. 1198–1199. Copyright © 1985. American Chemical Society, Washington, DC.

of the compound at RH $\leq 50\%$ will give a $P/P^o < 0.01$. By comparison, P/P^o rises to about 0.8 at RH $= 90\%$ with the same amount of *m*-dichlorobenzene in the soil, corresponding to nearly a two-order-of-magnitude increase in vapor concentration. These results are closely in agreement with the findings of Spencer et al. (1969) and Spencer and Cliath (1970) on the variation of dieldrin and lindane vapor concentrations with soil water content.

6-6 IMPLICATIONS FOR PESTICIDE AND CONTAMINANT ACTIVITY

In the preceeding sections, we have considered in some detail the dependence of sorption by soil on the properties of the compounds, the composition of soil or sediment, and ambient factors such as temperature and humidity. While a detailed account of the activity of pesticides and contaminants would involve other system parameters, the extent of sorption by soil or sediment is expected to be a primary factor affecting their activity and environmental behavior.

The proposed roles of soil organic matter and soil minerals in soil sorption and the influence of soil water on their functions provide an a priori basis for assessing the activity of soil-incorporated organic compounds and pesticides. In aqueous systems, the important properties for the sorption of the compounds are the organic matter content of the soil (sediment) and the water solubilities of the compounds. Given the concentration of a compound in water in equilibrium with a soil and the soil organic matter, one can make a reasonable estimate of the amount of chemical in the soil from its K_{om} value. The magnitude of K_{om} can in turn be estimated from its relationship with water solubility or with the octanol–water partition coefficient. Moreover, since the uptake by water-saturated soils is controlled mainly by partitioning into the organic matter (in which the solute has relatively small solubility), the concentration of a pesticide or a contaminant in soil or sediment should be much less than the organic matter content of the soil or sediment, except for possible cases in which the organic matter content is extremely small. Choi and Chen (1976), for instance, found that concentrations of chlorinated hydrocarbons (DDT, DDE, and PCBs) in sediments from a marine site bear a linear relationship with the organic C contents of the sediments; total chlorinated hydrocarbon contents of sediments ranged from 0.3 to 3.5 mg kg^{-1} whereas total organic C contents in the sediments ranged from 4.5 to 17 g kg^{-1} on a dry weight basis. Similarly, Goerlitz et al. (1985) found that sandy aquifer sediments in a creosote-contaminated groundwater site exhibited little retention of substituted phenols and polyaromatic hydrocarbons, whereas significant contaminant concentrations were found in groundwater.

The activity of a pesticide (or contaminant) may be expected to be most sensitively influenced by a changing soil-water content from above to below the saturation capacity in the field because this drying-wetting cycle would cause a sharp change in chemical activity of the pesticide, especially for soils low in organic matter content. The very top layer of surface soils supposedly undergoes a drying-wetting cycle to an extent depending on the change of ambient humidity as well as on agricultural practice. Under dry conditions, strong adsorption by soil minerals (particularly, clays) in addition to uptake by partitioning into organic matter lowers the activity (as measured by P/P^o) of the soil-applied pesticide. Upon wetting, the chemical activity would sharply rise as a result of the displacement by water of those species previously adsorbed on soil minerals (Goring, 1967; Chiou & Shoup, 1985). Volatile pesticides originally sorbed by dry soils and clays are thus readily released by the addition of water (Chisholm & Koblitsky, 1943; Schmidt, 1947; Stark, 1948; Hanson & Nex, 1953; Meggitt et al., 1954; Wade, 1954; Call, 1957; Gantz & Slife, 1960; Fang et al., 1961; Harris & Lichtenstein, 1961; Deming, 1963; Bowman et al., 1965; Gray & Weierich, 1965; Parochetti & Warren, 1966; Taylorson, 1966). While many pesticides and contaminants may not show significant vapor losses from soil, their chemical activities are similarly affected by the moisture content in soil as illustrated in Fig. 6–13 and 6–15.

Table 6-5. Influence of soil type and soil moisture on the toxicity of insecticides in soils to first instar nymphs of the common field cricket [*Gryllus pennsylvanicus* (Burmeister)] (reproduced from data of Harris, 1964).[†]

Insecticide	Soil type[‡]	LD$_{50}$, mg kg^{-1}	
		Moist	Dry
Heptachlor	Plainfield sand	0.068	0.53
	Muck	4.19	5.39
DDT	Plainfield sand	1.75	17.3
	Muck	67.2	99.8
Diazinon	Plainfield sand	0.26	34.1
	Muck	17.0	11.5
V-C 13	Plainfield sand	3.80	717
	Muck	279	165
Parathion	Plainfield sand	0.25	6.00
	Muck	22.6	9.10

† Reprinted by permission from *Nature*, Vol. 202, p. 724. Copyright © 1964. Macmillan Magazines, Ltd.
‡ Plainfield sand (mixed, mesic Typic Udipsamment).

Modifications of the chemical activity of pesticides in soil by soil-moisture content invariably lead to a change in the apparent pesticidal toxicity. Upchurch (1957) found that diuron was more toxic to cotton (*Gossypium hirsutum* L.) under moist than under dry soil conditions. Barlow and Hadaway (1955) observed that chlorinated hydrocarbon insecticides (lindane, DDT, and dieldrin) were inactivated by dry clay but were reactivated under high humidities for control of mosquitoes. Harris (1964) reported empirical correlations of the insecticide toxicities of heptachlor, DDT, diazinon, V-C 13 (dichlofenthion), and parathion on a sand (5.2 g organic matter kg^{-1} soil) and a muck (650 g organic matter kg^{-1} soil) with soil-moisture content. The results are shown in Table 6-5. Heptachlor was 7.8, DDT 9.9, parathion 24.4, diazinon 132, and V-C 13 189 times more toxic to the cricket when the sand was moist (5.5% water on a dry-weight basis) than when it was dry. By contrast, moisture in muck soil (162% water) had only a marginal effect on the insecticide toxicity. Comparison of insecticide toxicity between the two soil types yielded further striking results. At field moisture capacity DDT was 38.5, heptachlor 61.6, diazinon 65.7, V-C 13 73.3, and parathion 92.0 times less toxic in the muck as compared with the sand. In moist soil all the insecticides were found to be strongly inactivated by the organic content of the soil, with the extent of inactivation being dependent on the specific insecticide used; in dry soil there was no obvious correlation between organic content and toxicity of the two soils. Harris thus concluded that inactivation of the insecticides in moist soils was proportional to the organic content of the soil, while with dry soils inactivation was related to the adsorptive capacity of the mineral fraction.

In a similar study, Harris and Bowman (1981) compared the toxicities (LD$_{50}$) of soil-incorporated insecticides to crickets [*Gryllus pensylvanicus* (Burmeister)] in moist (20% water) and air-dried (2.5% water) sandy loam (39 g organic matter kg^{-1} soil and 8% clay) for fensulfothion, carbofuran, diazinon, fonofos, chlorpyrifos, phorate sulfoxide, phorate sulfone, phorate, terbufos sulfoxide, terbufos sulfone, and terbufos. The LD_{50} values are 2.8, 5.4, 32, 18, 26, 3.4, 4.3, 7.5, 4.8, 5.6, and 29 times greater (i.e., less toxic), respectively, for dry soil than for moist soil.

The observed magnitude of the effect of soil moisture on the apparent toxicity of soil-incorporated organic pesticides from those reported studies is very closely correlated with the recognized effect of soil moisture on the individual functions of soil minerals and organic matter in soil uptake. The pronounced effect of moisture on adsorption of the organic compound by soil minerals also appears to agree with our common experience that the air "in the field" gives a pleasant fragrance following a rain shower that succeeds a long period of drought. This phenomenon does not take place if the field has been wet prior to the rain shower. By these recognitions, we now have a better appreciation of the role of humidity in influencing the activity of organic compounds in the soil environment, as well as a deeper understanding of the distinct mechanisms through which the organic and mineral components of the soil function in sorption.

6-7 SUMMARY

This chapter gives consideration to the mechanistic functions of soil organic matter and minerals in sorption of nonionic organic compounds by soil to account for the soil behavior at varying system conditions. The soil is regarded as a dual sorbent, in which the mineral matter functions as a conventional solid adsorbent and the organic matter as a partition medium.

In aqueous suspensions, adsorption of organic compounds by mineral matter is suppressed by water due to strong polar interactions of water with minerals, and the soil sorption consists primarily of solute partitioning into soil organic matter. The presumed partition with organic matter is analyzed in relation to the equilibrium properties of organic compounds in solvent–water mixtures, and is supported by the observed linearity of soil–water isotherms, absence of competitive effects between solutes, low equilibrium heats, and difference in limiting sorption capacity for liquid and solid compounds. The theoretical basis underlying the empirical relationships of K_{om} to water solubility (S_w) and to octanol–water partition coefficient (K_{ow}) is explained. The importance of the organic matter content in soil and sediment in controlling the sorption coefficients of organic compounds in aqueous suspensions is emphasized.

The significance of mineral adsorption with dry and subsaturated soils is illustrated by data on soil sorption of organic compounds from nonpolar organic solvents and from the vapor phase. In these systems, the extent of adsorption of a compound by soil minerals in a soil decreases with an increase in moisture content and becomes minimal when the soil is fully saturated with water. The markedly greater sorption of a compound by the dry or subsaturated soil relative to that by the saturated soil is ascribed to mineral adsorption, which predominates over the simultaneous partition uptake by soil organic matter. As a result, a drying-wetting cycle in a normal soil causes a sharp change in the soil's sorption capacity and consequently the chemical activity of a compound in the soil. A change in soil moisture content below the saturation capacity would therefore have a strong impact on the apparent toxicity of soil-incorporated pesticides and on the movement of organic contaminants across the soil-air interface. Some specific examples are shown to demonstrate the influence of moisture in soil on the apparent toxicity of pesticides.

The assumed functions of soil minerals and organic matter as illustrated with dry and wet soils give a novel account of the noted fragrance in the field following a rain shower that succeeds a long period of drought; this phenomenon does not take place if the field has been wet prior to the rain shower.

ACKNOWLEDGMENT

The author wishes to thank Professor Milton Manes of Kent State University for valuable comments and discussions on the subject during the preparation of this chapter. The author also thanks Mrs. Caroline Hagelgans for her typing the manuscript in a timely manner.

REFERENCES

Adamson, A.W. 1967. Physical chemistry of surfaces. 2nd ed. Interscience Publishers, New York.

Bailey, G.W., and J.L. White. 1964. Review of adsorption and desorption of organic pesticides by soil colloids, with implications concerning pesticide bioactivity. J. Agric. Food Chem. 12:324–332.

Barlow, F., and A.B. Hadaway. 1955. Studies on aqueous suspensions of insecticides. V. The adsorption of insecticides by soils. Bull. Entomol. Res. 46:547–559.

Boucher, R.F., and G.F. Lee. 1972. Adsorption of lindane and dieldrin pesticides on unconsolidated aquifer sands. Environ. Sci. Technol. 6:538–543.

Bowman, M.C., M.S. Schechter, and R.L. Carter. 1965. Behavior of chlorinated insecticides in a broad spectrum of soil types. J. Agric. Food Chem. 13:360–365.

Boyd, S.A. 1982. Adsorption of substituted phenols by soil. Soil Sci. 134:337–343.

Briggs, G.G. 1973. A simple relationship between soil adsorption of organic chemicals and their octanol-water partition coefficients. p. 83–88. 7th British Insecticide and Fungicide Conf. The Boots Co., Nottingham.

Briggs, G.G. 1981. Theoretical and experimental relationships between soil adsorption, octanol-water partition coefficients, water solubilities, bioconcentration factors, and parachor. J. Agric. Food Chem. 29:1050–1059.

Browman, M.G., and G. Chesters. 1977. The solid-water interface: Transfer of organic pollutants across the solid-water interface. p. 49–105. In I.H. Suffet (ed.) Fate of pollutants in the air and water environments. Part I. John Wiley and Sons, New York.

Brunauer, S. 1945. The adsorption of gases and vapors. Princeton University Press, Princeton, NJ.

Call, F. 1957. The mechanism of sorption of ethylene dibromide on moist soils. J. Sci. Food Agric. 8:630–639.

Caron, G., I.H. Suffet, and T. Belton. 1985. Effect of dissolved organic carbon on the environmental distribution of nonpolar organic compounds. Chemosphere 14:993–1000.

Carringer, R.D., J.B. Weber, and T.J. Monaco. 1975. Adsorption-desorption of selected pesticides by organic matter and montmorillonite. J. Agric. Food Chem. 23:568–572.

Carter, C.W., and I.H. Suffet. 1982. Binding of DDT to dissolved humic material. Environ. Sci. Technol. 16:735–740.

Chiou, C.T. 1981. Partition coefficient and water solubility in environmental chemistry. p. 117–153. *In* J. Saxena and F. Fisher (ed.) Hazard assessment of chemicals: Current developments. Academic Press, New York.

Chiou, C.T. 1985. Partition coefficients of organic compounds in lipid-water systems and correlations with fish bioconcentration factors. Environ. Sci. Technol. 19:57–62.

Chiou, C.T., V.H. Freed, D.W. Schmedding, and R.L. Kohnert. 1977. Partition coefficient and bioaccumulation of selected organic chemicals. Environ. Sci. Technol. 11:475–478.

Chiou, C.T., D.E. Kile, T.I. Brinton, R.L. Malcolm, J.A. Leenheer, and P. MacCarthy. 1987. A comparison of water solubility enhancements of organic solutes by aquatic humic materials and commercial humic acids. Environ. Sci. Technol. 21:1231–1234.

Chiou, C.T., R.L. Malcolm, T.I. Brinton, and D.E. Kile. 1986. Water solubility enhancement of some organic pollutants and pesticides by dissolved humic and fulvic acids. Environ. Sci. Technol. 20:502–508.

Chiou, C.T., L.J. Peters, and V.H. Freed. 1979. A physical concept of soil-water equilibria for nonionic organic compounds. Science (Washington, DC) 206:831–832.

Chiou, C.T., L.J. Peters, and V.H. Freed. 1981. Soil-water equilibria for nonionic organic compounds. Science (Washington, DC) 213:684.

Chiou, C.T., P.E. Porter, and D.W. Schmedding. 1983. Partition equilibria of nonionic organic compounds between soil organic matter and water. Environ. Sci. Technol. 17:227–231.

Chiou, C.T., P.E. Porter, and T.D. Shoup. 1984. Reply to comment on "Partition equilibria of nonionic organic compounds between soil organic matter and water." Environ. Sci. Technol. 18:295–297.

Chiou, C.T., D.W. Schmedding, and M. Manes. 1982. Partitioning of organic compounds in octanol-water systems. Environ. Sci. Technol. 16:4–10.

Chiou, C.T., and T.D. Shoup. 1985. Soil sorption of organic vapors and effects of humidity on sorptive mechanism and capacity. Environ. Sci. Technol. 19:1196–1200.

Chiou, C.T., T.D. Shoup, and P.E. Porter. 1985. Mechanistic roles of soil humus and minerals in the sorption of nonionic organic compounds from aqueous and organic solutions. Org. Geochem. 8:9–14.

Chisholm, R.C., and L. Koblitsky. 1943. Sorption of methyl bromide by soil in a fumigation chamber. J. Econ. Entomol. 36:549–551.

Choi, W.-W., and K.Y. Chen. 1976. Associations of chlorinated hydrocarbons with fine particulates and humic substances in nearshore surficial sediments. Environ. Sci. Technol. 10:782–786.

Deming, J.M. 1963. Determination of volatility losses of C^{14}-CDAA from soil surfaces. Weeds 11:93–95.

Dzombak, D.A., and R.G. Luthy. 1984. Estimating adsorption of polycyclic aromatic hydrocarbons on soils. Soil Sci. 137:292–308.

Everett, D.H. 1983. Adsorption from solution. p. 1–29. *In* R.H. Ottewill et al. (ed.) Adsorption from solution. Academic Press, London.

Fang, S.C., P. Theisen, and V.H. Freed. 1961. Effects of water evaporation, temperature, and rates of application on the retention of ethyl-N,N-di-n-propylthiocarbamate in various soils. Weeds 9:569–574.

Felsot, A., and P.A. Dahm. 1979. Sorption of organophosphorus and carbamate insecticides by soil. J. Agric. Food Chem. 27:557–563.

Filonow, A.B., L.W. Jacobs, and M.M. Mortland. 1976. Fate of polybrominated biphenyls (PBB's) in soils. Retention of hexabromobiphenyl in four Michigan soils. J. Agric. Food Chem. 24:1201–1204.

Flory, P.J. 1942. Thermodynamics of high polymer solutions. J. Chem. Phys. 10:51–61.

Flory, P.J. 1953. Principles of polymer chemistry. p. 495–520. Cornell University Press, Ithaca, NY.

Frissel, M.J., and G.H. Bolt. 1962. Interaction between certain ionizable organic compounds (herbicides) and clay minerals. Soil Sci. 94:284–291.

Gantz, R.L., and F.C. Slife. 1960. Persistence and movement of CDAA and CDEC in soil and the tolerance of corn seedlings to these pesticides. Weeds 8:599–606.

Goerlitz, D.F., D.E. Troutman, E.M. Godsy, and B.J. Franks. 1985. Migration of wood-preserving chemicals in contaminated groundwater in a sand aquifer at Pensacola, Florida. Environ. Sci. Technol. 19:955–961.

Goring, C.A.I. 1967. Physical aspects of soil in relation to the action of soil fungicides. Annu. Rev. Phytopathol. 5:285–318.

Gray, R.A., and A.J. Weierich. 1965. Factors affecting the vapor loss of EPTC from soils. Weeds 13:141–147.

Gregg, S.J., and K.S.W. Sing. 1982. Adsorption, surface area and porosity. 2nd ed. Academic Press, London.

Gschwend, P.M., and S.-C. Wu. 1985. On the constancy of sediment-water partition coefficient of hydrophobic organic pollutants. Environ. Sci. Technol. 19:90–95.

Hamaker, J.W., and J.M. Thompson. 1972. Adsorption. p. 49–143. In C.A.I. Goring, and J.W. Hamaker (ed.) Organic chemicals in the soil environment. Vol. I. Marcel Dekker, New York.

Hance, R.J. 1965. Observations on the relationships between the adsorption of diuron and the nature of the absorbent. Weed Res. 5:108–114.

Hanson, W.J., and R.W. Nex. 1953. Diffusion of ethylene dibromide in soils. Soil Sci. 76:209–214.

Haque, R. 1975. Role of adsorption in studying the dynamics of pesticides in a soil environment. p. 97–114. In R. Haque and V.H. Freed (ed.) Environmental dynamics of pesticides. Plenum Press, New York.

Haque, R., and R. Sexton. 1968. Kinetic and equilibrium study of the adsorption of 2,4-D dichlorophenoxy acetic acid on some surfaces. J. Colloid Interface Sci. 27:818–827.

Harris, C.I., and G.F. Warren. 1964. Adsorption and desorption of herbicides by soil. Weeds 12:120–126.

Harris, C.R. 1964. Influence of soil type and soil moisture on the toxicity of insecticides in soils to insects. Nature (London) 202:724.

Harris, C.R., and B.T. Bowman. 1981. The relationship of insecticide solubility in water to toxicity in soil. J. Econ. Entomol. 74:210–212.

Harris, C.R., and E.P. Lichtenstein. 1961. Factors affecting the volatilization of insecticidal residues from soils. J. Econ. Entomol. 54:1038–1045.

Hartley, G.S. 1960. Physico-chemical aspects of the availability of herbicides in soil. p. 63–78. In E.K. Woodford and G.R. Sagar (ed.) Herbicides and the soil. Blackwell, Oxford.

Hassett, J.J., W.L. Banwart, S.G. Wood, and J.C. Means. 1981. Sorption of α-naphthol: Implication concerning the limits of hydrophobic sorption. Soil Sci. Soc. Am. J. 45:38–42.

Huggins, M.L. 1942. Thermodynamic properties of solutions of long-chain compounds. Ann. N.Y. Acad. Sci. 43:1–32.

Jurinak, J.J. 1957a. The effect of clay minerals and exchangeable cations on the adsorption of ethylene dibromide vapor. Soil Sci. Soc. Am. Proc. 21:599–602.

Jurinak, J.J. 1957b. Adsorption of 1,2-dibromo-3-chloropropane vapor by soils. J. Agric. Food Chem. 5:598–600.

Jurinak, J.J., and D.H. Volman. 1957. Application of the Brunauer, Emmett, and Teller equation to ethylene dibromide adsorption by soils. Soil Sci. 83:487–496.

Karickhoff, S.W. 1981. Semi-empirical estimation of sorption of hydrophobic pollutants on natural sediments and soils. Chemosphere 10:833–846.

Karickhoff, S.W. 1984. Organic pollutant sorption in aquatic systems. J. Hydraul. Eng. 110:707–735.

Karickhoff, S.W., D.S. Brown, and T.A. Scott. 1979. Sorption of hydrophobic pollutants on natural sediments. Water Res. 13:241–248.

Kenaga, E.E., and C.A.I. Goring. 1980. Relationship between water solubility, soil sorption, octanol-water partitioning, and concentration of chemicals in biota. p. 78–115. In J.C. Eaton et al. (ed.) Aquatic toxicology. Am. Soc. Testing and Materials. Philadelphia, PA.

Khan, S.U. 1974. Adsorption of 2,4-D from aqueous solution by a fulvic acid-clay complex. Environ. Sci. Technol. 8:236–238.

Kipling, J.J. 1965. Adsorption from solutions of nonelectrolytes. Academic Press, London and New York.

Knight, B.A.G., and T.E. Tomlinson. 1967. The interaction of paraquat (1:1'-dimethyl 4:4'-dipyridylium dichloride) with mineral soils. J. Soil Sci. 18:233–243.

Lambert, S.M. 1967. Functional relationship between sorption in soil and chemical structure. J. Agric. Food Chem. 15:572–576.

Lambert, S.M. 1968. Omega (Ω), a useful index of soil sorption equilibria. J. Agric. Food Chem. 16:340–343.

Leistra, M. 1970. Distribution of 1,3-dichloropropene over the phases in soil. J. Agric. Food Chem. 18:1124–1126.

MacIntyre, W.G., and C.L. Smith. 1984. Comment on "Partition equilibria on nonionic organic compounds between soil organic matter and water." Environ. Sci. Technol. 18:295.

Mackay, D. 1977. Comment on "Partition coefficient and bioaccumulation of selected organic chemicals." Environ. Sci. Technol. 11:1219.

Mackay, D., A. Bobra, W.Y. Shiu, and S.H. Yalkowsky. 1980. Relationships between aqueous solubility and octanol-water partition coefficients. Chemosphere 9:701–711.

Manes, M. 1980. The Polanyi adsorption potential theory and its applications to adsorption from water solution onto activated carbon. p. 43–63. In I.H. Suffet and M.J. McGuire (ed.) Activated carbon adsorption of organics from the aqueous phase. Vol. I. Ann Arbor Science Publ., Ann Arbor, MI.

McCall, P.J., D.A. Laskowski, R.L. Swann, and H.J. Dishburger. 1983. Estimation of environmental partitioning of organic chemicals in model ecosystems. Residue Rev. 85:231–244.

Means, J.C., S.G. Wood, J.J. Hassett, and W.L. Banwart. 1980. Sorption of polynuclear aromatic hydrocarbons by sediments and soils. Environ. Sci. Technol. 14:1524–1528.

Means, J.C., S.G. Wood, J.J. Hassett, and W.L. Banwart. 1982. Sorption of carboxy-substituted polynuclear aromatic hydrocarbons by sediments and soils. Environ. Sci. Technol. 16:93–98.

Meggitt, W.F., H.A. Borthwock, R.J. Aldrich, and W.C. Shaw. 1954. Factors affecting the herbicide action of dinitro-ortho-secondary-butyl phenol. Proc. Northeast. Weed Control Conf. 8:21.

Miller, M.M., S.P. Wasik, G.-L. Huang, W.-Y. Shiu, and D. Mackay. 1985. Relationships between octanol-water partition coefficient and aqueous solubility. Environ. Sci. Technol. 19:522–529.

Mills, A.C., and J.W. Biggar. 1969. Solubility-temperature effect on the adsorption of gamma- and beta-BHC from aqueous and hexane solutions by soil materials. Soil Sci. Soc. Am. Proc. 33:210–216.

Mingelgrin, U., and Z. Gerstl. 1983. Reevaluation of partitioning as a mechanism of nonionic chemicals adsorption in soils. J. Environ. Qual. 12:1–11.

Nkedi-Kizza, P., P.S.C. Rao, and J.W. Johnson. 1983. Adsorption of diuron and 2,4,5-T on soil particle-size separates. J. Environ. Qual. 12:195–197.

Parochetti, J.V., and G.F. Warren. 1966. Vapor losses of IPC and CIPC. Weeds 14:218–284.

Pierce, R.H., Jr., C.E. Olney, and G.T. Felbeck, Jr. 1974. p,p'-DDT adsorption to suspended particulate matter in sea water. Geochim. Cosmochim. Acta 38:1061–1073.

Plato, C., and R. Glasgow. 1969. Differential scanning calorimetry as a general method for determining the purity and heat of fusion of high purity organic chemicals. Application to 95 compounds. Anal. Chem. 41:330–336.

Saltzman, S., L. Kliger, and B. Yaron. 1972. Adsorption-desorption of parathion as affected by soil organic matter. J. Agric. Food Chem. 20:1224–1226.

Savage, K.E., and R.D. Wauchope. 1974. Fluometuron adsorption-desorption equilibria in soil. Weed Sci. 22:106–110.

Schellenberg, K., C. Leuenberger, and R.P. Schwarzenbach. 1984. Sorption of chlorinated phenols by natural sediments and aquifer materials. Environ. Sci. Technol. 18:652–657.

Schmidt, C.T. 1947. Dispersion of fumigants through soil. J. Econ. Entomol. 40:829–837.

Schwarzenbach, R.P., and J. Westall. 1981. Transport of nonpolar organic compounds from surface water to groundwater. Laboratory sorption study. Environ. Sci. Technol. 15:1360–1367.

Scott, R.L. 1949. The thermodynamics of high polymer solutions. IV. Phase equilibria in the ternary system: Polymer-liquid 1 − liquid 2. J. Chem. Phys. 17:268–279.

Sherburne, H.R., and V.H. Freed. 1954. Adsorption of 3(p-chlorophenyl)-1,1-dimethylurea as a function of soil constituents. J. Agric. Food Chem. 2:937–939.

Shin, Y.-O., J.J. Chodan, and A.R. Wolcott. 1970. Adsorption of DDT by soils, soil fractions, and biological materials. J. Agric. Food Chem. 18;1129–1133.

Shipinov, N.A. 1940. Sorption of hydrogen cyanide on soils during fumigation with cyanides. Bull. Plant Protection (U.S.S.R.) No. 1–2:192–199.

Spencer, W.F., and M.M. Cliath. 1970. Desorption of lindane from soil as related to vapor density. Soil Sci. Soc. Am. Proc. 34:574–578.

Spencer, W.F., M.M. Cliath, and W.J. Farmer. 1969. Vapor density of soil-applied dieldrin as related to soil-water content, temperature, and dieldrin concentration. Soil Sci. Soc. Am. Proc. 33:509–511.

Stark, F.L., Jr. 1948. Investigations of chloropicrin as a soil fumigant. New York (Cornell) Agr. Exp. Sta. Mem. 178:1–61.

Stevenson, F.J. 1982. Humus chemistry. Chap. 17. Wiley-Interscience, New York.

Swoboda, A.R., and G.W. Thomas. 1968. Movement of parathion in soil columns. J. Agric. Food Chem. 16:923–927.

Tahoun, S.A., and M.M. Mortland. 1966. Complexes of montmorillonite with primary, secondary, and tertiary amides. I. Protonation of amides on the surface of montmorillonite. Soil Sci. 102:248–254.

Taylorson, R.B. 1966. Influence of temperature and other factors on loss of CDEC from soil surfaces. Weeds 14:294–296.

Upchurch, R.P. 1957. The influence of soil moisture content on response of cotton to diuron. Weeds 5:112–120.

Wade, P.I. 1954. The sorption of ethylene dibromide by soils. J. Sci. Food Agric. 5:184–192.

Weber, J.B. 1966. Molecular structure and pH effects on the adsorption of 13 s-triazine compounds on montmorillonite clay. Am. Mineral. 51:1657–1670.

Weber, J.B., P.W. Perry, and R.P. Upchurch. 1965. The influence of temperature and time on the adsorption of paraquat, diquat, 2,4-D and prometone by clays, charcoal, and an anion-exchange resin. Soil Sci. Soc. Am. Proc. 29:678–688.

Weber, J.B., and S.B. Weed. 1974. Effects of soil on the biological activity of pesticides. p. 223–256. In W.D. Guenzi (ed.) Pesticides in soil and water. Am. Soc. Agron., Madison, WI.

Weed, S.B. 1968. The effect of adsorbent charge on the competitive adsorption of divalent organic cations by layer-silicate minerals. Am. Mineral. 53:478–490.

Weil, L., G. Dure, and K.-E. Quentin. 1974. Wasserlöslichkeit von insektiziden chlorierten Kohlenwasserstoffen und polychlorierten Biphenylen im Hinblick auf eine Gewässerbelastung mit diesen Stoffen. Z. Wasser Abwasser Forch. 7:169–175.

Wershaw, R.L., P.J. Burcar, and M.C. Goldberg. 1969. Interaction of pesticides with natural organic material. Environ. Sci. Technol. 3:271–273.

Yaron, B., and S. Saltzman. 1972. Influence of water and temperature on adsorption of parathion by soils. Soil Sci. Soc. Am. Proc. 36:583–586.

Yost, E.C., and M.A. Anderson. 1984. Absence of phenol adsorption on goethite. Environ. Sci. Technol. 18:101–109.

Chapter 7

Effects of Humic Substances on Plant Growth[1]

YONA CHEN AND TSILA AVIAD, *Seagram Center for Soil and Water Sciences, The Hebrew University of Jerusalem, Rehovot, Israel*

---

[1] Contribution from The Seagram Center for Soil and Water Sciences, Faculty of Agriculture, The Hebrew University of Jerusalem, Israel. Mailing address: P.O.B. 12, Rehovot, Israel 76100.

ABSTRACT

Studies of the effects of humic substances on plant growth, under conditions of adequate mineral nutrition, consistently show positive effects on plant biomass. Stimulation of root growth is generally more apparent than stimulation of shoot growth. Both increases in root length and stimulation of the development of secondary roots have been observed for humic substances in nutrient solutions. The typical response curve shows increasing growth with increasing humic substance concentration in nutrient solutions, followed by a decrease in growth at very high concentrations. Shoots generally show similar trends in growth response to humic substances but the magnitude of the growth response is less. Foliar sprays can also enhance both root and shoot growth. The stimulatory effects of humic substances has been correlated with enhanced uptake of macronutrients. Humic substances can complex transition metal cations, which can sometimes result in enhanced uptake and sometimes result in competition with the roots resulting in decreased uptake. A small fraction of lower molecular weight components in humic substances can be taken up by plants. These components seem to increase cell membrane permeability and may have hormone-like activity. In soils, addition of composts can stimulate growth beyond that provided by mineral nutrients presumably because of the effects of humic substances. Addition of Fe-enriched organic materials can alleviate high-lime chlorosis. Soil additions of prepared humic substances is not economical, but the response to foliar sprays has the potential to be economical because of the relatively small quantities needed.

Brief Historical Review

Man has realized for thousands of years that dark-colored soils are more productive than light-colored soils and that productivity was closely associated with decaying plant and animal residues. Bacon (1651) believed that plants absorbed a juice from the soil for their sustenance. Woodward (1699) demonstrated in the late 17th century that plant response to various sources of water was in the following order: soil water extract > river water > well water, an effect which was also correlated with the yellow color of these waters. In the early 19th century the direct role of humic substances in plant nutrition and growth was emphasized (Thaer, 1808). Thaer further suggested that humus was the only material which supplied nutrients to plants. Much later Grandeau (1872) still believed that humus was a major component of plant nutrition and provided plants with C and other nutrients (the "humus theory"). Evidence against the "humus theory" was assembled by many investigtors such as Sprengel (1832) and De Saussure (1804) who showed that plants synthesize organic substances from CO_2 and water. Liebig, in a number of publications (1841, 1856), supported the evidence against the "humus theory" and provided fundamental information on the role of minerals in plant nutrition.

Lawes and Gilbert (1905) working at Rothamsted, England, demonstrated after a long-term field study in the early 20th century that soil fertility may be maintained, for at least several years, by applying mineral fertilizers only (see also Russell, 1921, p. 1–29). However, the controversy between the humus and mineral theories was not ended by these experiments. Scientists have realized that more exact experimentation is required to determine the benefit of humus to plant growth and to determine possible synergistic effects of humic substances and minerals.

In the early 20th century, Bottomly (1914a, b; 1917; 1920) published a series of papers in which he showed that humic substances enhanced the growth of various plant species in mineral nutrient solutions. Bottomly believed humic substances acted as plant growth hormones and called them "auximones". Other investigators (Olsen, 1930; Burk et al., 1932) attributed the beneficial effects of humic substances on plant growth to the increased solubilization of some mineral ions such as Fe.

From this brief historical review, it is clear that when beneficial effects of humic substances on plant growth are to be studied, it is essential to provide these materials in the presence of sufficient mineral nutrients. This chapter will focus on studies in which this approach was applied. Evidence will also be provided from various field experiments in which humic substances seem to serve as plant growth promoting factors.

Nature of Soil Organic Matter

As organic materials in the soil decay, macromolecules of a mixed aliphatic and aromatic nature are formed. The term *humus* is widely accepted as synonymous for *soil organic matter*. It is defined as the total of the organic compounds in soil, exclusive of undecayed plant and animal tissues, their partial decomposition products, and the soil biomass (see chapter 1, this book).

The chemical and colloidal properties of soil organic matter can be studied only in the free state, that is, when freed of inorganic soil components. Thus, the first task of the researcher is to separate organic matter from the inorganic matrix. Alkali, usually 0.1 to $0.5M$ NaOH, has been a popular extractant of soil organic matter. Other methods of extraction have also been used by various researchers, but since this subject is beyond the objectives of this chapter, extraction procedures will, in general, be ignored. A detailed discussion of the extraction procedures can be found elsewhere (e.g., Stevenson, 1982).

Indirect Effects

It is generally accepted by soil scientists and plant physiologists that plant growth and yield are largely determined by mineral nutrition, water and air supply to the roots, and environmental conditions such as light and temperature. However, a number of studies in addition to those mentioned earlier suggest that soil organic matter also affects plant growth. Correlations between organic matter content of soils and plant yield are frequently reported in the literature (e.g., Scharpf, 1967; Agboola, 1978; Lykov, 1978; Ojenhiyi

& Agbede, 1980; Li et al., 1981; Pilus Zambi et al., 1982; Rebufetti & Labunora, 1982; Olsen, 1986).

It is well established that soil organic matter may affect soil fertility indirectly through the following mechanisms: (i) supply of minerals, mostly N, P, K and micronutrients to the roots; (ii) improved soil structure, thereby improving water-air ratios in the rhizosphere; (iii) an increase in the soil microbial population including beneficial microorganisms; (iv) an increase in the cation exchange capacity (CEC) and the pH buffering capacity of the soil; (v) supply of defined biochemical compounds to plant roots such as acetamide and nucleic acid (Hutchinson & Miller, 1912); and (vi) supply of humic substances that serve as carriers of micronutrients or growth factors.

This chapter will focus on the effects of humic substances on plant growth (item vi). Since there is a considerable overlap between the various mechanisms, however, the distinction between them may be difficult to establish.

Direct Effects

Direct effects are those which require uptake of organic macromolecules, such as humic substances, into the plant tissue resulting in various biochemical effects either at the cell wall, membrane level, or in the cytoplasm. A prerequisite for these activities is the decay of the bulk organic matter into humic materials.

Because the soil system is usually too complicated for studies on the direct effects of humic substances on plant growth, these problems were usually studied using extracted humic substances supplied to plants grown in nutrient solutions, with application either to the foliage or to the roots by addition to the nutrient solution. Both these methods of application and their effects will be discussed in this chapter.

7–1 RESPONSE OF PLANTS TO HUMIC SUBSTANCES

7–1.1 Germination and Seedling Growth

Effects of humic substances on seed germination and seedling development have been studied by a number of investigators. It was commonly assumed that since these growth stages are strongly responsive to humic substances, final yield would also be affected.

Smidova (1962) studied the effects of Na-humate on water imbibition and germination of winter wheat (*Triticum aestivum* L.). Increased water absorption, respiration and germination was observed in solutions of 100 mg L^{-1} Na-humate. Stimulated germination was attributed to the enhanced enzymatic activity in the seed tissue. Dixit and Kishore (1967) observed the stimulation of the germination of several varieties of crop seeds by humic (HA) and fulvic (FA) acids. Germination of corn (*Zea mays* L.) was stimulated by 60 mg L^{-1} HA or FA; 30 mg L^{-1} HA, and 45 mg L^{-1} FA enhanced germination of barley (*Hordeum vulgare* L.) and wheat. Also, coating of seeds with Ca-humate has been shown to be beneficial (Iswaran & Chonkar, 1971). Apparently, the treatments affected only the rate of germination and not the fraction of viable seeds (Pagel, 1960).

7-1.2 Root Initiation and Growth

Humic substances appear to have a greater effect on roots than on the aboveground parts of plants (Sladky, 1959a). Stimulated root growth and enhancement of root initiation have been commonly observed. Khristeva (1949) tested the response of several plants to Na-humate of various origins. Root length as well as stem length of winter wheat increased and in the optimal cases almost doubled. The optimal concentration of Na-humate was 60 mg L^{-1}, but at 600 mg L^{-1} plant growth resembled water-grown control plants. Secondary roots did not develop in water cultures but were strongly developed in the solutions with added humate. Kononova and Pankova (1950) compared root development of corn as affected by humic acids extracted from Podzol (Cryorthods, Fragiorthods, Haplorthods) and Chernozem (Cryoborolls) soils to growth in water. Root length and number was more than doubled in the 4 to 5 mg L^{-1} Na-humate solutions. Similar results were obtained by Ivanova (1965) using corn. Humic acid in concentrations of 1 to 10 mg L^{-1} applied to agar growth media containing a complete nutrient solution enhanced root growth whereas a concentration of 100 mg L^{-1} HA inhibited growth.

Sladky (1959a) applied humic acid, fulvic acid, and an alcoholic extract of organic matter at concentrations of 50, 50, and 10 mg L^{-1}, respectively, to tomato (*Lycopersicon esculentem* L.) plants grown in nutrient solution. The three fractions of soil organic matter significantly stimulated root length and weight as compared to a pure nutrient solution. Sanchez-Conde and Ortega (1968) used nutrient solutions containing 8, 80, or 160 mg L^{-1} HA to grow pepper (*Capsicum annuum* L.) plants in a greenhouse and found enhanced growth in all three treatments. A maximum increase in root weight of 56.1% over the control was reported. The same concentrations of humic acid were reported to increase growth of shoots and roots of sugar beet (*Beta vulgaris* L.) (Sanchez-Conde et al., 1972). Fernandez (1968) conducted a series of experiments on roots of corn to test the effects of humic substances originating from decomposing organic matter of two sources: (i) farmyard manure humified for 1 yr, and (ii) vegetable residue compost produced after a number of years of composting. Farmyard manure humic acid increased root weight whereas vegetable compost humic acid inhibited growth. The dependence of effects on source of humate is somewhat unusual in the literature on root response to humic substances, but it may be related to the young age and possible difference in chemical structure of the tested humic acids.

Lee and Bartlett (1976) studied the effects of Na-humate on corn roots and found that root proliferation is enhanced at optimum concentration of 8 mg L^{-1}. Tan and Nopamornbodi (1979) extracted humic acid from a clayey soil and studied its influence on root growth of corn seedlings after

5 d growth in Hoagland's solution with or without humic acid. At concentrations of 640 mg L^{-1} HA and 1600 mg^{-1} HA, roots were significantly longer whereas a higher concentration of 3200 mg L^{-1} HA was less effective. The dry weight of the seedling after 16 d showed a similar pattern. Mylonas and McCants (1980a, b) tested both soil humic and fulvic acid effects on tobacco (*Nicotiana tabacum* L.) seedlings grown on filter paper in Petri dishes and water with Hoagland's solution with or without humic substances at various concentrations. Root length and numbers were measured 10, 14, and 17 d after germination. Humic substance concentrations of 10 to 100 mg L^{-1} were optimal for root growth. At concentrations of 1000 mg L^{-1} or more of humic substances, root length was shorter than that of the control. The effects of the humic substances became more prominent with time after germination.

A typical response curve of plant roots to increasing concentrations of a fulvic acid is shown in Fig. 7-1 which presents data from a study on cucumber (*Cucumis sativus* L.) plants by Rauthan and Schnitzer (1981). Addition of fulvic acid to Hoagland's solution was beneficial up to concentrations of 300 mg L^{-1} FA. The increase in weight and length was significant at concentrations of 100 to 300 mg L^{-1}.

In contrast to the results of many others, Fortun and Lopez-Fando (1982) reported that additions of 100, 250, or 500 mg L^{-1} HA did not increase dry weight of corn roots.

There are a considerable number of studies showing effects of humic substances on the pattern of root proliferation. Only a few will be reviewed here. O'Donnell (1973) was especially interested in studying the effects of humic substances on root proliferation and the formation of lateral roots, and in comparing humic substances from various sources. Geranium (*Pelargonium hortorum* L.) cuttings were placed in aqueous solutions containing one of the following organic substances, all originating from leonardite (a coal-like substance from the lignite group): (i) 500 mg L^{-1} FA, (ii) 500 mg

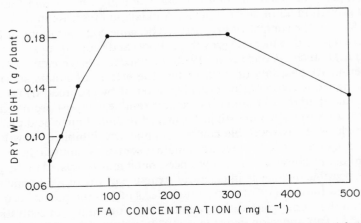

Fig. 7-1. Influence of fulvic acid concentration on dry weight of cucumber roots (after Rauthan & Schnitzer, 1981).

L^{-1} HA, (iii) 5000 mg L^{-1} leonardite, and (iv) 500 mg L^{-1} Na-humate. Plant response was compared to that of indole 3-butyric acid (IBA) at 100 mg L^{-1} and tap water. The humic substances were found to affect not only root growth rate but also proliferation of stout, healthy roots. The highest number of lateral roots was observed in the leonardite solution, which also contained the highest concentration of humic substances. The effects of the various humic substances were similar and resembled the response to the IBA treatment. It should be noted, however, that in O'Donnell's study the humic substances were added to water and not to nutrient solutions.

Alexandrova (1977) extracted humic substances from humified clover (*Trifolium alexandrium* L.) leaves and added them at concentrations up to 85 mg L^{-1} to nutrient solutions containing N, P, and K (NPK) at various levels, and grew millet (*Pennicitum* sp. L.) and corn in the solutions. Roots were longer in all the solutions containing humic substances. Moreover, the humic substances reduced negative effects that were observed when the nutrient solution concentration exceeded the optimum. Kononova and Pankova (1950) reported that excised roots of various plants placed in 7 mg L^{-1} HA solution produced new root initials whereas roots placed in water did not. Linehan (1976) studied the effects of 50 mg L^{-1} FA solutions on excised tomato roots. The optimal concentration for both root elongation and formation of root initials was 25 mg L^{-1} although a significant enhancement was also found at 50 mg L^{-1}. In addition, Linehan (1976) compared the effects of fulvic acids extracted from a number of soils and found them to be similar.

Reports by Schnitzer and Poapst (1967) and Poapst and Schnitzer (1971) clearly show beneficial effects of fulvic acid on root growth and the number of root initials in beans (*Phaseolus vulgaris* L.). The optimal concentration of fulvic acid reported in these studies significantly differs from those described earlier. Concentrations of 3000 to almost 6000 mg L^{-1} FA produced the greatest response. This discrepancy is probably explained by the differences in the mode of application. Cotyledons of bean seedlings were dipped for 3 to 3.5 h in solutions of $NaHCO_3$ containing fulvic acid at various concentrations up to 6000 mg L^{-1} and then transferred to a pot containing perlite previously drenched with a nutrient solution. After 6 d in perlite, the plants were removed and roots counted. When the fulvic or humic acids were added directly to the nutrient solution, concentrations ranging from 25 to 100 mg L^{-1} were optimum promoters for root growth.

Tomato seedlings have often been used to study effects of humic substances both on root and seedling growth. Some results showing beneficial effects of humic acid, fulvic acid, and an alcohol extract of organic matter on growth are presented in Table 7–1 (Sladky, 1959a). The tomato seedlings in this study were grown in a complete nutrient solution (Sachs) with or without the addition of humic substances. The beneficial effects both on roots and stem weight were significant.

Foliar application can also affect root growth. Sladky (1959b) applied humic materials as a foliar spray on begonia (*Begonia semperflorens* L.) plants and found enhanced root growth. The plants were grown in nutrient solu-

Table 7-1. Effect of soil organic matter fractions on tomato seedling growth (Sladky, 1959a).

Treatment	Stem length	Root length	Fresh weight		Dry weight	
			Stem	Roots	Stem	Roots
	cm		g			
Control	20.9	13.1	6.4	1.6	0.52	0.05
Alcohol extract (10 mg L^{-1})	32.4	17.0	8.9	3.3	0.62	0.09
HA (50 mg L^{-1})	51.5	20.2	14.9	3.2	1.07	0.23
FA (50 mg L^{-1})	56.8	14.0	17.5	5.4	1.60	0.24

tions and the concentrations of humic and fulvic acids in the sprays were 100 mg L^{-1}. Similar observations were obtained by Sladky (1965) with sugar beets grown in distilled water sprayed with solutions containing NPK and NPK + 300 mg L^{-1} HA. The foliar spray of NPK + humic acid enhanced root as well as shoot growth compared to NPK alone.

To summarize this section on interactions of humic substances and root growth, the following conclusions can be drawn:

1. Humic substances of various origins enhance root growth with either nutrient solution or foliar application.
2. Both elongation and formation of root initials are affected.
3. Results of some experiments indicate that fulvic acids have a slightly stronger effect than humic acids.
4. The concentration of the humic material is important, and generally the response decreases at high concentrations.

7-1.3 Shoot Development

In most of the studies that were reported in the previous section of this chapter, the effects of humic substances are often more prominent on roots than on shoots (e.g., Table 7-1), but the response of shoots has more often been reported. Many of the publications on shoot growth enhancement are limited to young plants grown in pots or in nutrient solutions.

Sladky and Tichy (1959) compared the effects of foliar or nutrient solution applications of humic substances on roots and shoots. When tomato plants were sprayed with a solution of 300 mg L^{-1} HA, both fresh and dry weight of shoots was increased. Young leaves responded to a greater extent than older ones. Higher application rates inhibited growth and caused leaf deformation. Application of foliar spray on begonia plants (Sladky, 1959b) yielded similar results and indicated that fulvic acid is slightly more effective than humic acid. Sugar beets responded similarly to foliar spray of 300 mg L^{-1} HA solution containing NPK (Sladky, 1965). Poapst et al. (1970) examined the effects of direct applications of high fulvic acid concentrations (1000 to 8000 mg L^{-1}) on the elongation of excised bean stems and reported inhibition effects at extremely high concentrations. Lee and Bartlett (1976) studied stimulation of corn seedling growth and found optimum response at about 8 mg L^{-1} Na-humate applied with the nutrient solution. The in-

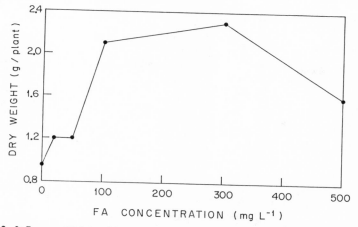

Fig. 7-2. Influence of fulvic acid concentration on dry weight of cucumber shoots (after Rauthan & Schnitzer, 1981).

crease in seedling growth was 30 to 50% in nutrient solution or low organic matter soil. The optimum level given in their report was quite low, probably because of the plant type and a short growth period. Rauthan and Schnitzer (1981) studied the stimulation of cucumber plants in fulvic acid and Hoagland's solutions and extended their study to the flowering stage that occurred after 6 wk. The resulting plant response curve, shown in Fig. 7-2, indicates optimal shoot growth at 100 to 300 mg L^{-1} FA, which overlaps the optimum for root growth (Fig. 7-1).

In most of the studies plants were grown in nutrient solutions but some research has been performed in sand or soil cultures. Lee and Bartlett (1976) found that corn plant response in low organic matter loamy soil was similar to that in nutrient solutions. Tan and Tantiwiramanond (1983) found stimulatory effects in sand cultures of applications of both fulvic and humic acids on growth of soybeans (*Glycine max* L.), peanuts (*Arachis hypogea* L.) and clover plants. In general, shoot, root, and nodule dry weights showed a tendency to increase in response to treatments with 100 to 400 mg FA or HA kg^{-1} soil. Optimum application levels were 400 to 800 mg kg^{-1} soil. These concentrations are reported on a per kilogram–soil basis and should not be directly compared with concentrations in nutrient solutions.

In conclusion, the studies discussed in this section indicate that fulvic and humic acids may stimulate shoot growth of various plants when applied either as foliar spray at concentrations of 50 to 300 mg L^{-1}, or when applied in nutrient solutions at concentrations of 25 to 300 mg L^{-1}. The stimulatory effect of shoot growth usually correlates to root response regardless of the mode of application. A summary of the effects of humic substances on seeds, roots, and shoots of higher plants and relevant concentration of humic materials is presented in Table 7-2.

Table 7-2. Effects of humic substances on seeds, roots and shoots of plants.

Plant organ response	Material	Concentration range, mg L^{-1}	Effect
Seed germination	HA	0–100	Enhanced rate. Accelerated water uptake.
Root initiation and elongation	HA, FA	50–300	Stimulated root initiation. Lateral roots development.
Excised root elongation	HA	5–25	Enhanced growth. Cell elongation.
Intact plant growth	HA, FA	0–500 optimum 50–300	Enhanced growth of shoots and roots

7-2 MECHANISMS INVOLVED IN PLANT GROWTH STIMULATION

7-2.1 Uptake of Macroelements

The stimulatory effect of humic substances on plant growth has been commonly related to enhanced uptake of macronutrients. In the introduction to this chapter, the historical argument whether humus, mineral nutrition, or possibly both, are required for optimal plant growth, was described. This argument has not been resolved and the effects of humic substances on the uptake of elements of nutritional significance is still the focus of many studies. Some of these will be briefly described in this section.

Gaur (1964) observed enhanced uptake of N, P, and K and a decrease in the uptake of Ca in ryegrass (*Lolium perenne* L.) grown in pots in a soil amended with humic acid extracted from compost. Sanchez-Conde and Ortega (1968) irrigated pepper plants with solutions containing 8, 80, and 100 mg L^{-1} HA and found an increase in the uptake of N, P, and Mg and a decrease in the uptake of K, Ca, and Na. Moreover, they also observed toxic accumulation of N, P, and Mg in the roots. Fernandez (1968) reported that humic acid originating from manure increased the uptake rate of N at all the application rates studied, whereas humic acid extracted from peat showed a similar activity only at low application rates.

Dormaar (1975) extracted humic acids from three soils and added 1, 5, 10, 20, and 50 mg L^{-1} to *Festuca scabrella* Torr. plants grown in nutrient solutions. Nitrogen uptake increased in solutions containing 20 and 50 mg L^{-1}, but uptake of P, K, Na, Ca, and Mg was not affected. In a thorough study performed on cucumber plants, Rauthan and Schnitzer (1981) grew their plants in Hoagland's solution containing up to 2000 mg L^{-1} FA. The treatments enhanced the uptake of N, P, K, Ca, and Mg to the shoots and N to the roots. Maximum uptake of all these elements as well as maximum growth level were obtained at concentrations of 100 to 300 mg L^{-1} FA.

Several investigators have concluded that the growth enhancement effect of humic substances in the rooting medium is due to an increase in the uptake of P. Lee and Bartlett (1976) reported that the addition of Na-humate, at concentrations up to 85 mg L^{-1}, to nutrient solutions resulted in an increased uptake of P. At concentrations higher than 85 mg L^{-1} Na-humate, a yield decline was reported but P uptake increased further. Similar effects on P uptake were also observed in soils. Jelenic et al. (1966) added ^{32}P-labeled superphosphate plus Na-humates to two soils at levels of 2, 4, 8, and 12 mg Na-humate kg^{-1} soil and found enhanced uptake of both soil- and superphosphate-P by corn. Maximum uptake was achieved with the additions of 3 mg and 4 to 8 mg Na-humate per kg of soil in a pseudogley and a Brown Forest (Eutrochrepts) soil, respectively. Various fractions of humic materials that were extracted from lignite differed in their effect on the uptake of P, but, in general, all the fractions enhanced P uptake.

Studies on the effects of humic materials on P uptake were also performed on excised roots and cell cultures of plants using radioactively labeled P (Vaughan et al., 1978). Humic acid was added at levels of 5, 50, and 500 mg L^{-1} to nutrient solutions containing ^{32}P in which excised corn roots were cultured. The lower concentrations of 5 and 50 mg L^{-1} HA enhanced ^{32}P uptake whereas 500 mg L^{-1} HA inhibited uptake. Vaughan and McDonald (1971) observed an increase in the rate of ^{32}P uptake by discs of beet root tissue although the total quantity taken up was not affected.

The fact that humic substances may affect the uptake of some elements by infuencing their rate of release from the soil mineral component should not be overlooked. For example, Tan (1978) has shown that both humic and fulvic acids are capable of releasing fixed K from illite or montmorillonite. Obviously, these effects may be important in soils, but do not contribute to the response of plants grown in nutrient solutions.

7–2.2 Uptake of Microelements

Complexation of transition metals, such as Cu, Zn, Fe, Mn, and others by humic substances, has been the focus of a large number of publications. This topic was recently reviewed by Chen and Stevenson (1986) in relation to plant growth and, therefore, will not be discussed in great detail in this chapter. Solubilization of micronutrients from their inorganic forms may be the major factor in the promotion of plant growth in soils by humic substances. The same situation may apply to nutrient solutions in which solubility of most of the micronutrients is limited. The presence of humic substances in either a nutrient or soil solution may contribute to improved availability of elements.

Iron has drawn the attention of researchers more than any other microelement. Dekock (1955) concluded that lignite-derived humic substances maintained Fe in solution in both nutrient solutions and plant tissues even at high phosphate concentrations. Untreated chlorotic plants contained high Fe concentration in their roots probably due to precipitation as ferric phosphate. Humic substances not only increased the solubility of Fe in solution but also affected Fe translocation from roots to shoots (Dekock, 1955). Aso and Sakai (1963) found that in nutrient solutions at pH 7 rice (*Oryza sativa* L.) and barley were chlorotic unless ammonium nitro–humic acid (NH$_4$–NHA) was added. Addition of Fe(III)-humic substance complexes significantly reduced chlorosis severity, while unferrated humic substances alone were ineffective. Dyakonova and Maksimova (1967) reported that soluble Fe-humic acid complexes occurred in natural peat and prevented chlorosis in plants. Lee and Bartlett (1976) found that 5 mg L^{-1} Na-humate in a nutrient solution enhanced yield of corn plants as well as increased the Fe concentration in roots and shoots. Linehan and Shepherd (1979) compared effects of fulvic acid to those of polymaleic acid and other polycarboxylates. Addition of fulvic acid to nutrient solutions, at concentrations up to 25 mg L^{-1}, enhanced Fe uptake to shoots of wheat seedlings. Polymaleic acid and other polycarboxylates showed similar effects.

In a number of publications, Chen and collaborators (Barak & Chen, 1982; Chen et al., 1982a, b; Chen & Barak, 1983; Bar-Ness & Chen, 1990a, b) have shown that Fe-enriched organic materials such as peat or manure could serve as a remedy to lime-induced chlorosis in soils. The corrective effect was attributed to complexation of Fe by humic substances in the organic materials.

The effect of humic substances on the uptake of Zn and Cu was also investigated on intact plants and on plant tissues. Vaughan and McDonald (1976) studied Zn uptake by beet root tissue cut into discs, and found that the addition of humic acid slightly inhibited Zn uptake by aged discs when concentrations exceeded 25 mg L^{-1} HA. Lower concentrations did not affect Zn uptake. On intact plants, Jalali and Takkar (1979) reported that Fe, Cu, and Zn uptake by rice plants was enhanced by increased levels of organic matter.

Soil humus can reduce the plant uptake of some metal ions by strong adsorption from solution. White and Chaney (1980) followed the uptake of Zn, Cd, and Mn in two soils amended with Zn and Cd. The soils contained 1.2% and 3.8% organic matter [Sassafras (Typic Hapludult) and Pocomoke (Typic Umbraqualt) soils, respectively]. Toxic effects on plants were reduced in the high organic matter soil due to increased binding of metals to insoluble soil components, possibly fractions of soil organic matter. Reduced uptake of Cd resulting from peat amendments to sand cultures of soybeans was reported by Strickland et al. (1979). The plants were grown in sand mixed with 0.5 to 8% of peat (w/w). Up to 20 mg of Cd kg^{-1} soil were added and yield and Cd uptake recorded. Yields increased with increasing levels of peat. Cadmium concentrations in roots and shoots increased with low additions of peat to the medium (up to 0.5%) but decreased at high additions.

Apparently, the solubility of added organic matter fractions is an important factor determining whether uptake enhancement or inhibition will occur. Another important factor is the occurrence of specific uptake mechanisms for micronutrients from complexes at the root surfaces, such as those known for Fe (Marschner et al., 1986). Clearly, low concentrations of soluble complexes of humic substances with Fe, Zn, Cu, or Mn will enhance their uptake by plants, thereby improving the nutritional status of the plant (e.g., Rauthan & Schnitzer, 1981).

Table 7–3 is presented to conclude the sections describing the beneficial effects of humic substances on plant growth under conditions in which a com-

Table 7-3. The effect of 50 mg L^{-1} of humic acid on growth of wheat in water or Hoagland's nutrient solution (after Vaughan & Malcolm, 1985).

Culture medium	Plant organ	Fresh weight, mg/plant	Stimulation, %
Water	Root	93	0
	Shoot	185	0
Water + HA	Root	146	57.5
	Shoot	252	36.2
Hoagland	Root	182	96.3
	Shoot	342	84.9
Hoagland + HA	Root	203	119.0
	Shoot	390	110.8

plete nutrient solution was supplied. The data clearly show growth enhancement in water by additions of humic acid, and further stimulation of growth in Hoagland's solution. The stimulation in the presence of humic acid exceeds that of Hoagland's solution alone by about 25%, which provides evidence for a synergistic effect of combined applications of mineral nutrition and humic substances.

7-2.3 Uptake of Humic Substances and Biochemical Effects

7-2.3.1 Uptake of Humic Substances

The assertion that humic substances can have a direct effect on plant growth implies that these materials are taken up by plants. Initially, researchers determined uptake by observation of staining of the plant tissue with black- or brown-colored materials. More recent investigations have used ^{14}C-labeled humic substances. Aso and Sakai (1963) immersed mulberry (*Moorus* sp. L.) seedlings in NH$_4$-humic acid, homogenized the plant parts, and observed staining of the tissue. Prat (1963) allowed shoots of several plant species to stand with their cut ends in a humic acid solution. Microscopic evidence of staining due to humic acid uptake to the shoots was obtained. The uptake of humic acid by roots was not tested in either of these studies.

Using ^{14}C-labeled material, Prat and Pospisil (1959) showed that humic acid accumulated in the roots of sugar beets and corn. Only a small fraction of radioactivity was transported from the roots to the shoots. Similar observations were also reported by other workers (Führ & Sauerbeck, 1967b; Vaughan & Linehan, 1976). Some investigations showed that fulvic acid is transported to the shoot to a greater extent than is humic acid (Führ & Sauerbeck, 1967a; Führ, 1969). Vaughan and Linehan (1976) found that labeled humic acid was taken up by wheat roots and about 5% was transported to the shoots. Führ and Sauerbeck (1967b) demonstrated that whereas much of the absorbed radioactivity from ^{14}C-humic acid was incorporated into the epidermis of sunflower (*Helianthus annus* L.), radish (*Raphanus sativus* L.) and carrot (*Daucus carota* L.) roots, a substantial amount of activity originating from low molecular weight components of the humic materials also entered the stele.

Vaughan and McDonald (1976) studied the uptake of ^{14}C-humic acid by subcellular components of beet roots. The greatest amount of radioactivity was associated with cell walls and smaller levels with mitochondria and ribosomes. It was suggested that only low molecular weight fractions of the aromatic core are biologically active. In a later study on excised pea (*Pisum sativum* L.) roots, Vaughan and Ord (1981) showed that the ratio of uptake of fulvic to humic acids increased with incubation time, indicating preferred uptake of the low molecular weight substances. Their study also indicated that low molecular weight humic acid fractions are taken up both actively and passively, whereas humic acid of molecular weight > 50 000 daltons is taken up only passively. Although Vaughan et al. (1985) concluded that mostly lower molecular weight fractions of humic substances were taken up actively by plants, the chemical nature of these fractions remains to be clarified.

It appears likely that fulvic acid may be somewhat more biologically active than humic acid. This fact is corroborated by results of the studies mentioned earlier. Since humic acids often contain low molecular weight fractions, however, their potential for biological activity should not be underestimated.

7-2.3.2 Biochemical Effects

Sections 7-2.1 and 7-2.2 summarize a substantial amount of literature about indirect effects of humic substances on plant growth via macro- and microelements supply. The fact that humic substances are also taken up by plants encouraged scientists to investigate possible direct effects on cell membranes and biochemical cycles in plant cells. Since this subject has been reviewed recently (Vaughan et al., 1985) and because it is beyond the scope of this chapter, this topic will only be briefly summarized here and some typical examples will be presented.

7-2.3.2.1 Effects on Membranes. The stimulation of ion uptake by treatments with humic materials led many investigators to propose that these materials affect membrane permeability (Vaughan & McDonald, 1971; Vaughan & McDonald, 1976). Prozorovskaya (1936) demonstrated that the exoosmosis of sugars from bulb scales was increased in the presence of humic acid and concluded that humic acid increased the permeability of cell membranes, resulting in an increase of nutrient uptake. Heinrich (1964) reported an increase in urea uptake by epidermal cells of *Gentiana rochelli* in the presence of humic acid.

The mode of action of humic substances on membranes is not clear, but it is probably related to the surface activity of humic substances (Chen & Schnitzer, 1978) resulting from the presence of both hydrophilic and hydrophobic sites. Thus, the humic substances may interact with the phospholipid structures of cell membranes and react as carriers of nutrients through them.

7-2.3.2.2 Energy Metabolism. Both photosynthesis and respiration rates of plants were enhanced by the presence of humic substances in the

Table 7-4. Effect of humic substance fractions on respiration and chlorophyll levels in tomato plants (% of control) (Sladky, 1959a).

Treatment	Oxygen uptake		Chlorophyll
	Leaves	Roots	
	% of control		
Control	100	100	100
Alcohol extract (10 mg L^{-1})	110	176	130
HA (50 mg L^{-1})	124	123	163
FA (50 mg L^{-1})	130	138	169

nutrient solutions in which they were grown (Table 7-4 and Fig. 7-3). Tomato plants grown in nutrient solutions containing either humic acid, fulvic acid, or an alcoholic extract of soil organic matter produced higher concentrations of chlorophyll. Their oxygen consumption increased as compared to control plants (Table 7-4). The effect of fulvic acid was greater than that of humic acid. Humic acid was shown to enhance respiration of beet root slices (Fig. 7-3). This experiment (Vaughan, 1967) was conducted under axenic conditions to eliminate the possibility that the enhanced respiration was due to microbial degradation of the humic materials. Boiling the humic acid in 6M HCl prior to the experiment ruled out the possibility that carbohydrates and/or proteins were responsible for the respiration enhancement since these materials are removed during the boiling process. When leaves of begonia were sprayed with aqueous solutions of humic acid, a marked increase in O_2 uptake was recorded (Sladky, 1959b; Sladky & Tichy, 1959). This change shows that both foliar and nutrient solution applications of humic materials can affect respiration. Clearly, humic and fulvic acids can have a direct effect on respiration. This effect was also observed for a model synthetic humic acid used by Flaig (1968).

Many investigations have shown an increase in chlorophyll contents resulting from applications of humic substances in nutrient solutions or foliar spray. An example is presented for tomatoes in Table 7-4 (Sladky, 1959a).

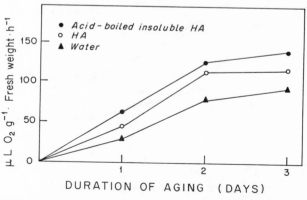

Fig. 7-3. Effect of humic acid on the development of respiration in beet root slices (after Vaughan, 1967).

Sladky and Tichy (1959) concluded that the dry weight increase in begonia plants sprayed with humic acid solution was correlated with the observed increase in chlorophyll contents which in turn, affected photosynthesis. Production of chlorophyll alone, however, does not necessarily result in higher yields.

7-2.3.2.3 Protein and Nucleic Acid Synthesis. Interactions of humic substances with the synthesis of nucleic acid have been reported. Changes in ribonucleic acid (RNA) synthesis were observed for excised pea roots by Vaughan and Malcolm (1979b). The authors concluded that humic acid influences the production of m-RNA which is essential for many biochemical processes in the cell. Many publications, of which only a few will be mentioned here, report the influence of humic substances on protein, especially enzyme synthesis and development.

Bukvova and Tichy (1967) found that humic acid affects the development of phosphorylase in wheat plants grown in sand cultures. Low concentrations of 10 mg L^{-1} HA enhanced the enzyme synthesis in the root, but higher concentrations of 100 mg L^{-1} were inhibitory. Humic substances influenced the development of catalase, o-diphenoloxidase, and cytochrome in tomatoes (Stanchev et al., 1975), and invertase and peroxidase in beets (Vaughan, 1967; Vaughan et al., 1974).

7-2.3.2.4 Enzyme Activity. The notion that humic substances may act as growth hormones led many scientists to investigate their influence on enzyme activity, particularly effects on indole-3-acetic acid (IAA) metabolism.

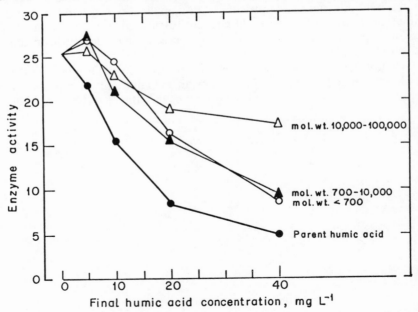

Fig. 7-4. Effect of soil humic acid and its molecular weight fractions on indoleacetic acid destruction by lentil root extract (after Mato et al., 1972a).

Mato and collaborators have shown in a series of publications (e.g., Mato et al., 1971; Mato et al., 1972a, b) that humic acid, fulvic acid, and fractions of humic substances inhibit IAA-oxidase, thereby hindering IAA destruction. Unfractionated humic acid was more effective than any of the three fractions at suppressing IAA destruction (Fig. 7–4). Obviously, maintenance of high activity of IAA will positively affect plant growth.

Humic substances have been shown to affect the activities of other enzymes such as phosphatase (Malcolm & Vaughan, 1979a, b, c), invertase (Vaughan, 1967; Malcolm & Vaughan, 1979b), choline esterase (DeAlmeida et al., 1980) and peroxidase (Vaughan & Malcolm, 1979a). Generally, inhibition of these enzymes occurred in positive, direct correlation with humic concentration. Ladd and Butler (1971), using purified enzymes, showed that humic acid inhibited carboxypeptidase A, chymotrypsin A, pronase, and trypsin activities, but stimulated papin and ficin.

The mechanisms by which humic substances affect enzyme activities are not completely understood. Several hypotheses have been suggested. Butler and Ladd (1969) proposed that humic acid may inhibit pronase activity by either competing with the substrate for the catalytically active sites on the enzyme or by causing conformational changes in the enzyme. Vaughan and Malcolm (1979a) suggested that the humic material inhibited peroxidase activity by competitive chelation of Fe. Apparently, there are a number of different mechanisms that are related to the reactivity of functional groups on the humic materials and which vary depending on the specific enzyme.

It may be concluded from the above evidence that the net direct effect of humic materials on growth probably involves interactions of a series of biochemical stimulations and inhibitions, thereby partially explaining the dependence on the humic substance concentration and the type and degree of effects observed on plants.

7–3 EFFECTS ON PLANT GROWTH RELATED TO HUMIC SUBSTANCES IN GREENHOUSE SUBSTRATES AND IN FIELD TRIALS

7–3.1 Mobilization of Fe in Soils and Potting Media

In contrast to the substantial literature on the activity of humic substances in stimulation of plant growth in nutrient cultures, information on the activity of humic substances in the field is scarce. The use of humic materials to help overcome Fe and Zn deficiencies, however, has been studied in field trials and has been reviewed recently (Chen & Stevenson, 1986). Naturally occurring organic materials rich in humic substances, such as peat and decomposed manure, have been used as fertilizers after the employment of enrichment processes with Fe or Zn (e.g., Barak & Chen, 1982; Bar-Ness & Chen, 1990a, b). Figure 7–5 shows the effect of increasing application rates of Fe-enriched peat on chlorophyll production in peanuts grown on a highly calcareous soil (Xerorthent, 63% $CaCO_3$). Chen and Barak (1983) used the

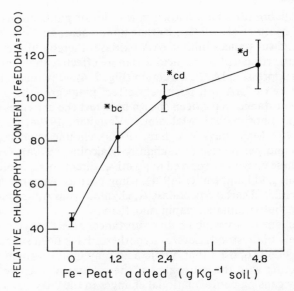

Fig. 7–5. Response of relative chlorophyll content to iron-enriched peat (Fe-Peat) added to an Xerorthent soil (Mitzpeh Massua) (average ± SE). Lower-case letters indicate results of a Newman-Keuls multiple range test. Asterisk indicates treatment not significantly different from Fe-EDDHA ($\alpha = 0.05$) (after Barak & Chen, 1982).

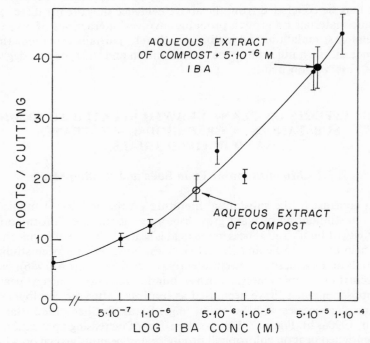

Fig. 7–6. Adventitious root formation in mung bean cuttings as affected by auxin (IBA) and by an aqueous extract of composted manure (Raviv et al., 1983).

same peat preparation for the remedy of Fe deficiency in peanuts grown in the field. Yields increased from 4740 kg ha^{-1} in control plants to 5540 kg ha^{-1} and 5810 kg ha^{-1} when fertilized with Fe-enriched peat and FeEDDHA (at a level considered adequate to provide a complete remedy), respectively. Bar-Tal et al. (1988) have shown that fulvic acid maintains $10^{-3.5}$ mM Zn in solution in the presence of Ca-montmorillonite at pH 7.5 whereas Zn levels decreased to $10^{-5.5}$ mM in the absence of fulvic acid. These and many other studies show that humic substances can play an important role in the micronutrient nutrition of plants under field conditions.

Chen and Katan (1980) observed an increased growth response in a number of crops in fields that were solar heated under transparent plastic mulch and wet soil conditions. As a result of the treatment, soluble soil organic matter, mostly consisting of fulvic acid, increased from 20 to 80 mg L^{-1}. Since no significant changes in the composition of the soluble organic matter were detected, Chen and Katan (1985, unpublished) attributed the increased growth response to the observed increase in fulvic acid concentration.

Composts originating from plant residues or manure are known for their high levels of humic substances. Additions of composts to potting media mixtures such as with peat, perlite, vermiculite, etc. often result in significant yield increases (Raviv et al., 1986; Chen et al., 1988) which can be attributed to humic substances, nutritional elements and microbial activity. Water extracts from composts were shown to contain an auxin-like activity (Raviv et al., 1983). This activity was characterized by its effect on mung bean (*Vigna mungs* L.) root formation. These results as summarized in Fig. 7–6 show a prominent effect of a water extract of compost on root formation resembling that of $5 \times 10^{-6} M$ indole butyric acid (IBA). This influence showed an additive response of the extract to that of IBA (Fig. 7–6).

Reports on the response of agricultural crops to applications of humic substances are scarce in the scientific literature. Reports by commercial companies on beneficial effects of various products containing humic substances often lack statistical analysis and should, therefore, be interpreted with caution.

7–3.2 Foliar Sprays

Xudan (1986) studied effects of foliar application of fulvic acid on water use, nutrient uptake, and yield of wheat in pot experiments and in the field. Fulvic acid reduced the stomatal conductance of well-watered plants in pots from 0.80 to 0.25 cm S^{-1}. The stomatal conductance of control plants fell continuously from 0.85 cm S^{-1} to almost zero over a 9-d drying cycle. Plants sprayed with fulvic acid at the beginning of the cycle maintained stomatal conductance of 0.30 cm S^{-1} for the whole period. Spraying with fulvic acid resulted in a higher level of chlorophyll in the leaves and of greater uptake of ^{32}P by the roots. When control wheat plants were subjected to drought stress at head development stage, grain yield was depressed by 30%. Spraying with fulvic acid increased the yield of the plants grown at dry conditions to 97% of the irrigated control. Field trials on wheat conducted in north China demonstrated that when fulvic acid was used to decrease water stress imposed by hot dry winds during head development, grain yield increased by 7.3 to 18.0%.

Brownell et al. (1987) tested the response of various field crops to application of two extracts from leonardite. No details on the chemical properties were presented because of the commercial origin of the products. Positive yield and hormone-like responses on a number of field crops were observed. Field trials on processing tomatoes produced average yield increases of 10.5% compared to untreated controls. Trials on cotton (*Gossypium hirsutum* L.) during the same year, with and without other growth controlling chemicals, produced an average yield increase of 11.2%. Unreplicated large field trials on grape vines (*Vitis vinifera*) of various cultivars produced a range of increases in total yield from 3 to 70%, with an average of 25%. In most of these studies, one of the products that was rich in humic substances was used as an early season soil treatment, while the other was used as a post-emergence foliar spray. Based on their observations, the investigators hypothesized that the tested products, when used singly or in tandem, triggered a flowering response in many species of plants. The most pronounced effect was obtained by Brownell et al. (1987) using a combination of early season soil treatment with a post-emergence foliar spray.

7-4 CONCLUSIONS AND FUTURE PROSPECTS

Favorable effects of humic substances on plant growth under laboratory conditions have been demonstrated for the following parameters: length and fresh and/or dry weights of shoots and roots, number of lateral roots, root initiation, seedling growth after germination, microbial population and activity, nutrient uptake, and flowering.

Favorable effects were also observed under field conditions, either directly or indirectly, as a response to application of humic substances. These effects on agricultural crops were related to increases in fulvic acid concentration in the soil solution, which induced increased growth response or increased concentrations of soluble micronutrients.

The modes of action of humic substances on plant growth can be divided into direct (requiring uptake by the plant tissue) and indirect effects as follows:

Direct effects:
1. Effects on membranes resulting in improved transport of nutritional elements.
2. Enhanced protein synthesis.
3. Plant-hormone like activity.
4. Enhanced photosynthesis.
5. Effects on enzyme activity.

Indirect effects:
1. Solubilization of microelements (e.g. Fe, Zn, Mn) and some macroelements (e.g. K, Ca, P).

2. Reduction of active levels of toxic elements.
3. Enhancement of microbial populations.

It should be noted that effects of organic amendments, such as enhanced nutrient supply, soil structure improvement, increased cation exchange capacity, increased water retention, and enhanced microorganism populations due to increased C and N sources, should be distinguished from those of the specific effects of humic substances. This comparison is not always easy to achieve because of the considerable overlap between various sources of humic substances (i.e. manure, compost, soil organic matter) and their effects on the plant environment.

To assist in evaluating the prospects of finding favorable effects under field conditions, some calculations can be made. In the following calculations, foliar and soil applications were compared on the basis of concentrations that were frequently reported as those concentrations of humic substances that are required to affect plant growth in growth chambers.

a. *Soil Application:*
Assumptions: 1. Plow layer weight: 2500 Mg ha^{-1}
2. Water content at field capacity (% by weight): 30
3. Increase required in humic substances (HS) concentration: 100 mg L^{-1}
For 1 ha the following quantity of HS is required:

$$2500 \text{ Mg} \times 0.3 \text{ m}^3 \text{ Mg}^{-1} \times 0.1 \text{ kg HS m}^{-3} = 75 \text{ kg.}$$

b. *Foliar Spray:*
Assumptions: 1. Required volume of spray: 2000 L ha^{-1}
2. Required concentration: 250 mg L^{-1}
For 1 ha the following quantity of HS is required:

$$2000 \text{ L} \times 0.25 \text{ g L}^{-1} = 500 \text{ g.}$$

These calculations, based on extrapolation of numerous laboratory studies, show that the amount of humic materials required for an effective soil application is about 75 kg ha^{-1}. Foliar spray may be effective at an application rate that is at least 100 times lower. Because of the relatively high cost of preparations of humic substances, it seems that future prospects for economical use in agriculture of these products are much better for foliar spray application. Soil and foliar application of complexes of humic materials with micronutrients such as Fe and Zn require separate evaluation due to their specific effects under deficiency conditions prevailing in calcareous soils.

In fertile soils of humid climates in which soluble organic matter may reach levels up to 400 mg L^{-1} (Chen & Schnitzer, 1978), beneficial effects due to application of humic substances are not likely to occur. In soils of semiarid or arid zones, however, in which soluble organic matter does not exceed 20 to 30 mg L^{-1} (e.g. Chen & Katan, 1980), beneficial effects may be observed at sufficient application rates of humic substances. The possibility of obtaining increases of soluble humic substances by applications of manure or compost should, however, be considered as an economical alternative.

To conclude, it seems that commercial humates applied to normally productive agricultural soils at rates recommended by their promoters would not appear to contain sufficient quantities of the necessary ingredients to produce the claimed beneficial effects. Yield increases, if any, from the use of such products would appear to be insufficient to offset increased production costs to the farmer.

ACKNOWLEDGMENT

This research was jointly supported by the U.S.–Israel (Binational) Agricultural Research and Development Fund (BARD), and by the Niedersachsischen Ministry of Science and Technology.

REFERENCES

Agboola, A.A. 1978. Influence of soil organic matter on cow pea's response to N fertilizer. Agron. J. 70:25–28.

Alexandrova, I.V. 1977. Soil organic matter and the nitrogen nutrition of plants. Sov. Soil Sci. (Engl. Transl.) 9:293–301.

Aso, S., and I. Sakai. 1963. Studies on the physiological effects of humic acid. 1. Uptake of humic acid by crop plants and its physiological effects. Soil Sci. Plant Nutr. (Tokyo) 9:85–91.

Bacon, F. 1651. Sylva sylvarum, London.

Bar-Ness, E., and Y. Chen. 1990a. Manure and peat based iron-organo complexes: I. Characterization and enrichment. *In* Y. Chen and Y. Hadar (ed.) Iron nutrition and interactions in plants. Martinus Nijhof. Dordrecht. The Netherlands (in press).

Bar-Ness, E., and Y. Chen. 1990b. Manure and peat based iron-enriched complexes: II. Transport in soils. *In* Y. Chen and Y. Hadar (ed.) Iron nutrition and interactions in plants. Martinus Nijhof. Dordrecht. The Netherlands (in press).

Bar-Tal, A., B. Bar-Yosef, and Y. Chen. 1988. Effects of fulvic acid and pH on zinc sorption on montmorillonite. Soil Sci. 146:367–373.

Barak, P., and Y. Chen. 1982. The evaluation of iron deficiency using a bioassay-type test. Soil Sci. Soc. Am. J. 46:1019–1022.

Bottomley, W.B. 1914a. Some accessory factors in plant growth and nutrition. Proc. R. Soc. London, B 88:237–247.

Bottomley, W.B. 1914b. The significance of certain food substances for plant growth. Ann. Bot. (London) 28:531–540.

Bottomley, W.B. 1917. Some effects of organic growth-promotion substances (auximones) on the growth of *Lemna minor* in mineral cultural solutions. Proc. R. Soc. London, B 89:481–505.

Bottomley, W.B. 1920. The effect of organic matter on the growth of various plants in culture solutions. Ann. Bot. (London) 34:353–365.

Brownell, J.R., G. Nordstrom, J. Marihart, and G. Jorgensen. 1987. Crop responses from two new Leonardite extracts. Sci. Total Environ. 62:492–499.

Bukvova, M., and V. Tichy. 1967. The effect of humus fractions on the phosphorylase activity of wheat (*Triticum aestivum* L.). Biol. Plant. 9:401–406.

Burk, D., H. Lineweaver, and C.K. Horner. 1932. Iron in relation to the stimulation of growth by humic acid. Soil Sci. 33:413–435.

Butler, J.H.A., and J.N. Ladd. 1969. The effect of methylation of humic acids on their influence on proteolytic enzyme activity. Austr. J. Soil Res. 7:263–268.

Chen, Y., and P. Barak. 1983. Iron-enriched peat and lignite as iron fertilizer. p. 195–202. *In* K.M. Schallinger (ed.) Proc. Int. Symp. Peat Agric. Hort., 2nd, 9–14 Oct. 1983. Volcani Center, Bet-Dagan, Israel.

Chen, Y., Y. Inbar, and Y. Hadar. 1988. Composted agricultural wastes as potting media for ornamental plants. Soil Sci. 145:298–303.

Chen, Y., and J. Katan. 1980. Effect of solar heating of soils by transparent polyethylene mulching on their chemical properties. Soil Sci. 130:271–277.

Chen, Y., J. Navrot, and P. Barak. 1982a. Remedy of lime-induced chlorosis with iron-enriched muck. J. Plant Nutr. 5:927–940.

Chen, Y., and M. Schnitzer. 1978. The surface tension of aqueous solutions of soil humic substances. Soil Sci. 125:7–15.

Chen, Y., B. Steinitz, A. Cohen, and Y. Elber. 1982b. The effect of various iron-containing fertilizers on growth and propagation of *Gladiolius grandiflorus*. Sci. Hortic. (Amsterdam) 18:169–175.

Chen, Y., and F.J. Stevenson. 1986. Soil organic matter interaction with trace elements. p. 73–116. *In* Y. Chen and Y. Avnimelech (ed.) The role of organic matter in modern agriculture. Martinus Nijhoff Publ., Dordrecht.

De Almeida, R.M., F. Pospisil, K. Vackova, and M. Kutacek. 1980. Effect of humic acids on the inhibition of pea choline esterase and choline a-cyltransferase with malathion. Biol. Plant. 22:167–175.

De Saussure, Th. 1804. Recherches chimiques sur la vegetation. Paris.

Dekock, P.C. 1955. The influence of humic acids on plant growth. Science (Washington, DC) 121:473–474.

Dixit, V.K., and N. Kishore. 1967. Effect of humic acid and fulvic acid fraction of soil organic matter on seed germination. Indian J. Sci. Ind. Sec. A 1:202–206.

Dormaar, J.F. 1975. Effects of humic substances from chernozemic Ah horizons on nutrient uptake by *Phaseolus vulgaris* and *Festuca scabrella*. Can. J. Soil Sci. 55:111–118.

Dyakonova, K.V., and A.E. Maksimova. 1967. Humic substances of the most active part of organic fertilizers and their influence on plants. p. 79–85. Trans. Jt. Meet. Comm. 2 and 4, Int. Soc. Soil Sci. 1966. Dokuchaev Soil Inst., Moscow.

Fernandez, V.H. 1968. The action of humic acids of different sources on the development of plants and their effect on increasing concentration of the nutrient solution. Pont. Acad. Sci. Scr. Varia 32:805–850.

Flaig, W. 1968. Uptake of organic substances from soil organic matter by plants and their influence on metabolism. Pont. Acad. Sci. Scr. Varia 32:1–48.

Fortun, C., and C. Lopez-Fando. 1982. Influence of humic acid on the mineral nutrition and the development of maize roots cultivated in normal nutrient solutions and lacking Fe and Mn. Anales de Edafologiay Arbogiologia XLI:335–349.

Führ, F., and D. Sauerbeck. 1967a. The uptake of colloidal organic substances by plant roots as shown by experiments with ^{14}C-labelled humus compounds. p. 73–82. *In* Report FAO/IAEA Meeting, Vienna, Pergamon Press, Oxford.

Führ, F., and D. Sauerbeck. 1967b. The uptake of straw decomposition products by plant roots. p. 317–327. *In* Report FAO/IAEA Meeting, Vienna, Pergamon Press, Oxford.

Gaur, A.C. 1964. Influence of humic acid on growth and mineral nutrition in plants. Bull. Assoc. Fr. Itude Sol. 35:207–219.

Grandeau, L. 1872. Recherches sur le role des matieres organiques du sol dans les phenomenes de la nutrition des vegetaux. Comptes rendus hebdomadaire seances de l'academie des sciences. Paris.

Heinrich, G. 1964. Huminsaeure und Permeabilitat. Protoplasma 58:402–425.

Hutchinson, H.B., and N.H.J. Miller. 1912. The direct assimilation of inorganic and organic forms of nitrogen by higher plants. J. Agric. Sci. 4:282–302.

Iswaran, V., and P.K. Chonkar. 1971. Action of sodium humate and dry matter accumulation of soybean in saline alkali soil. *In* B. Novak et al. (ed.) Humus et Planta. V:613–615. Prague.

Ivanova, L.V. 1965. Influence of humic substances on growth of excised maize roots. Dokl. Akad. Nauk. BSSR 9:255–257.

Jalali, V.K., and P.N. Takkar. 1979. Evaluation of parameters for simultaneous determination of micronutrient cations available to plants from soils. Indian J. Agric. Sci. 49:622–626.

Jelenic, D.B., M. Hajdukovic, and Z. Aleksic. 1966. The influence of humic substances on phosphate utilization from labelled superphosphate. p. 85–88. *In* The use of isotopes in soil organic matter studies. FAO/IAEA Tech. Meet., Pergamon Press, Oxford.

Khristeva, L. 1949. Nature of the effect of humic acids on the plant. Dokl. Vses. Akad. Skh. Nauk im. V.I. Lenina. Vol. 7.

Kononova, M.M., and N.A. Pankova. 1950. The action of humic substances on the growth and development of plants. Doklady Akad. Nauk SSSR 73:1069–1071.

Ladd, J.M., and J.H.A. Butler. 1971. Inhibition and stimulation of proteolytic enzyme activities by soil humic acids. Austr. J. Soil Res. 7:253–261.

Lawes, J.B., and J.H. Gilbert. 1905. Collected papers. In W.H. Hall (ed.) The Book of the Rothamsted Experiments, London.

Lee, Y.S., and R.J. Bartlett. 1976. Stimulation of plant growth by humic substances. Soil Sci. Soc. Am. J. 40:876–879.

Li, C.H., G.H. Xu, and Z.W. Feng. 1981. Essential properties of woodland soil in major China fir producing areas and their relationship with growth of China fir. T'u Jang T'ung Pao 4:1–6.

Liebig, J.V. 1841. Organic chemistry in its applications to agriculture and physiology. Translated by J.W. Webster, and J. Owen, Cambridge.

Liebig, J.V. 1856. On some points of agricultural chemistry. J. Royal Agric. Soc. 17:284–326.

Linehan, D.J. 1976. Some effects of a fulvic acid component of soil organic matter on the growth of cultivated excised tomato roots. Soil Biol. Biochem. 8:511–517.

Linehan, D.J., and H. Shepherd. 1979. A comparative study of the effects of natural and synthetic ligands on ion uptake by plants. Plant Soil 52:281–289.

Lykov, A. 1978. The effect of the organic matter of derno podzolic soil on the yields of field crops. Problemy Zemledeliya. Referentivnyi Zhurnal Seriya. 5:195–202.

Malcolm, R.E., and D. Vaughan. 1979a. Comparative effects of soil organic matter fractions on phosphatase activities in wheat roots. Plant Soil 51:117–126.

Malcolm, R.E., and D. Vaughan. 1979b. Effects of humic acid fractions on invertase activities in plant tissues. Soil Biol. Biochem. 11:65–72.

Malcolm, R.E., and D. Vaughan. 1979c. Humic substances and phosphatase activities in plant tissues. Soil Biol. Biochem. 11:253–259.

Marschner, H., V. Romheld, and M. Kissel. 1986. Different strategies of higher plants in mobilization and uptake of iron. J. Plant Nutr. 9:695–714.

Mato, M.C., R. Fabregas, and J. Mendez. 1971. Inhibitory effect of soil humic acids on indoleacetic acid oxidase. Soil Biol. Biochem. 3:285–288.

Mato, M.C., M.G. Olmedo, and J. Mendez. 1972a. Inhibition of indoleacetic acid oxidase by soil humic acids fractionated in Sephadex. Soil Biol. Biochem. 4:469–473.

Mato, M.C., L.M. Gonzalez-Alonso, and J. Mendez. 1972b. Inhibition of enzymatic indoleacetic acid oxidation by fulvic acids. Soil Biol. Biochem. 4:475–478.

Mylonas, V.A., and C.B. McCants. 1980a. Effects of humic and fulvic acids on growth of tobacco. 1. Root initiation and elongation. Plant Soil 54:485–490.

Mylonas, V.A., and C.B. McCants. 1980b. Effects of humic and fulvic acids on growth of tobacco. 2. Tobacco growth and ion uptake. J. Plant Nutr. 2:377–393.

O'Donnell, R.W. 1973. The auxin-like effects of humic preparations from Leonardite. Soil Sci. 116:106–112.

Ojeniyi, S.O., and O.O. Agbede. 1980. Soil organic matter and yield of forest and tree crops. Plant Soil 57:61–67.

Olsen, C. 1930. On the influence of humus substances on the growth of green plants in water culture. Comptes-rendus du Laboratoire Carlsberg 18:1–16.

Olsen, S.R. 1986. The role of organic matter and ammonium in producing high corn yields. p. 29–70. In Y. Chen and Y. Avnimelech (ed.) The role of organic matter in modern agriculture. Martinus Nijhoff Publ., Dordrecht.

Pagel, H. 1960. Ueber den Einfluss von Humusstoffen auf das Pflanzenwachstum. 1. Einfluss von Humusstoffen auf Ertrag und Naehrstoffaufnahme. Albercht-Thaer-Arch. 4:492–506.

Pilus Zambri, M., M. Yaacob, A.J.M. Kamal, and S. Paramananthan. 1982. The determination of soil factors on growth of cashew on bri soils. Part I. Pertanika 5:200–206.

Poapst. P.A., C. Genier, and M. Schnitzer. 1970. Effect of soil fulvic acid on stem elongation in peas. Plant Soil 32:367–372.

Poapst, P.A., and M. Schnitzer. 1971. Fulvic acid and adventitious root formation. Soil Biol. Biochem. 3:215–219.

Prat, S. 1963. Permeability of plant tissues to humic acids. Biol. Plant. 5:279–283.

Prat, S., and F. Pospisil. 1959. Humic acids with ^{14}C. Biol. Plant. 1:71–80.

Prozorovskaya, A.A. 1936. The effect of humic acid and its derivatives on the uptake of nitrogen, phosphorus, potassium and iron by plants. Tr. NIUIFa. 127.

Febufetti, A., and D. Lubunora. 1982. Wheat yield in north eastern Uruguay in relation to NPK fertilizers, soil organic matter content and climatic conditions. p. 117–122. *In* C.C. Cerri (ed.) Regional colloquium on soil organic matter studies. 18–22 Oct. 1982, Sao Paulo, Brazil. Promocet, Sao Paulo.

Rauthan, B.S., and M. Schnitzer. 1981. Effects of soil fulvic acid on the growth and nutrient content of cucumber (*Cucumis sativus*) plants. Plant Soil. 63:491–495.

Raviv, M., Y. Chen, Z. Geler, S. Medina, E. Putievski, and Y. Inbar. 1983. Slurry produced by methanogenic fermentation of cow manure as a growth medium for some horticultural crops. Acta Hort. 150:563–573.

Raviv, M., Y. Chen, and Y. Inbar. 1986. Peat and peat substitutes as growth media for container-grown plants. p. 257–287. *In* Y. Chen and Y. Avnimelech (ed.) The role of organic matter in modern agriculture, Martinus Nijhoff Publ., Dordrecht.

Russell, E.J. 1921. Soil conditions and plant growth. Longmans, Green, & Co., London.

Sanchez-Conde, M.P., and C.B. Ortega. 1968. Effect of humic acid on the development and the mineral nutrition of the pepper plant. p. 745–755. *In* Control de la Fertilizacion de las plantas cultivadas, 2° Cologuio Evr. Medit. Cent. Edafol. Biol. Aplic. Cuarto, Sevella, Spain.

Sanchez-Conde, M.P., C.B. Ortega, and M.I. Perz Brull. 1972. Effect of humic acid on sugar beet in hydroponic culture. Arales de edafologia y'Agrobiologia 31:319–331.

Scharpf, H. 1967. Relationships between the humus content of soil and crop yields in a long term fertilizer trial. Albrecht-Thaer-Arch. 11:133–141.

Schnitzer, M., and P.A. Poapst. 1967. Effects of a soil humic compound on root initiation. Nature (London) 213:598–599.

Sladky, Z. 1959a. The effect of extracted humus substances on growth of tomato plants. Biol. Plant. 1:142–150.

Sladky, Z. 1959b. The application of extracted humus substances to overground parts of plants. Biol. Plant. 1:199–204.

Sladky, Z. 1965. Anatomic and physiological alternations in sugar beet receiving foliar applications of humic substances. Biol. Plant. 7:251–260.

Sladky, Z. and V. Tichy. 1959. Applications of humus substances to overground organs of plants. Biol. Plant. 1:9–15.

Smidova, M. 1962. Effect of sodium humate on swelling and germination of plant roots. Biol. Plant. 4:112–118.

Sprengel, C. 1832. Ueber Pflanzenhumus, Humussaeure und Humussaure Salz. Chemie fur Landwirthe, Forstmaenner und Cameralisten, Goettingen.

Stanchev, L., Z. Tanev, and K. Ivanov. 1975. Humus substances as suppressors of biuret phytotoxicity. *In* Humus et Planta VI:373–381.

Stevenson, F.J. 1982. Humus chemistry: Genesis, composition, reactions. Wiley-Interscience, New York.

Strickland, R.C., W.R. Chaney, and R.J. Lamoreaux. 1979. Organic matter influences phytotoxicity of cadmium to soybeans. Plant Soil 52:393–402.

Tan, K.H. 1978. Effects of humic and fulvic acids on release of fixed potassium. Geoderma 21:67–74.

Tan, K.H., and V. Nopamornbodi. 1979. Effect of different levels of humic acids on nutrient content and growth of corn (*Zea mays* L.). Plant Soil 51:283–287.

Tan, K.H., and D. Tantiwiramanond. 1983. Effect of humic acids on nodulation and dry matter production of soybean, peanut and clover. Soil Sci. Soc. Am. J. 47:1121–1124.

Thaer, A.D. 1808. Grundriss der Chemie fur Landwirte. Berlin.

Vaughan, D. 1969. The stimulation of invertase development in aseptic storage tissue slices by humic acid. Soil Biol. Biochem. 1:15–28.

Vaughan, D., M.V. Cheshire, and C.M. Mundie. 1974. Uptake by beetroot tissue and biological activity of ^{14}C-labelled fractions of soil organic matter. Biochem. Soc. Trans. 2:126–129.

Vaughan, D, and D.J. Linehan. 1976. The growth of wheat plants in humic acid solutions under axenic conditions. Plant Soil 44:445–449.

Vaughan, D., and R.E. Malcolm. 1979a. Effect of soil organic matter on peroxidase activity of wheat roots. Soil Biol. Biochem. 11:57–63.

Vaughan, D., and R.E. Malcolm. 1979b. Effect of humic acid on invertase synthesis in roots of higher plants. Soil Biol. Biochem. 11:247–272.

Vaughan, D., and R.E. Malcolm. 1985. Influence of humic substances on growth and physiological processes. p. 37–75. *In* D. Vaughan and R.E. Malcolm (ed.) Soil organic matter and biological activity. Martinus Nijhoff/Dr. W. Junk Publ., Dordrecht.

Vaughan, D., R.E. Malcolm, and B.G. Ord. 1985. Influence of humic substances on biochemical processes in plants. p. 77–108. *In* D. Vaughan and R.E. Malcolm (ed.) Soil organic matter and biological activity. Martinus Nijhoff/Dr. W. Junk Publ., Dordrecht.

Vaughan, D., and I.R. McDonald. 1971. Effects of humic acid on protein synthesis and ion uptake in beet discs. J. Exp. Bot. 22:400–410.

Vaughan, D., and I.R. McDonald. 1976. Some effects of humic acid on the cation uptake by parenchyma tissue. Soil Biol. Biochem. 8:415–421.

Vaughan, D., and B.G. Ord. 1981. Uptake and incorporation of ^{14}C-labelled soil organic matter by roots of *Pisum sativum* L. J. Exp. Bot. 32:679–687.

Vaughan, D., B.G. Ord, and R.E. Malcolm. 1978. Effect of soil organic matter on some root surface enzymes of and uptake into winter wheat. J. Exp. Bot. 29:1337–1344.

White, M.C., and R.L. Chaney. 1980. Zinc, cadmium and manganese uptake by soybean from two zinc and cadmium amended coastal plain soils. Soil Sci. Soc. Am. J. 44:308–313.

Woodward, J. 1699. Thoughts and experiments on vegetation. Philos. Trans. R. Soc. London, B. 21:382–398.

Xudan, X. 1986. The effect of foliar application of fulvic acid on water use, nutrient uptake and wheat yield. Aust. J. Agric. Res. 37:343–350.

Chapter 8

Ecological Aspects of Soil Organic Matter in Tropical Land Use

WOLFGANG ZECH,
LUDWIG HAUMAIER,
AND REINHOLD HEMPFLING, *Institute of Soil Science, University of Bayreuth, Federal Republic of Germany*

ABSTRACT

Carbon-13 nuclear magnetic resonance (NMR) spectroscopy and pyrolysis field ionization mass spectrometry have been used to evaluate the differences in organic matter composition of a very fertile, man-made Amazon soil (*Terra Preta Do Indio*) and a relatively barren Oxisol derived from the same parent material. The most pronounced difference found is the high aromaticity of the *Terra Preta* humus which is considered to be due to mineralization and humification of large amounts of nutrient-rich organic materials. Carbon-13 NMR spectroscopy also has been applied to the examination of changes in organic matter composition due to clearing of forests and cultivation. Five neighboring soils in Yucatan, Mexico, differing only in land use have been investigated. The results indicate that intensive cultivation leads to a strong increase in the aromaticity of the organic matter of these soils.

Soil organic matter strongly affects many of the chemical and physical properties of soils that are essential to permanent soil fertility. Many soils in the tropics are rich in kaolinite and iron and aluminum oxides, but relatively poor in organic matter. Rapid mineralization of the organic matter occurs after the clearing of forests, resulting in a further decrease of organic matter content. This process is well known and explains some aspects of the low fertility levels of many soils in the tropics under permanent cropping systems.

In this chapter, two examples are described that demonstrate in detail the ecological significance of soil organic matter in tropical land use. The objectives are: (i) to show how the organic matter composition of a very fertile, obviously anthropogenic soil differs from that of a relatively barren soil derived from the same parent material, and (ii) to show how land use decisively influences soil organic matter composition.

8-1 MATERIALS AND METHODS

8-1.1 Soils from the Brazilian Amazon Region

The first example deals with Amazon soils in Brazil. Oxisols and Ultisols predominate in this region, but in addition a peculiar, black earth-like soil occurs in small areas rarely exceeding 2 ha (Fig. 8–1). These are the so-called *Terra Preta Do Indio,* or Red Indian Black Earth soils. Because of the similarity of their texture to that of immediately surrounding soils (from more or less sandy to very clayey), and because of ceramics that are nearly always found in the upper horizons, these soils are considered to be man-made (Sombroek, 1966).

We studied the less common clayey variant of *Terra Preta* soils located on the *planalto* (plateau) about 40 km southwest of Santarém (Fig. 8–1), 165 m above sea level and 105 m above the watertable. In addition, Oxisols nearby were investigated for comparison (Bechtold, 1982; Pabst, 1985). Both soil types are derived from the so-called Belterra clay. They have the same mineralogical properties with a predominance of kaolinite in the clay fraction, and rutile, zircon, and tourmaline as heavy minerals in the 100- to 200-μm fraction (Bechtold, 1982). The *Terra Preta,* situated immediately at the edge of the planalto, is not of alluvial origin and shows no hydromorphic features. It is classified as a Haplohumox; the Oxisol about 9 km away is a Haplorthox. The vegetation is secondary rain forest (capoeira) with remnants of an abandoned rubber (*Hevea brasiliensis*) plantation. No pronounced differences in vegetation cover between the two sites could be detected. Mean annual precipitation and temperature are about 2050 mm and 25 °C, respectively. The rainy season lasts from December to July (Bechtold, 1982; Pabst, 1985). Some characteristics of the two soil profiles investigated are given in Table 8–1.

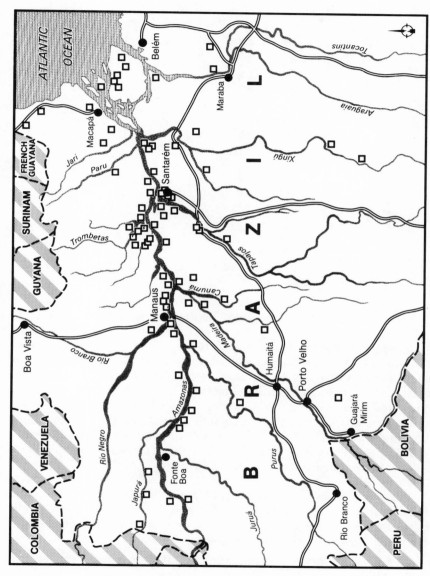

Fig. 8-1. The basin of the Amazon River in Brazil. The squares indicate the occurrence of *Terra Preta Do Indio* (Bechtold, 1982).

Table 8-1. Some characteristics of *Terra Preta* (Haplohumox) and Oxisol (Haplorthox) near Santarém, Brazil.

Depth	Sand	Silt	Clay	pH (CaCl$_2$)	Organic C	Total N	C/N ratio	CEC	Exchange- able Ca
cm	———— % ————			— g kg^{-1} —			— cmol$_c$ kg^{-1} —		
				Terra Preta					
3	14	23	63	6.3	92	6.8	13.5	44	39
12	10	22	68	6.3	66	4.4	15.0	41	32
20	7	23	70	5.6	60	3.9	15.4	44	20
30	7	18	75	5.4	43	2.4	17.9	29	11
50	7	12	81	5.1	28	1.6	17.5	22	5
70	5	19	76	4.7	22	1.2	18.3	19	3
95	6	14	80	4.5	13	0.7	18.6	13	1
120	3	17	80	4.6	8	0.6	13.3	11	1
150	4	29	67	4.7	6	0.4	15.0	9	1
				Oxisol					
3	8	9	83	3.7	30	2.7	11.1	15	t†
11	4	8	88	3.8	15	1.4	10.7	9	t
22	4	8	88	4.0	10	1.0	10.0	9	t
35	3	8	89	4.1	8	0.8	10.0	7	t
55	3	6	91	4.0	6	0.6	10.0	7	t
75	3	11	86	4.2	5	0.6	8.3	7	t
100	3	8	89	4.1	4	0.5	8.0	6	t
125	3	8	89	4.2	4	0.5	8.0	7	t
150	2	13	85	4.2	3	0.4	7.5	6	t

† t = trace.

8-1.2 Soils from Yucatan, Mexico

The second example concerns soils located 10 km south of Oxkutzcab in Yucatan, Mexico (average annual precipitation about 1000 mm; average annual temperature about 27 °C; rainy season from May to November). The parent material consists of tertiary limestone. As a result of the hilly topography and intensive land use, which dates back to the ancient Maya, soil catenas have developed in this region with Rendolls on the hills and upper slopes and with deep and clay-rich Vertisols on lower slopes and in depressions. The soils lying at the bottom of the hills are classified as Typic Chromuderts. Some properties of a representative profile of these soils are given in Table 8-2.

In order to investigate the effects of cultivation on soil organic matter composition, samples from homogeneous neighboring soils, differing only in land use, were examined. Samples of the upper 10 cm of the mineral soils were taken from the following locations:

1. *Secondary forest:* The trees (*acacia* and *vitex* species) are approximately 15 yr old.
2. *Young tree garden, 3 yr after clearing of the secondary forest:* Cassava, beans, corn, and vegetables are cultivated amongst young citrus, mango, papaya, and avocado plantations. The canopy does not cover the floor yet. Slight manual preparation of the surface soil

Table 8–2. Some characteristics of a representative soil profile (Typic Chromudert) in depressions near Oxkutzcab, Yucatan, Mexico.

Horizon	Depth	Sand	Silt	Clay	pH (CaCl$_2$)	Organic C	Total N	C/N ratio	Cation exchange capacity	Base saturation	Bulk density	Pore volume
	cm	%				g kg^{-1}			cmol$_c$ kg^{-1}	%	Mg m^{-3}	%
Ah	15	3	39	58	7.1	27.5	2.4	11.4	36.4	98	0.85	66
AB	17	2	25	73	5.8	12.7	1.3	9.8	25.6	72	1.06	60
B1	18	2	22	76	5.4	8.3	1.0	8.6	24.9	68	--	--
B2	25+	2	21	77	5.3	5.0	0.8	6.3	22.4	67	1.16	56

Table 8-3. Characterization of the surface soils (0-10 cm) near Oxkutzcab, Yucatan, Mexico.

Land use	Sand	Silt	Clay	pH (CaCl$_2$)	Organic C	Pore volume
		%			g kg^{-1}	%
Secondary forest	4	46	50	6.4	28.7	67
Young tree garden	3	41	56	6.9	17.7	62
Mature tree garden	4	41	55	7.5	24.8	62
Corn field	4	39	57	7.1	22.0	63
Compost bed	5	39	56	7.3	28.7	69

is done. No fertilizers are applied. This type of cropping represents a permanent system of tropical land use. In contrast to shifting cultivation, the vegetation consists of three floors (herbs, bushes, trees). Nutrient leaching is reduced, and fallow periods are not necessary.

3. *Mature tree garden, approximately 25 yr after clearing of the secondary forest:* Mango, avocado, banana, and citrus trees (8–10 m in height) densely cover the floor. Pineapples, beans, corn, and cassava are cultivated under the canopy. Chickens and pigs mix the surface soil layers and enrich them by their excrement.

4. *Corn field, 10 yr after clearing of the secondary forest:* There is no tree or bush vegetation. The soil is intensively cultivated by tractors. Mineral fertilizers are applied.

5. *Compost bed:* About 2 yr before sampling, compost (8000 kg ha^{-1}) was applied and the soil was turned over to a depth of 50 cm by hand.

Table 8-3 gives information about the surface soil properties of these five experimental plots (five samples were taken per location, mixed, and then analyzed).

8-1.3 Methods

Organic C was determined by dry combustion with a Woesthoff Carmhomat 8-ADG, total N by a Kjeldahl procedure, and total P by the molybdenum blue method after digestion with HF–HClO$_4$. The pH was measured in 0.01M CaCl$_2$ at a soil-to-solution ratio of 1:3 (w/v). Cation exchange capacity (CEC) was determined by treatment of the soil with 0.1M BaCl$_2$ buffered with triethanolamine (pH 8.2) (Amazon soils) or 0.1M BaCl$_2$ without buffer (Mexican soils) and re-exchange of Ba^{2+}, with 0.1M MgCl$_2$. Texture was determined according to de Leenheer et al. (1955), and bulk density and pore volume according to Hartge (1971).

^{13}C-NMR spectra were obtained on a Bruker CXP 300 spectrometer. Cross polarization magic angle spinning (CPMAS) ^{13}C-NMR spectroscopy (see chapter 10, this book): spectrometer frequency, 75.5 MHz; spectral width, 31 250 Hz; number of scans, 6500 to 29 000; spinning speed, 3.5 and 4.5 kHz; contact time, 0.7 and 1.0 ms. No attempt was made to reduce spinning side bands. The liquid-state ^{13}C-NMR spectra were obtained by the method described by Wilson and Goh (1983) (spectrometer frequency 75.5 MHz; spectral width, 25 000 Hz; number of scans, 11 000 and 30 000). Bond assignments were made according to Preston and Schnitzer (1984), Breitmaier and Voelter (1979), and Kalinowski et al. (1984).

For thermally-programmed, time-resolved pyrolysis mass spectrometry, 100 to 200 μg of the whole soil sample was degraded in the field ion source of a double-focusing mass spectrometer (Finnigan MAT 731). The direct introduction system with an aluminum crucible was used. The sample was heated linearly from 50 to 500 °C at a rate of 0.4 °C s^{-1} (for more details see Haider & Schulten, 1985).

8–2 PROPERTIES AND GENESIS OF THE *TERRA PRETA DO INDIO* (BRAZIL)

8–2.1 Chemical Properties

According to Sombroek (1966) the *Terra Preta* is very fertile, and after clearing of forests the soils are not immediately exhausted as the Oxisols are. Rubber trees growing on patches of *Terra Preta* allegedly are more resistant to diseases. These properties of the *Terra Preta* can be explained by greater stability of its soil organic matter. Carbon mineralization rates during incubation of the samples from Santarém confirm this idea. The CO_2-production of the *Terra Preta* Ah horizon is significantly lower than that of the Oxisol Ah (Zech et al., 1979), despite the fact that the organic C content in the surface horizon of the *Terra Preta* is 92 g kg^{-1} and that of the Oxisol is only 30 g kg^{-1} (Table 8–1). Organic C contents decrease with increasing soil depth from 92 to 6 g kg^{-1} in the *Terra Preta,* and from 30 to 3 g kg^{-1} in the Oxisol. A similar depth function is found for total N (Table 8–1; *Terra Preta:* 6.8 to 0.4 g kg^{-1}; Oxisol: 2.7 to 0.4 g kg^{-1}). The C/N ratios are narrow in the Oxisol (7.5 to 11.1) and relatively narrow in the *Terra Preta* (13.3–18.6). Total N reserves are high: 11 200 kg ha^{-1} (1-m depth) for the Oxisol and 17 500 kg ha^{-1} (1-m depth) for the *Terra Preta.*

Total P contents in the *Terra Preta* are very high (up to 2.00 g kg^{-1} in the upper 0 to 20 cm, decreasing to 0.15 g kg^{-1} at a depth of 150 cm). Phosphorus contents in the Oxisol are very low, varying from 0.22 g kg^{-1} in the Ah to 0.11 g kg^{-1} at a depth of 150 cm (for more details see Zech et al., 1979). The two soils differ significantly in pH (Table 8–1; *Terra Preta* Ah: 6.3; Oxisol Ah: 3.7). A close correlation exists between pH, CEC, and the nature of the exchangeable cations. The CEC of the Oxisol decreases from 15 cmol$_c$ kg^{-1} in the surface layer to 6 cmol$_c$ kg^{-1} at a soil depth of 150 cm. Base saturation is only 6% in the Ah, and exchangeable Al and H are dominant. In contrast, the *Terra Preta* surface horizons are characterized by a high CEC of 41 to 44 cmol$_c$ kg^{-1}, high base saturation, and dominance of exchangeable Ca (39 cmol$_c$ kg^{-1}). Cation exchange capacity and exchangeable Ca decrease to 9 cmol$_c$ kg^{-1} and 1 cmol$_c$ kg^{-1}, respectively, with increasing soil depth (Table 8–1). Because the clay fraction consists nearly exclusively of kaolinite in both soils, the great differences in CEC must be due to the differences in organic matter content.

The characterization of the organic matter of *Terra Preta* and Oxisol Ah horizons by spectrometric methods indicates that *Terra Preta* is more humified, richer in high molecular weight organics but poorer in mobile (oxalate extractable) humic substances than the Oxisol (Zech et al., 1979). Gas chromatography/mass spectroscopy (GC/MS) studies of methylated permanganate oxidation products, however, reveal no differences between *Terra Preta* and Oxisol humic substances. Therefore, other modern analytical techniques for the characterization of soil organic matter such as carbon-13 nuclear magnetic resonance (^{13}C-NMR) spectroscopy (cf. chapter 10, this book) and pyrolysis field ionization mass spectrometry were employed.

8–2.2 Results of ^{13}C-NMR Spectroscopy

Figure 8–2 shows the solid-state ^{13}C-NMR spectra of the *Terra Preta* and Oxisol surface soils. In both spectra the dominant peak has a chemical

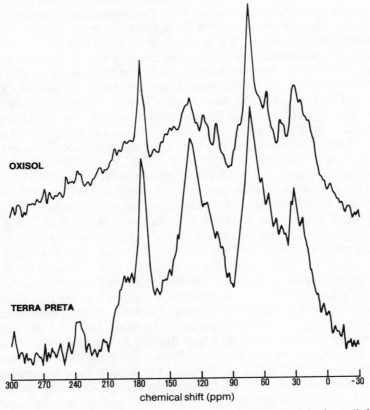

Fig. 8–2. CPMAS ^{13}C-NMR spectra of the *Terra Preta* and Oxisol Ah horizons (0–20 cm).

shift of about 70 ppm that is due to polysaccharides. Signals at 175 ppm and near 30 ppm are due to carboxyl groups, amides, esters, and aliphatic structures, respectively. The main differences, however, are in the aromatic C signals near 130 ppm. Obviously, the organic matter of the *Terra Preta* is characterized by high contents of aromatic structures. Their degree of oxygen substitution seems to be rather low because signals in the 150-ppm region are weak. Hydrogen- and carbon- (e.g., carboxyl) substituted or fused aromatic rings apparently predominate.

Better resolved spectra can be obtained using solution-state NMR spectroscopy (Fig. 8–3). The NaOH-soluble humic substances of the *Terra Preta* Ah are richer in carboxyl groups, amides and esters (175 pm), carbon- and hydrogen-substituted aromatic carbons, and carbons of fused aromatic rings (around 130 ppm) than those of the Oxisol. Also, there are differences in the signals between 0 and 56 ppm. In the *Terra Preta* spectrum the peak due to methoxy groups (56 ppm) is accompanied by three signals at 54, 52, and 50 ppm that are attributed to propylic side chain carbons of lignin phenylpropane units (Lüdemann & Nimz, 1973) and/or amino acids.

The signals between 0 and 50 ppm are due to aliphatic carbons such as CH_2 carbons in saturated hydrocarbon chains, C, CH, CH_2 carbons

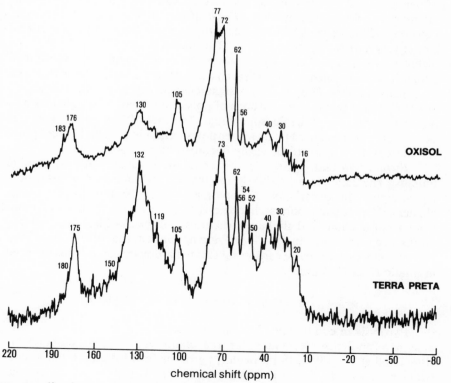

Fig. 8–3. ^{13}C-NMR spectra of NaOH-soluble substances of *Terra Preta* and Oxisol Ah horizons (0–20 cm).

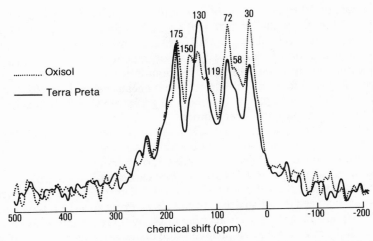

Fig. 8-4. CPMAS ^{13}C-NMR spectra of *Terra Preta* and Oxisol humins, insoluble in NaOH.

bound to aromatic rings,CH_2 carbons in alicyclic groups, aliphatic carbons of peptides, and methyl carbons (González-Vila et al., 1983). These signals are more pronounced in the *Terra Preta* spectrum. There is little evidence, however, of long chains of saturated hydrocarbons that would give rise to a relatively narrow signal at 30 ppm. The chains seem to be rather short or highly branched. The Oxisol is particularly rich in OH, O- (and N-) substituted aliphatic carbons such as in polysaccharides (and amino acids) with maximum peaks at 62 to 77 ppm. This assignment is in agreement with the signal at 105 ppm due to anomeric carbons of polysaccharides. The higher carbon mineralization rate of the Oxisol compared to that of *Terra Preta* (see section 8-2.1) may be caused by its higher polysaccharide content because polysaccharides are known to be easily degraded by microorganisms.

Terra Preta and Oxisol also differ in the nature of alkali-insoluble humins. The solid-state ^{13}C-NMR spectra of the humins (Fig. 8-4) show typical signals due to carboxyl (175 ppm), aromatic (119, 130, and 150 ppm), aliphatic (30 ppm), and polysaccharide carbon (72 ppm). In the Oxisol humin, contents of phenolic and aliphatic structures, polysaccharides, and methoxy groups are higher than in the *Terra Preta* humin. On the other hand, *Terra Preta* humin contains more carbon-substituted aromatic compounds or fused aromatic rings (130 ppm).

8-2.3 Results of Pyrolysis Mass Spectrometry

Pyrolysis in combination with mass spectrometry has become a valuable new analytical technique for the characterization of biopolymers (Bracewell et al., 1989). In particular, pyrolysis field ionization mass spectrometry (Py-FIMS) has been recognized as a powerful tool in the evaluation of the structural components of lignin, humic compounds, and whole soil material (Haider & Schulten, 1985; Schulten, 1987). Figure 8-5 shows

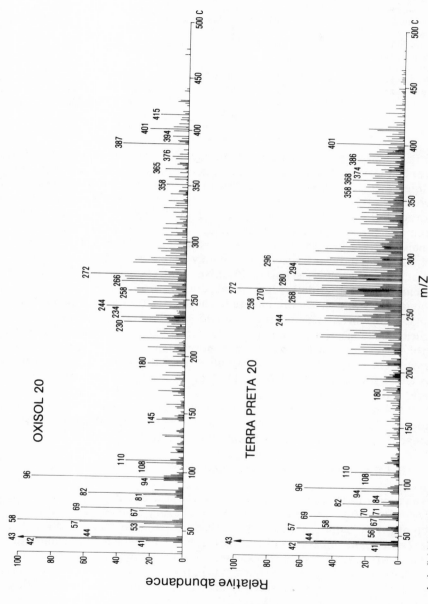

Fig. 8-5. Pyrolysis field ionization mass spectra of the *Terra Preta* and Oxisol Ah horizons (whole soil samples).

Table 8-4. Field ionization mass signals of typical pyrolysis products from organic soil constituents.

Compound class	(m/z)
Polysaccharides	72, 82, 84, 96, 98, 110, 112, 114, 126, 132, 144, 162
Proteins	67, 79, 81, 93, 95, 117, 131, 145
Phenols	94, 108, 120, 122
Aliphatic hydrocarbons	42, 43, 44, 56, 57, 58, 69, 70, 71
Lignin monomers	
Guaiacyl	124, 152, 166, 178, 180
Syringyl	168, 192, 194, 196, 208, 210
Lignin dimers	270, 272, 274, 284, 286, 296, 298, 300, 302, 310, 312, 314, 316, 326, 328, 330, 332, 340, 342, 344, 346, 356, 358, 372, 374, 376, 386, 388

the field ionization mass spectra of the pyrolysis products from the *Terra Preta* and Oxisol Ah horizons. Similar patterns of mass signals differing only in intensities are observed for the two samples. The mass signals indicate thermal degradation products of polysaccharides, proteins, phenols, aliphatic hydrocarbons, and lignin (Table 8-4). The high intensities of signals due to furan and furfural products (m/z 82, 96, 110) compared to those of other signals characteristic for pentose and hexose units in polysaccharides (m/z 114, 126, 144) indicate a precursor that may be related to, but does not consist of any known polysaccharide. Bracewell et al. (1989) term this precursor "pseudo-polysaccharide". Mass signals due to pseudo-polysaccharide and proteins are somewhat more intense in the spectrum of the Oxisol sample.

Distinct differences between the two samples become obvious in the higher mass range of the spectra (m/z 230 to m/z 400). Mass signals of dimeric pyrolysis products of lignin are much more intense in the spectrum of the *Terra Preta* sample. In addition, signals due to demethylated and demethoxylated lignin dimers (m/z 242, 244, 246, 256, 258, 260, 268, 282) prevail there, too.

In summary, Py-FIMS corroborates the findings from ^{13}C-NMR spectroscopy: The *Terra Preta* humus is richer in lignin-derived aromatics and poorer in polysaccharide constituents than the Oxisol humus.

8-2.4 Genesis and Ecological Aspects of the *Terra Preta* Soils

There is little doubt that *Terra Preta* soils are of anthropogenic origin. Presumably, they are the results of long-lasting cultivation by the pre-Columbian Indian tribes. Sombroek (1966, p. 176) states that "apparently, the *Terra Preta* soil is a kind of 'kitchen-midden', which has acquired its specific fertility, notably much calcium and phosphorus, from dung, household garbage, and the refuse (bones) of hunting and fishing." Humus-rich soils similar to *Terra Preta* recently have been found by the senior author around former settlements in West Africa (Benin, Liberia). However, no data are available yet.

The detailed mechanisms by which *Terra Preta* humus gains its stability and special properties are still subject to speculation. As a whole, the decisive factor seems to be the input of large quantities of organic materials rich in N, P, and Ca. Stimulation of microbial activity by these inputs may lead to accelerated mineralization of substances less stable to degradation (such as polysaccharides and proteins), and thus to a relative enrichment of the stable humic fractions rich in aromatic constituents. On the other hand, special pathways of humification may be induced that promote selective preservation or microbial synthesis of aromatic compounds. The existence of *Terra Preta,* man-made or not, proves that infertile Oxisols in principle can be transformed to permanently fertile and stable *Terra Preta.* Such a transformation cannot be achieved solely by replenishing the mineral nutrient supply, however, because the soil organic matter is of prime importance for the prevention of nutrient leaching in these freely draining soils.

8-3 INFLUENCE OF LAND USE ON ORGANIC MATTER COMPOSITION OF HUMIC SURFACE LAYERS OF VERTISOLS IN YUCATAN (MEXICO)

Some chemical properties of these soils are given in Table 8–2 and Table 8–3. The CPMAS ^{13}C-NMR spectrum of the surface horizon under a 15-year-old secondary forest (Fig. 8–6) shows a broad peak with a maximum near 70 ppm due to polysaccharides. In the aliphatic region of the spectrum the polymethylene signal near 30 ppm is relatively sharp. The peaks due to aromatic carbons derive from lignin (150, 130, 119 ppm). The distinct signal at 175 ppm arises from carboxyl, ester, and amide carbons.

Three years after clearing the forest and establishing a tree garden most of the soil surface is still bare and exposed to the sun. The organic C content is only 17.7 g kg^{-1} compared to 28.7 g kg^{-1} in the forest soil (Table 8–3). The CPMAS ^{13}C-NMR spectrum (Fig. 8–6, young tree garden) indicates relatively reduced polysaccharide and increased aromatic C. In particular, the signal around 130 ppm is more pronounced. Ten years after clearing and establishing a mechanized cropping system for corn production, the organic C content is 22 g kg^{-1}. Aromaticity is considerably higher, whereas peaks due to polysaccharides and aliphatics are further decreased (Fig. 8–6, corn field).

Twenty-five years after clearing and establishing a tree garden with high biomass production and nutrient inputs by animal manure, and with mixing of the surface soils by chickens and pigs, organic C content is 24.8 g kg^{-1}. The pH and total P content have increased from 6.4 and 0.64 g kg^{-1} in the forest soil to 7.5 and 1.19 g kg^{-1}, respectively, in the soils of the mature tree garden. The distribution of the C species in the organic matter of the surface layer is nearly the same as that of the soil of the corn field. The most pronounced peaks in the ^{13}C-NMR spectrum (Fig. 8–6, mature tree garden) are due to aromatic structures (maximum near 130 ppm) and to carboxyl groups, amides, and esters (175 ppm). The signals of the polysaccharides and aliphatic structures are surprisingly low in intensity. Presumably, polysaccharides and aliphatics accumulate due to litter fall only. However, perma-

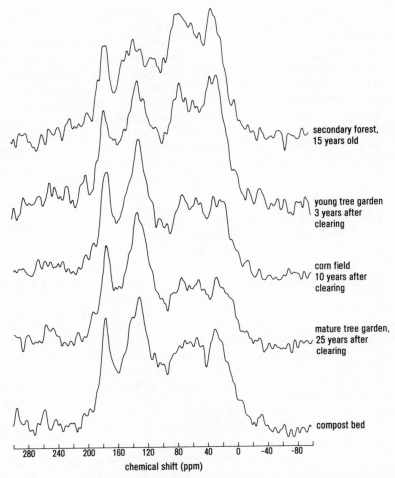

Fig. 8–6. CPMAS ^{13}C-NMR spectra of surface soils under different land use in Yucatan, Mexico.

nent mixing of mineral soil and litter by chickens and pigs, and nutrient rich inputs (animal manure), stimulate humification and mineralization. Humification may promote the generation of aromatic compounds, whereas mineralization could be responsible for relative enrichment of aromatic structures because of accelerated breakdown of polysaccharides and aliphatics. According to this idea, fresh compost prepared from tree litter should be relatively rich in aliphatic and O-alkyl substances but relatively poor in aromatic components. The compost bed in Yucatan, however, is rich in old, extensively decomposed materials. In the CPMAS ^{13}C-NMR spectrum (Fig. 8–6) peaks due to carboxyl groups and aromatics are dominant, and those due to polysaccharides and aliphatics are somewhat more intense than in the spectrum of the mature tree garden soil.

In summary, clearing of the forest and establishing of cropping systems leads to an initial decrease in organic C content of the soil (28.7 g kg^{-1} in the forest stand and 17.7 g kg^{-1} in the young tree garden) due to mineralization of components easily accessible to microorganisms. The consequence is a relative enrichment of residues, which are more resistant to biodegradation. Permanent cultivation results in a slow but steady increase in organic C content (24.8 g kg^{-1} in the mature tree garden) because of the continual input of root and litter biomass. Enhanced microbial activity due to nutrient inputs and permanent mixing of the soil prevents accumulation of easily degradable substances. Only organic materials resistant to microbial attack (predominantly aromatics) remain.

8-4 CONCLUSIONS

The *Terra Preta* phenomenon provides evidence that Oxisols in principle can be converted to fertile black earth-like soils. These soils are characterized by high contents of aromatic structures with a low degree of oxygen substitution. The dominance of these aromatic building blocks is considered to be responsible for increased stability. The composition of the *Terra Preta* organic C seems to be the result of intensive humification and rapid mineralization of large quantities of organic materials rich in N, P, and Ca incorporated into the mineral soil. Changes in the organic matter composition of the Yucatan soils under different land use clearly show that cultivation can cause considerable increase in aromaticity of humus. Transformation of Oxisols to *Terra Preta* cannot be achieved on wide areas but only in small gardens. The improvement of such gardening systems seems to be a key to higher food production in the tropics.

ACKNOWLEDGMENT

The authors are indebted to Dr. Falesi, formerly EMBRAPA, Belem, and Dr. Neugebauer (GTZ), for supporting the field work, which was partly done by E. Pabst and G. Bechtold. Transportation of further soil samples was carried out by Dr. Burger and M. Denich (GTZ). ^{13}C-NMR spectra were obtained with the generous help of Dr. Förster (Bruker), Prof. Dr. Lüdemann, and R. Fründ (University of Regensburg). Prof. Dr. Schulten (Fresenius Institute, Wiesbaden) obtained the pyrolysis field ionization mass spectra. This study has been carried out with financial support by the Deutsche Forschungsgemeinschaft, Bonn (SFB 137).

REFERENCES

Bechtold, G. 1982. Terra Preta Do Indio: Anorganisch-chemische Kennzeichnung eines brasilianischen Anthrohumox. Thesis. Univ. of Bayreuth/Univ. of Munich, FRG.

202   ZECH ET AL.

Bracewell, J.M., K. Haider, S.R. Larter, and H.-R. Schulten. 1989. Thermal degradation relevant to structural studies of humic substances. p. 181–222. *In* M.H.B. Hayes et al. (ed.) Humic substances. II: In search of structure. John Wiley & Sons, Ltd., Chichester, UK.

Breitmaier, E., and W. Voelter. 1978. ^{13}C-NMR spectroscopy. Verlag Chemie, Weinheim, FRG.

González-Vila, F.J., H.D. Lüdemann, and F. Martin. 1983. ^{13}C-NMR structural features of soil humic acids and their methylated, hydrolyzed and extracted derivatives. Geoderma 31:3–15.

Haider, K., and H.-R. Schulten. 1985. Pyrolysis field ionization mass spectrometry of lignins, soil humic compounds and whole soil. J. Anal. Appl. Pyrolysis 8:317–331.

Hartge, K.H. 1971. Die physikalische Untersuchung von Böden. Enke, Stuttgart.

Kalinowski, H.-O., S. Berger, and S. Braun. 1984. ^{13}C-NMR-Spektroskopie. Thieme, Stuttgart.

de Leenheer, L., M.v. Ruymbeke, and L. Maes. 1955. Die Kettenaräometer-Methode für die mechanische Bodenanalyse. Z. Pflanzenernaehr., Dueng., Bodenk. 68:10–19.

Lüdemann, H.D., and H. Nimz. 1973. Carbon-13 nuclear magnetic resonance spectra of lignins. Biochem. Biophys. Res. Commun. 52:1162–1169.

Pabst, E. 1985. *Terra Preta Do Indio:* chemische Kennzeichnung und ökologische Bedeutung einer brasilianischen Indianerschwarzerde. Thesis. Univ. of Bayreuth/Univ. of Munich, FRG.

Preston, C.M., and M. Schnitzer. 1984. Effects of chemical modifications and extractants on the carbon-13 NMR spectra of humic materials. Soil Sci. Soc. Am. J. 48:305–311.

Schulten, H.-R. 1987. Pyrolysis soft ionization mass spectrometry of aquatic/terrestrial humic substances. J. Anal. Appl. Pyrolysis 12:149–187.

Sombroek, W.G. 1966. Amazon soils. A reconnaissance of the soils of the Brazilian Amazon region. Versl. Landbouwkd. Onderz. no. 672.

Wilson, M.A., and K.M. Goh. 1983. N.M.R. spectroscopy of soils: structure of organic material in sodium deuteroxide extracts from Patua loam, New Zealand. J. Soil Sci. 34:305–313.

Zech, W., E. Pabst, and G. Bechtold. 1979. Analytische Kennzeichnung der Terra Preta Do Indio. Mitt. Dtsch. Bodenkd. Ges. 29:709–716.

Chapter 9

Humic and Fulvic Acid Fractions from Sewage Sludges and Sludge-amended Soils[1]

STEPHEN A. BOYD, *Michigan State University, East Lansing, Michigan*

LEE E. SOMMERS, *Colorado State University, Fort Collins, Colorado*

ABSTRACT

Humic and fulvic acid fractions obtained from sewage sludges show several distinguishing chemical and structural characteristics when compared to soil-derived humics including: (i) higher N contents, (ii) lower C/N ratios, (iii) higher H/C ratios indicating a higher fraction of aliphatic components, and (iv) lower carboxyl group acidity. Infrared spectra of sludge humic and fulvic acids show the presence of associated protein and aliphatic materials like fats and waxes. Several nonhumic components, accounting for 30 to 55% by weight, have also been identified in sludge fulvic acid fractions. These include: (i) amino acids, (ii) hexosamines, (iii) neutral

---

[1] Contribution of the Michigan Agricultural Experiment Station, East Lansing, MI 48824.

sugars, and (iv) anionic surfactants. The presence of sodium lauryl sulfate and other organic ester sulfate detergents (anionic surfactants) results in a much higher S content in sludge fulvic acid fractions as compared to soil fulvic acid fractions. Humic and fulvic acids from sewage sludges have complex metal binding sites involving oxygen-containing chelating groups and mixed nitrogen/oxygen ligand systems. Humic and fulvic acid fractions obtained from sludge-amended soils show characteristics of sludge humic and fulvic acid fractions. These include (i) higher S contents; (ii) presence of lauryl sulfate type surfactants; and (iii) increased N contents, higher H/C ratios, and lower acidity. The association of sludge-derived protein and aliphatic materials with humic materials from sludge-amended soils can be seen clearly in their infrared spectra. With time after sludge application, the effects of sludge amendment on soil humics become less apparent.

The application to agricultural soils of residues originating from agricultural, municipal, and industrial activities has been practiced for many years. One waste management option that has been receiving increased attention in recent years has been the application of municipal sewage sludges to agricultural cropland. In addition to waste disposal considerations, the main advantage of applying sludges to agricultural land is that sludges contain N, P, K, and trace elements and thus can serve as a substitute for conventional fertilizer materials. Sludges are also an organic waste that can contribute to desirable improvements in soil physical characteristics. For an agricultural system, the amount of sludge applied should be consistent with the N needs of the crop in order to minimize leaching of nitrate into groundwater.

Additional constraints associated with sludge use on agricultural land include the potential presence of pathogens, Cd and other metals, soluble salts, and organic chemicals originating from industrial activities. The above constraints associated with sludge use on cropland have been documented in conference proceedings (Page et al., 1983) and design manuals (U.S. EPA, 1983). Metals receiving the greatest attention from a human health standpoint are Cd and Pb while Zn, Cu, and Ni may be phytotoxic at elevated soil levels particularly in soils at pH < 5.5. The characterization of metals in sludges and sludge-amended soils has been recently summarized by Lake et al. (1984). The U.S. Environmental Protection Agency and many state regulatory agencies have developed regulations and guidelines governing the application of sludge on land.

The chemical composition of sludges is a function of inputs to the sewage treatment plant and the processes used for sludge treatment and handling. The types of treatment processes commonly used to treat sewage at the wastewater treatment plant are summarized in Fig. 9-1. However, the particular configuration of the treatment plant varies significantly at different locations. Sewage sludge originates from the material which settles out of raw sewage (wastewater) in the primary sedimentation tank (Fig. 9-1). One common method of sludge treatment prior to disposal is anaerobic digestion (Fig. 9-1). The anaerobically-digested sludge may then be applied to agricultural cropland. In general, a typical sludge contains from 20 to 30%

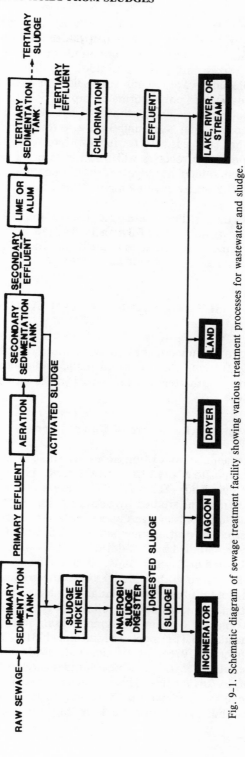

Fig. 9–1. Schematic diagram of sewage treatment facility showing various treatment processes for wastewater and sludge.

organic C (Sommers, 1977). The organic C includes the various constituents of sewage plus intermediate and end products of decomposition. The sludge applied to soil contains a mixture of compounds including proteins, polysaccharides, fats, waxes, and greases. In addition, discharge of industrial effluents into municipal sewage systems can result in elevated levels of specific organic compounds such as polychlorinated biphenyls, phthalic acid esters, and a variety of solvents (see Overcash, 1983; U.S. EPA, 1983, 1985). Organic compounds containing N, P, and S are also found in sludges and will be added to soil when sludge is used for fertilization. After addition to soils, microbial decomposition processes will result in mineralization of part of the C, N, P, and S contained in sludges. Also, it is possible that specific constituents found in the sludge may be incorporated into the humic components of the soil.

The objectives of this chapter are (i) to describe the chemical and structural characteristics of humic and fulvic acid fractions obtained from sewage sludge, and (ii) to evaluate changes in the humic and fulvic acid fractions of soil resulting from sludge amendments to soils.

9–1 SEWAGE SLUDGE FULVIC ACIDS

A considerable fraction of the C extracted from sewage sludges with strong base occurs in the fulvic acid fraction. In a study of 10 sewage sludges, Riffaldi et al. (1982) found that the mean fulvic acid content was 8.9% of the total C of sludge dry matter. The individual values varied widely ranging from 2.4 to 21.9%.

9–1.1 Elemental Composition

The mean elemental composition of sewage sludge fulvic acid fractions from two studies (Sposito et al., 1976; Riffaldi et al., 1982) and mean values for soil-derived fulvic acids (Schnitzer, 1978) are given in Table 9–1. The sludge fulvic acid fractions studied by Sposito et al. (1976) were extracted from anaerobically-digested sewage sludges that were collected from drying ponds at three different locations in southern California (Holtzclaw et al., 1976). The 10 sludges studied by Riffaldi et al. (1982) were from wastewater treatment plants located in Tuscany, Italy. All of the plants primarily received domestic wastewater. Two of these sludges were anaerobically digested. The procedure used to obtain the sewage sludge fulvic acid fractions consisted of NaOH extraction and acidification, followed by centrifugation. The fulvic acid fractions obtained were not fractionated using adsorption chromatography as recommended by Thurman and Malcolm (1981). Thus, the materials studied by Sposito et al. (1976) and Riffaldi et al. (1982) may have contained certain nonhumic substances (e.g., polysaccharides and low molecular weight acids) in addition to fulvic acids. This is a general problem with investigations of fulvic acids (see chapter 2 by Malcolm, this book). It is important to understand that humic and fulvic acids are high molecular weight, poly-

Table 9-1. Mean values of the elemental composition[†] and atomic ratios of fulvic acid fractions from sewage sludges and soils.

Analysis	Sludge fulvic acids		Soil fulvic acids[¶]
	A[‡]	B[§]	
C	4.08	3.26	4.57
H	0.66	--	0.54
N	0.28	0.44	0.21
O	4.23	--	4.48
S	0.82	--	0.19
C/N, mole ratio	17	8.8	25.4
H/C, mole ratio	2.06	1.84	1.42

[†] g kg^{-1} on an ash- and moisutre-free basis.
[‡] From Sposito et al. (1976); means of three anaerobic sludges.
[§] From Riffaldi et al. (1982); means of 10 sludges.
[¶] From Schnitzer (1978); means of all data.

acidic, brown to yellow colored materials formed from secondary synthesis reactions (Stevenson, 1982). The fulvic acid fraction contains fulvic acids, as well as nonhumic substances that are co-extracted with the fulvic acids (see chapter 2).

Several significant observations can be made from the data (Table 9–1) concerning the sludge fulvic acid fractions:

1. They show significant variability in elemental content.
2. They have very high S contents, 20 to 40 times greater than that for "average" soil fulvic acid.
3. The H/C ratios were higher than those of the soil fulvic acids indicating that the sludge-derived fulvic acids have a higher fraction of aliphatic components than soil fulvic acids.
4. The C/N ratios were lower than for soil fulvic acids; this has been attributed to the high content of protein decomposition products in the sludge fulvic acid fraction.

The high S content of the sludge fulvic acid fraction, which was its most distinguishing feature, was characterized by Schaumberg et al. (1982). They found that the majority (69%) of the S was present as organic ester sulfate; about 25% of the ester S occurred as sodium lauryl sulfate and related detergents, but 70% was unidentified. Linear alkyl benzene sulfonates were also present and accounted for ca. 7 to 12% of the total S.

9-1.2 Total Acidity, Carboxyl Group Content, and E_4/E_6 Ratios

The mean total acidity of 10 sewage sludge fulvic acid fractions (Riffaldi et al., 1982) was considerably lower than the mean value for soil fulvic acid fractions (Schnitzer, 1978). These values are given in Table 9–2. The carboxyl group contents of sludge and soil fulvic acid fractions are also given in Table 9–2. It appears that the carboxyl (COOH) acidity for sludge-derived fulvic acids is significantly lower than that for soil fulvic acids (Schnitzer, 1978; Holtzclaw & Sposito, 1979; Riffaldi et al., 1982). The values in Table

Table 9-2. Mean values of carboxyl group contents and E_4/E_6 ratios of fulvic acid fractions from sewage sludges and soils.

Analysis	Sludge fulvic acids		Soil fulvic acids	
	A†	B‡	Altamont†	C§
Total acidity (mol$_c$ kg^{-1})	--	2.7	--	10.3
CO$_2$H (mol kg^{-1})				
Standard	9.47	1.7	12.52	8.2
Modified	0.68	--	4.52	--
E_4/E_6	5.9	2.0	--	9.6

† From Sposito et al. (1976); Holtzclaw and Sposito (1979); means of three sludges. Altamont soil (fine, montmorillonitic, thermic Typic Chromoxerert).
‡ From Riffaldi et al. (1982); means of 10 sludges.
§ From Schnitzer (1978); means of all data.

9-2 for the sludge fulvic acids reported by Sposito and coworkers (samples denoted A) and the Altamont soil fulvic acid fraction were determined using a modification of the standard calcium acetate method (Holtzclaw & Sposito, 1979). This modified method was used to eliminate very significant positive interferences due to sulfonic acid (SO$_3$H) groups in the sludge fulvic acid fraction, and to acidic OH groups. Positive interferences from acidic OH groups might also be expected in soil fulvic acids, which would lower the mean value reported in Table 9-2. It is clear, however, that soil fulvic acids have a significantly higher carboxyl content than sludge fulvic acids. The magnitude of this difference can be assessed by comparing the mean values for soils to those for the sludges studied by Riffaldi et al. (1982) (samples denoted B).

The ultraviolet-visible spectra of sludge fulvic acid fractions are qualitatively similar to soil fulvic acid spectra in that they are essentially featureless (Sposito et al., 1976). The E_4/E_6 ratios (Table 9-2) calculated from these spectra were lower than for soil fulvic acids and were close to the upper limit of E_4/E_6 values of humic acids (Table 9-2). The conventional interpretation based on lower E_4/E_6 ratios is that the sludge fulvic acids have more condensed structures than soil fulvic acids. This does not seem likely, however, because the H/C ratios (Table 9-1) indicate that the soil fulvic acids are in fact more condensed.

9-1.3 Infrared Spectra

Infrared spectra of sludge fulvic acid fractions (Sposito et al., 1976; Riffaldi et al., 1982) resembled a Type III spectrum (Stevenson & Goh, 1971) of soil fulvic acids (Fig. 9-2). The common features are: C–H stretch at 2940 cm^{-1} due to aliphatic C–H; bands at 1720 and 1650 cm^{-1} due to the C=O of carboxyl and amide groups; and a 1540 cm^{-1} band due to the C–O stretch of polysaccharides. The primary distinguishing features of sludge fulvic acids are bands at 1360, 1160, 880, and 580 cm^{-1}. The 1360, 1160, and 580 cm^{-1} bands were assigned as originating from sulfonyl units (Sposito et al., 1976; 1982; Sposito & Holtzclaw, 1980). Sposito et al. (1978) have also recorded ^{13}C- and ^{1}H-NMR spectra of sludge fulvic acid fractions.

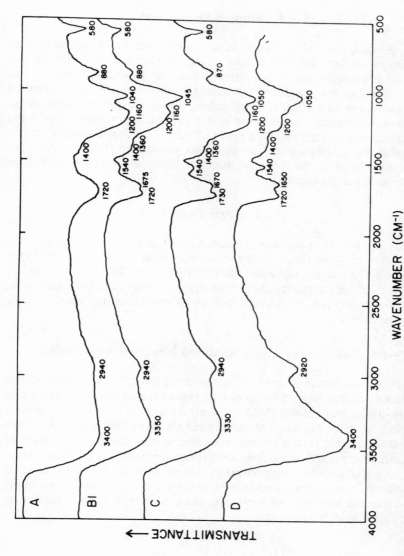

Fig. 9–2. Infrared spectra of the fulvic acid fractions from three anaerobic sewage sludges A, B1, and C (Sposito et al., 1976). The spectrum of fulvic acid extracted from Rifle peat (Type III spectrum) is labeled D (from Stevenson & Goh, 1971, Fig. 5).

Although these spectra were not especially intense or well resolved, they did provide structural information that indicated the presence of both aliphatic and aromatic components as well as degradation products originating from proteins and polysaccharides.

9-1.4   Nitrogen Distribution

In general, the N-distribution in anaerobic sludge and soil fulvic acid fractions is similar (Table 9-3, and references therein). The NH_3-N and amino acid-N contents and the distribution of acidic, neutral, and basic amino acids are similar, but the sludge fulvic acid fraction has a higher content of S-containing amino acids. The most striking difference is in the amino sugar content which is much higher in the sludge fulvic acid, primarily due to the high galactosamine content. As a result of this, the glucosamine/galactosamine ratio was 10 times lower in the sludge fulvic acid fraction. The high galactosamine content has been attributed to the presence of anaerobic bacteria in sludge digesters.

9-1.5   Sugar Content

The neutral sugar contents of sludge fulvic acid fractions reported by Holtzclaw et al. (1980) are similar to the value reported by Oades (1967) for the carbohydrate content of a soil fulvic acid (Table 9-4). There is considerable variability in the sugar contents among the sludge fulvic acid fractions, which presumably is a reflection of the type of waste entering the wastewater treatment plants.

9-1.6   Summary of Major Identified Nonhumic Components

The major identified nonhumic components present in the sewage sludge fulvic acid fractions studied by Sposito and co-workers (Table 9-5) are amino acids (6-9%), hexosamines (5-14%), neutral sugars (10-20%), and anionic surfactants (10%). Together, these account for 30 to 55% by weight of the purified fulvic acid fraction from anaerobic sewage sludge. The nature of the association between these components and the humified fulvic acid component is not clear. Some components such as proteins and hexosamines could be covalently bound whereas others such as the anionic surfactants may simply be co-extracted and have no particular association with other components in the fulvic acid fractions.

9-1.7   Polyacidic Nature

From titration studies, Sposito and Holtzclaw (1977) and Sposito et al. (1977) have determined that sludge-derived fulvic acid fractions contain a continuum of classes of functional groups ranging in acidity from very strong to very weak (Table 9-6). These workers have identified four separate classes

Table 9-3. N-Distribution in the fulvic acid (FA) fractions of three anaerobic sewage sludges and a soil.

| | Source of sewage sludge FA[†] | | | Soil FA | |
N Fraction	A	B1	C	Typical[‡]	Elliot silt loam[§]
Total N[¶]	0.177	0.306	0.367	0.228	--
Unhydrolyzed N[#]	0.822	0.701	0.916	0.44	1.28
Hydrolyzed N[#]	9.178	9.299	9.084	9.56	8.72
Unidentified N[#]	0.110	0.150	0.101	3.05	0.79
NH_3-N[#]	2.775	3.279	2.826	2.03	4.99
Amino acid-N[#]	3.995	2.974	3.124	4.27	2.18
Amino sugar-N[#]	2.298	2.896	3.033	0.22	0.76
Glucosamine[#]	0.203	0.395	0.401	--	--
Galactosamine[#]	2.095	2.501	2.632	--	--
Gluc./Galact. ratio	0.10	0.16	0.15	1.3	--

† Holtzclaw et al. (1980).
‡ Kowalenko (1978).
§ Stevenson (1960). Elliot silt loam (Fine, illitic, mesic Aquic Argiudoll).
¶ g kg^{-1} of ash- and moisture-free fulvic acid.
# g kg^{-1} of total N in fulvic acid.

Table 9-4. Neutral sugar carbon content of fulvic acid fractions from three anaerobic sewage sludges and a soil.

Source of fulvic acid	Carbon[§]
A[†]	12.9
B1[†]	22.6
C[†]	16.7
Soil[‡]	19.6

† Holtzclaw et al. (1980).   ‡ Oades (1967).   § Percent of fulvic acid carbon.

Table 9-5. Major identified nonhumic components of sewage sludge fulvic acid (FA) fractions from three anaerobic sludges.

Fraction	FA weight, %
Amino acids[†]	6-9
Hexosamines[†]	5-14
Neutral sugars[†]	10-20
Anionic surfactants[‡]	
Lauryl sulfate type	4.5
Linear alkyl benzene sulfonate type	5.3

† From Holtzclaw et al., 1980.
‡ Compiled from Sposito et al. (1976, 1982); Holtzclaw and Sposito (1978, 1979).

Table 9-6. Polyacidic nature of anaerobic sewage sludge fulvic acids.[†]

Type of group	pH for ionization	Type of acidity	Class
SO_3H	<3	Very strong	I
COOH	3-5	Strong	II
Weakly acidic COOH	5-8	Weak	III
N-containing groups	5-8	Weak	III
Phenolic-OH	>8	Very weak	IV
SH	>8	Very weak	IV

† From Sposito and Holtzclaw (1977) and Sposito et al. (1977).

of functional group acidity, designated as classes I, II, III, and IV. The relatively numerous strongly acidic groups (Class I) which ionized at pH < 3 were probably sulfonic acid groups. Type II acidity was attributed to primarily carboxyl groups. Type III groups, which titrated around pH 7, were identified as weakly acidic carboxyls and N-containing groups (e.g., $-NH_3^+$). Those titrating in the alkaline range (Type IV) were identified as phenolic OH and thiol (SH) groups. The net result was that the pH for proton dissociation ranged from < 3 to > 8, which resulted in very complex behavior in solution, and titration curves distinctly different from those of soil fulvic acids.

9-1.8 Stability of Sludge–Organic Metal Complexes

Several studies have been conducted to evaluate the stability of complexes formed between fulvic acid extracted from sewage sludge–soil mixtures and various metal ions. Emphasis has been placed on the fulvic acid fraction because of its water-soluble nature and the potential to mediate metal uptake by plants. Potentiometric titrations of fulvate with $Cu(ClO_4)_2$ at pH 5 indicated that a relatively strong 1:1 complex (log K_1 = 3.9) is formed between Cu and the more weakly acidic functional groups (Sposito et al., 1979). The same experimental conditions were used to show that two classes of complexes exist between sludge fulvate and Cd^{2+} and Pb^{2+} (Sposito et al., 1981). These authors also used linear correlation analysis to predict the stability constants (K_1 and K_2) for complexes of sludge-derived fulvic acid with the divalent cations of Mg, Ca, Mn, Fe, Ni, and Zn. The log K_1 values for the fulvate–divalent metal ion complexes (in parenthesis) decreased in the order Pb (4.22) > Fe (3.96) > Mn (3.93) > Cu (3.88) > Ni (3.81) > Zn (3.54) > Ca (3.12) > Cd (3.04) > Mg (2.71).

An alternative to using measured stability constants is the development of a model to predict metal complexation with sludge-derived fulvate. Mattigod and Sposito (1979) found that a mixture of simple organic acids could be used to simulate the titration properties of sludge-derived fulvate and, more importantly, the stability constants for each metal complex were known. The organic acids used in this "mixture model" were benzenesulfonic, salicylic, phthalic, citric, maleic, ornithine, lysine, valine, and arginine. A potential advantage of the "mixture model" is that speciation calculations can be performed for a greater number of metal ions since stability constants are known for many metals with the simple organic acids employed in the mixture model. Speciation calculations for soil solutions obtained from sludge-amended soils were used to evaluate the mixture model. This comparison revealed that the mixture model could be used to predict Cd(II) speciation in the aqueous phase of sludge-amended soils (Sposito et al., 1982). This model, however, did not accurately predict speciation of Cu(II) and is therefore not uniformly applicable to all metals found in soils and sludges.

Table 9-7. Elemental and carboxyl content, total acidity, and E_4/E_6 ratios for humic acid fractions from soil, sludges, and sludge-amended soil.

Analysis	Average sludge[†]	Chicago sludge[‡]	Composted municipal refuse[§]	Chalmers soil[‡]	Chicago sludge-amended soil (2 mo)[‡]	Chicago sludge-amended soil (14 mo)[‡]
Elemental content¶						
C	4.46	5.20	5.26	4.92	5.034	4.905
H	--	0.721	0.623	0.486	0.530	0.502
N	0.61	0.878	0.660	0.394	0.502	0.454
H/C, mole ratio	--	1.7	1.4	1.2	1.3	1.2
C/N, mole ratio	8.53	6.9	9.3	14.8	11.7	12.6
CO_2H (mol kg^{-1})	2.3	0.95#	1.8	2.6#	2.20	2.50
Total acidity (mol$_c$ kg^{-1})	3.6	2.4	6.2	7.2	4.0	5.1
E_4/E_6	2.8	3.8	6.5	5.0	5.1	5.2

† From Riffaldi et al. (1982); means of 10 sludge humic acids.
‡ From Boyd et al. (1980). Times shown in parentheses indicate sampling time after sludge application. Chalmers soil (fine-silty, mixed, mesic Typic Haplaquoll).
§ From Gonzalez-Vila and Martin (1985); mean of three compost humic acids.
¶ g kg^{-1} of humic acid weight on an ash- and moisture-free basis.
# Modified calcium acetate method, Holtzclaw and Sposito (1979).

9-2 SEWAGE SLUDGE HUMIC ACIDS

There is somewhat less information on humic acids derived from sludge than on sludge fulvic acids. In one study Boyd et al. (1980) examined humic acid fractions extracted from anaerobically-digested sewage sludge (Chicago), and also from a soil and a sludge-amended soil. In another study Riffaldi et al. (1982) characterized humic acids extracted from 10 different sludges. They found that humic materials accounted for 17% of the sludge organic C, which, on average, was distributed equally between humic and fulvic acids; the mean ratio of humic acid to fulvic acid was 1.3. However, in most samples fulvic acid C was greater than humic acid C. The humic acid to fulvic acid C ratios ranged from 0.3 to 3.0. Table 9-7 shows data on the elemental composition, total acidity, carboxyl group content, and E_4/E_6 ratios for the sludge and soil humic acid fractions. Compared to the soil humic acid fraction, the sludge humic acid fraction contained about twice as much N and as a result the C/N ratio was half. The high sludge N was probably due to associated protein decomposition products, and infrared spectra of the sludge humic acid fraction supported this contention (see below). The sludge humic acid fraction also had a higher H/C ratio indicating a less condensed structure. The sludge humic acid fraction had a much lower carboxyl group content, total acidity and E_4/E_6 ratios than the soil humic acid. As discussed above, it is very doubtful that the E_4/E_6 ratio has any significant relationship to degree of humification. Thus, when compared to the soil-derived materials, both sludge fulvic and humic acid fractions had higher N con-

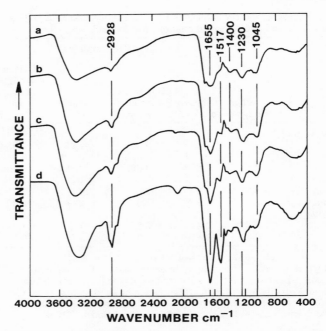

Fig. 9-3. Infrared spectra of humic acids extracted from the (a) Chalmers soil, (b) sludge-amended soil 14 mo after application, (c) sludge-amended soil 2 mo after application, and (d) Chicago anaerobic sludge (Boyd et al., 1980).

tents, lower C/N ratios, higher H/C ratios, lower carboxyl group contents, and lower E_4/E_6 ratios.

Infrared (IR) spectra of sludge and soil humic acid fractions are shown in Fig. 9-3. The sludge humic acid spectrum shows prominent amide I and II bands at 1650 and 1520 cm^{-1} originating from proteinaceous materials. Other prominent features in the IR spectrum of the sludge humic acid are the intense bands in the 2850 and 2950 cm^{-1} region due to C–H stretching vibrations. The IR spectrum strongly suggests that the sludge humic acid fraction has a relatively high content of associated protein and aliphatic materials like fats and waxes. Similar IR spectra were reported by Riffaldi et al. (1982) and Gonzalez-Villa and Martin (1985); these IR spectra of sludge humic acids resemble type III spectra of soil humic substances (Stevenson & Goh, 1971). ^{13}C-NMR spectra also indicate that humic acids extracted from composted municipal refuse are highly aliphatic, and in this regard, different from soil humic acids (Gonzalez-Villa & Martin, 1985).

9-3 HUMIC AND FULVIC ACIDS IN SLUDGE-AMENDED SOILS

Because sewage sludge is approximately 50% organic matter, its application to soil could affect the humic and fulvic acid fractions of soil organic matter. Sposito et al. (1982) studied fulvic acid from Domino soil that had

Table 9-8. Properties of fulvic acid fractions from soil, sludge, and sludge-amended soil.[†]

Fulvic acid	C	H	N	S	LS[‡]	LAS[‡]	Mole ratio H/C	C/N	COOH
				g kg^{-1}					mol$_c$ kg^{-1}
Domino (sludge-amended) soil	3.60	0.79	0.47	0.60	0.345	0.097	2.6	8.9	7.9
Average sludge	4.08	0.66	0.28	0.82	0.45	0.53	1.9	17.0	0.68
Average soil§	4.57	0.54	0.21	0.19	--	--	1.4	25.3	8.2

† From Sposito et al. (1982). Domino soil (fine-loamy, mixed, thermic Xerollic Calciorthid).
‡ LS = anionic surfactants of the lauryl sulfate type; LAS = anionic surfactants of the linear alkyl benzene sulfonate type.
§ From Schnitzer (1978).

received a total of 245 Mg ha^{-1} yr^{-1} (Table 9-8). The sludge application dates were: winter 1975, summer 1976, winter 1976, summer 1977, winter 1977, and summer 1978. The soil was sampled in September 1978. This material was compared to an "average" soil fulvic acid (Schnitzer, 1978) and an "average" sludge fulvic acid (compiled from Sposito et al., 1976; Holtzclaw & Sposito, 1978, 1979). The impact of sludge application was apparent in the fulvic acid fraction. One obvious difference is the high S content, due to the presence of significant quantities of unaltered lauryl sulfate type surfactants. Another difference is the low glucosamine to galactosamine ratio (Table 9-9). The proton titration curve, was, however, very similar to that of a sludge fulvic acid fraction. The curve clearly showed three classes of weakly acidic functional groups (Class II, III, IV, Table 9-6). On the other hand, the carboxyl group content of the sludge-amended soil fulvic acid fraction was much higher than what is typical for a sludge fulvic acid fraction but within the normal range for a soil fulvic acid (Table 9-8). Presumably, this was the result of partial oxidation of the sludge organic matter after its application to soil. In general, it was concluded that the fulvic acid fraction from the sludge-amended soil more resembled a sludge fulvic acid fraction than a soil fulvic acid fraction.

Table 9-9. N-distribution in fulvic acid fractions from anaerobic sludge and sludge-amended soil.

Fraction	Fulvic acid Sludge-amended soil[†]	Average sludge[‡]
Total N§	0.467	0.28
Unhydrolyzed N¶	0.957	0.82
Hydrolyzed N¶	9.043	9.18
Unidentified N¶	0.104	0.12
NH$_3$-N¶	5.681	2.96
Amino acid-N¶	2.307	3.36
Amino sugar-N¶	0.951	2.74
Glucosamine¶	0.103	0.33
Galactosamine¶	0.848	2.41
Glucosamine/galactosamine ratio	0.12	0.14

† Sposito et al. (1982). Domino soil.   § g kg^{-1} of ash- and moisture-free weight.
‡ Holtzclaw et al. (1980).   ¶ g kg^{-1} of total N.

Boyd et al. (1980) examined changes in the humic acid fraction of soil resulting from sludge application. That study compared an unamended Chalmers (Typic Argiaquoll, fine, mixed mesic) soil to a sludge-amended soil (100 Mg ha^{-1}, single application), 2 and 14 mo after application. The effects of sludge application on the elemental content, total acidity, and carboxyl content are shown in Table 9-7. As described above, the sludge humic acid fraction had more N, a higher H/C ratio, and a lower total acidity and carboxyl content than the soil humic acid. Humic acid from the sludge-amended soil showed these same trends. With time, however, the effects of sludge application became less apparent. For example, the carboxyl content was reduced somewhat by sludge application, but with time after application the carboxyl content increased.

The effect of sludge application could be seen clearly in the infrared spectra (Fig. 9-3). The most dramatic changes could be seen soon after application (2 mo, spectrum C) and showed up in the aliphatic C–H stretching region (2928 cm^{-1}) and the 1650 cm^{-1} region, which is the amide I band arising from proteinaceous materials. Again with increasing time after application (14 mo, spectrum B), the infrared spectra more resembled the spectrum of the soil humic acid.

The data from the application of sludge to the Chalmers soil showed that sludge organic matter was recovered in the humic fraction of sludge-amended soil for up to 14 mo after application. Generally, the humic acid fraction from the sludge-amended soil more resembled the soil humic acid than the sludge humic acid fraction. With increasing time after application, the sludge effects were diminished.

The fulvic acid fraction extracted by Sposito et al. (1982) from a sludge-amended soil more resembled a sludge fulvic acid fraction than a soil fulvic acid fraction. The primary difference between the studies of Boyd et al. (1980) and Sposito et al. (1982) was the magnitude of the sludge effect. This can be attributed to the fact that in the latter case (Sposito et al., 1982) a soil that had received three annual applications of sludge was used, and was sampled soon after the third application. In the former case (Boyd et al., 1980) the soil used had received a single application of sludge and was sampled up to 14 mo after the sludge addition. In addition, the humic acid fraction studied by Boyd et al. (1980) may be subject to smaller and slower changes as compared to the fulvic acid fraction studied by Sposito et al. (1982).

9-4 ESR STUDIES ON Cu(II) BINDING BY SLUDGE-DERIVED HUMIC AND FULVIC ACIDS

Electron spin resonance (ESR) has been used to study the structures of complexes formed between various transition metal cations and sludge humic and fulvic acid fractions. The Cu(II) complexes have been most widely studied primarily because Cu(II) gives an ESR spectrum from which a great deal of information can be derived. In this sense, Cu(II) has been used as a probe to determine specific structural details about the metal binding sites of humic and fulvic acids. The types of questions that have been dealt with include (i) the type and number of ligand donor atoms (O, N and S), (ii) formation of chelate vs. nonchelate complexes, (iii) stereochemistry of the binding sites, and (iv) the heterogeneity of the binding sites.

The two basic parameters that are obtained from an ESR spectrum are the g and A values. In the case of Cu(II), the spectra are often anisotropic because the coordination of Cu(II) in humic materials forms a distorted octahedron with one axis longer than the other two due to Jahn-Teller distortion. As a result, g and A values can be obtained in both the parallel and perpendicular regions, $g_{\parallel}$, $A_{\parallel}$, $g_{\perp}$, and $A_{\perp}$. The g values of Cu(II) are given by the equation.

$$g = 2.0023\ (1 + r\lambda/\delta)$$

where r and λ are constants and δ is the energy level splitting between the d-orbitals of Cu(II) (Knowles et al., 1976). Various ligand donor atoms produce different degrees of ligand splitting so that the g values clearly give information on the type and strength of ligand interactions with Cu(II). The practical result is that there is a correlation between the kind of atom coordinating to Cu(II) and the Cu(II) g values, and this can be used to help determine the structures of Cu(II) complexes with humic and fulvic acids in soils and sludge. As it turns out, for CuX_4 chromophores (where X_4 are the four equatorial donor atoms of complexed Cu(II)) the $g_{\parallel}$ values decrease going from X = O ($g_{\parallel}$ = 2.3–2.4) to X = N ($g_{\parallel}$ = 2.21–2.24) to X = S (Boyd et al., 1981). For example, a Cu(II) complex with four O donor atoms in the equatorial plane will give a $g_{\parallel}$ value between 2.3 to 2.4; four N donor atoms a $g_{\parallel}$ value between 2.21 to 2.24. Mixed complexes give intermediate values.

Boyd et al. (1983) used this relationship to study the structures of a Cu(II)–sludge–humic acid complex. From these studies it was concluded that Cu(II) formed at least two equatorial bonds with humic acid O donor atoms in a chelate type complex. This structure is shown in Fig. 9–4. The Cu(II)

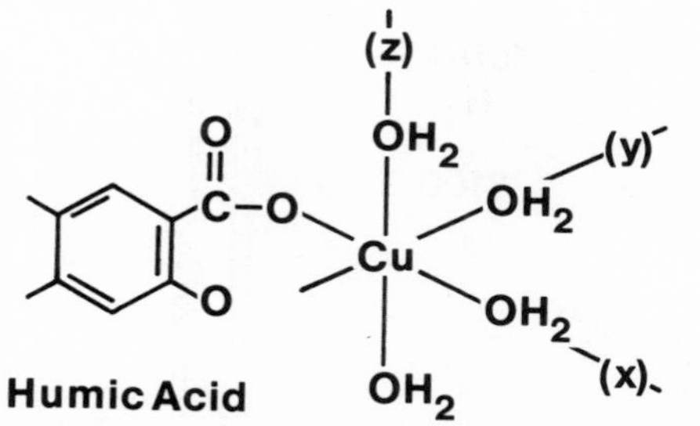

Fig. 9–4. Complexation of Cu(II) by sludge-derived humic acid. Note that Cu(II) is chelated by the sludge humic acid. The Cu(II) also appeared to form axial bonds (along Z-axis) with humic acid ligands originating from proteinaceous materials associated with the sludge humic acid fraction (Boyd et al., 1983).

Table 9-10. ESR spectral data (77 K) of Cu(II)–sludge fulvic acid (FA) complexes in aqueous solution.†

Cu/FA mole ratio	$g_\parallel$	$g_\perp$	Lines at $g_\perp$	FA ligand atoms
0.08:1	2.307	2.055		4 O
	2.278	2.040	2 × 5	2 O, 2 N
0.8:1	2.300	2.057		4 O‡
	2.277	2.038	5	2 O, 2 N‡
	2.250		7	1 O, 3 N
8:1	2.300	2.055		4 O‡
	2.279	2.037	5	2 O, 2 N‡
	2.250		7	1 O, 3 N
80:1	2.300	2.056		4 O‡
	2.277	4.041	5	2 O, 2 N‡
	2.251		7	1 O, 3 N

† Fulvic acid from the Ontario sludge (Senesi et al., 1985).
‡ Main ligand system.

ion also appeared to form axial bonds (along the Z axis, Fig. 9–4) with ligands originating from proteinaceous material associated with the sludge humic acid.

ESR studies on Cu(II) complexes with sludge-derived fulvic acids have also been reported (Senesi & Sposito, 1984; Senesi et al., 1985). Both residual Cu(II)–fulvic acid complexes and those prepared at different Cu/fulvic acid ratios were characterized. The data on Cu(II)–fulvic acid complexes prepared at different Cu/fulvic acid ratios (Table 9–10) indicated two "well-resolved" components in the parallel region of all spectra as indicated by $g_\parallel$ values of 2.30 and 2.28, corresponding to binding sites of either a four oxygen (4 O) or a mixed two oxygen/two nitrogen (2 O, 2 N) ligand donor

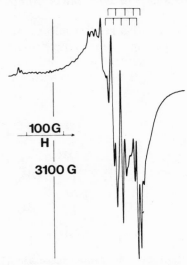

Fig. 9-5. ESR spectrum [scan range, 1 kG (1 gauss = 10^{-4} tesla)] of a Cu-Ontario fulvic acid (FA) in aqueous solution at 77 K after resin treatment and at Cu/FA = 9 (gain, 8 × 10^4; time constant, 200 ms; scan time, 500 s) (Senesi et al., 1985).

system. These results suggested primarily equatorial coordination of O and N donor ligands corresponding to at least two macroscopic classes of poly-dentate complexes between sludge fulvic acid and Cu(II). The strongest bind-ing site was suggested to involve sulfonic acid groups together with N groups. The $g_\perp$ region of the ESR spectra reported by Senesi et al. (1985) was high-ly structured (Fig. 9–5). This was attributed to the superhyperfine splitting of Cu(II) by a number of N-ligand nuclei, and was further evidence for bind-ing sites involving N-donor atoms in sludge fulvic acid.

9–5 SUMMARY

Humic and fulvic acid fractions from sludges have several distinguish-ing chemical/structural features. As compared to average soil humic and ful-vic acids, they have higher N contents, lower C/N ratios, higher S contents, a higher fraction of aliphatic components (higher H/C ratios), higher galac-tosamine contents, lower total acidity, lower amounts of carboxyl groups, associated anionic surfactants (detergents), and complex binding sites involv-ing both N and O donor ligand systems. Although the properties of soil and sludge humic and fulvic acids are in many ways similar, the environment of the sludge digester produces significant differences in structure. When sludge is added to soil, it also impacts the humic and fulvic acid fraction for at least several years after application. Over time the effects of sludge amendments become less apparent.

REFERENCES

Boyd, S.A., L.E.Sommers, and D.W. Nelson. 1980. Changes in the humic acid fraction of soil resulting from sludge application. Soil Sci. Soc. Am. J. 44:1179–1186.

Boyd, S.A., L.E. Sommers, D.W. Nelson, and D.X. West. 1981. The mechanism of copper(II) binding by humic acid: An electron spin resonance study of a copper(II)-humic acid com-plex and some adducts with nitrogen donors. Soil Sci. Soc. Am. J. 45:745–749.

Boyd, S.A., L.E. Sommers, D.W. Nelson, and D.X. West. 1983. Copper(II) binding by humic acid extracted from sewage sludge: An electron spin resonance study. Soil Sci. Soc. Am. J. 47:43–46.

Gonzalez-Villa, F.J., and F. Martin. 1985. Chemical structural characteristics of humic acids extracted from composted municipal refuse. Agric. Ecosyst. Environ. 14:267–278.

Holtzclaw, K.M., G.D. Schaumberg, C.S. LeVesque-Madore, G. Sposito, J.A. Heick, and C.T. Johnston. 1980. Analytical properties of the soluble, metal-complexing fractions in sludge-soil mixtures: V. Amino acids, hexosamines, and other carbohydrates in fulvic acid. Soil Sci. Soc. Am. J. 44:736–740.

Holtzclaw, K.M., and G. Sposito. 1978. Analytical properties of the soluble, metal-complexing fractions in sludge-soil mixtures: III. Unaltered anionic surfactants in fulvic acid. Soil Sci. Soc. Am. J. 42:607–611.

Holtzclaw, K.M., and G. Sposito. 1979. Analytical properties of the soluble, metal-complexing fractions in sludge-soil mixtures: IV. Determination of carboxyl groups in fulvic acid. Soil Sci. Soc. Am. J. 43:318–323.

Holtzclaw, K.M., G. Sposito, and G.R. Bradford. 1976. Analytical properties of the soluble, metal-complexing fractions in sludge-soil mixtures. I. Extraction and purification of ful-vic acid. Soil Sci. Soc. Am. J. 40:254–258.

Knowles, P.F., D. Marsh, and H.W.E. Rattle. 1976. Magnetic resonance of biomolecules. John Wiley & Sons, New York.

Kowalenko, C.G. 1978. Organic nitrogen, phosphorus, and sulfur in soils. p. 95–136. In M. Schnitzer and S.U. Khan (ed.) Soil organic matter. Elsevier, Amsterdam.

Lake, D.L., P.W.W. Kirk, and J.N. Lester. 1984. Fractionation, characterization, and speciation of heavy metals in sewage sludge and sludge-amended soils: A review. J. Environ. Qual. 13:175–183.

Mattigod, S.V., and G. Sposito. 1979. Chemical modeling of trace metal equilibria in contaminated soil solutions using the computer program GEOCHEM. p. 837–856. In E.A. Jenne (ed.) Chemical modeling in aqueous systems. ACS Symp. Ser. no. 93. American Chemical Society, Washington, DC.

Oades, J.M. 1967. Carbohydrates in some Australian soils. Aust. J. Soil Res. 5:103–115.

Overcash, M.R. 1983. Land treatment of municipal effluent and sludge: Specific organic compounds. p. 199–231. In A.L. Page et al. (ed.) Utilization of municipal wastewater and sludge on land. Univ. of California, Riverside.

Page, A.L., T.L. Gleason, III, J.E. Smith, Jr., I.K. Iskandar, and L.E. Sommers (ed.). 1983. Utilization of municipal wastewater and sludge on land. Univ. of California, Riverside.

Riffaldi, R., F. Sartori, and R. Levi-Miniz. 1982. Humic substances in sewage sludge. Environ. Pollut. Ser. B 3:139–146.

Schaumberg, G.D., K.M. Holtzclaw, C.S. LeVesque, and G. Sposito. 1982. Characterization of sulfur in fulvic acids extracted from anaerobically digested sewage sludge. Soil Sci. Soc. Am. J. 46:310–314.

Schnitzer, M. 1978. Humic substances: Chemistry and reactions. p. 1–64. In M. Schnitzer and S.U. Khan (ed.) Soil organic matter. Elsevier, Amsterdam.

Senesi, N., and G. Sposito. 1984. Residual copper (II) complexes in purified soil and sewage sludge fulvic acids: Electron spin resonance study. Soil Sci. Soc. Am. J. 48:1247–1253.

Senesi, N., D.F. Bocian, and G. Sposito. 1985. Electron spin resonance investigation of copper (II) complexation by fulvic acid extracted from sewage sludge. Soil Sci. Soc. Am. J. 49:119–126.

Sommers, L.E. 1977. Chemical composition of sewage sludges and analysis of their potential as fertilizers. J. Environ. Qual. 6:225–239.

Sposito, G., F.T. Bingham, S.S. Yadav, and C.A. Inouye. 1982. Trace metal complexation by fulvic acid extracted from sewage sludges: II. Development of chemical models. Soil Sci. Soc. Am. J. 46:51–56.

Sposito, G., and K.M. Holtzclaw. 1977. Titration studies on the polynuclear, polyacid nature of fulvic acid extracted from sewage sludge-soil mixtures. Soil Sci. Soc. Am. J. 41:330–336.

Sposito, G., and K.M. Holtzclaw. 1980. Interpretation of the infrared spectrum of fulvic acid extracted from sewage sludge. Soil Sci. Soc. Am. J. 44:177–178.

Sposito, G., K.M. Holtzclaw, and J. Baham. 1976. Analytical properties of the soluble, metal-complexing fractions in sludge-soil mixtures: II. Comparative structural chemistry of fulvic acid. Soil Sci. Soc. Am. J. 40:691–697.

Sposito, G., K.M. Holtzclaw, and D.A. Keech. 1977. Proton binding in fulvic acid extracted from sewage sludge-soil mixtures. Soil Sci. Soc. Am. J. 41:1119–1125.

Sposito, G., K.M. Holtzclaw, and C.S. LeVesque-Madore. 1979. Cupric ion complexation by fulvic acid extracted from sewage sludge-soil mixtures. Soil Sci. Soc. Am. J. 43:1148–1154.

Sposito, G., K.M. Holtzclaw, and C.S. LeVesque-Madore. 1981. Trace metal complexation by fulvic acid extracted from sewage sludge: I. Determination of stability constants and linear correlation analysis. Soil Sci. Soc. Am. J. 45:465–468.

Sposito, G., K.M. Holtzclaw, C.S. LeVesque-Madore, and C.T. Johnson. 1982. Trace metal chemistry in arid-zone field soil amended with sewage sludge: II. Comparative study of the fulvic acid fraction. Soil Sci. Soc. Am. J. 46:265–270.

Sposito, G., G.D. Schaumberg, T.G. Perkins, and K.M. Holtzclaw. 1978. Investigation of fulvic acid derived from sewage sludge using carbon-13 and proton NMR spectroscopy. Environ. Sci. Technol. 12:931–934.

Stevenson, F.J. 1960. Chemical nature of the nitrogen in the fulvic fraction of soil organic matter. Soil Sci. Soc. Am. J. 24:472–477.

Stevenson, F.J. 1982. Humus chemistry. Chapt. 2. Wiley-Interscience, New York.

Stevenson, F.J., and K.M. Goh. 1971. Infrared spectra of humic acids and related substances. Geochim. Cosmochim. Acta 35:471–483.

Thurman, E.M., and R.L. Malcolm. 1981. Preparative isolation of aquatic humic substances. Environ. Sci. Technol. 15:463–466.

U.S. EPA. 1983. Process design manual. Land application of municipal sludge. EPA-625/1-83-016. Municipal Environmental Research Lab., Cincinnati, OH.

U.S. EPA. 1985. Summary of environmental profiles and hazard indices for constituents of municipal sludge: Methods and results. Office of Regulations and Standards, EPA, Washington, DC.

Chapter 10

Application of Nuclear Magnetic Resonance Spectroscopy to Organic Matter in Whole Soils

MICHAEL A. WILSON, *CSIRO Division of Fossil Fuels, North Ryde NSW, Australia*

ABSTRACT

Basic principles of solid state nuclear magnetic resonance (NMR) spectroscopy relevant to the study of the structure of organic matter of whole soils or solid fractions of soils are outlined. These include cross polarization, magic angle spinning, relaxation phenomena, and decoupling. There are problems in obtaining quantitative data by cross polarization techniques since there is more than one spin lattice relaxation time in the rotating frame. Selective relaxation can be used, however, to identify specific functional groups in soil. Spectra of wood, cellulose, lignin, and their decomposition products can be related to spectra obtained from soil organic matter. The most notable feature about ^{13}C-NMR spectra of whole soils is their variability, and it is shown that this is due to soil-forming factors such as climate and vegetation. The isotopes ^{31}P and ^{15}N in soils can also be studied by NMR spectroscopy and can be used to trace the incorporation of these elements into soils.

Nuclear magnetic resonance spectroscopy (NMR) is a nondestructive technique which can provide information about the types of forms of C, P, N, and other elements in materials and hence has great potential in soil science. The application of some aspects of NMR spectroscopy to the analysis of soil organic matter is the subject of this chapter. A full description of the NMR technique is beyond the scope of this chapter and the reader is referred elsewhere for a more detailed treatment (Becker, 1980; Fukushima & Roeder, 1981; Fyfe, 1983; Shaw, 1984), in particular to the text specifically written for geochemists (Wilson, 1987). However, a fairly lengthy description is necessary here in order that the reader who has little NMR background can follow the text. Thus, the chapter has a secondary function: that is to act as an NMR spectroscopy primer for the practicing scientist. A list of symbols and definitions is shown in the Appendix.

10-1 THE NMR EXPERIMENT

Many nuclei spin about their axis, that is, they have angular momentum. The total angular momentum of a nucleus is given by $[h/2\pi] [I(I + 1)]^{1/2}$ where h is Planck's constant and I is the spin quantum number. The spin quantum number may have values of 0, 1/2, 1, 3/2, 5/2 depending on the nucleus in question. For hydrogen (^1H) and carbon of atomic mass 13 (^{13}C), $I = 1/2$ whereas for carbon of atomic mass 12 (^{12}C), $I = 0$. Since all nuclei are charged they create a magnetic field when they spin. Thus the nucleus can be regarded as a small bar magnet with a dipole, the nuclear magnetic moment.

If a spinning nucleus is placed in a magnetic field, H_o, it may take up one of $2I + 1$ orientations with respect to the direction of the magnetic field. For hydrogen, $I = 1/2$ and hence there are two ($2I + 1 = 2$) possible orientations. Each orientation has a different energy state. It is possible to induce transitions (resonance) between energy states using electromagnetic radiation in the radio frequency range, where the frequency, ν, is given by

$$\nu = \gamma H_o/2\pi \qquad [1]$$

where γ, the gyromagnetic ratio, is a constant for a particular nuclear type, and H_o is the strength of the magnetic field.

Not every nucleus exhibits this phenomenon. For ^{12}C, $I = 0$ and there is only one ($2I + 1 = 1$) orientation that the nucleus can adopt in a magnetic field. Consequently, it is not possible to induce nuclear spin resonance in the case of ^{12}C. If the element carbon is to be studied, only the ^{13}C isotope ($I = 1/2$), which is naturally less abundant, can be observed.

Table 10-1. NMR properties of important elements in the geosphere available for study by NMR spectroscopy.

Isotope	Spin	Natural abun-dance N	Magnetic moment	Gyro-magnetic ratio	Quadrupole moment	Frequency	Relative detectability
		%	μ	$\gamma/10^7$ rad T^{-1} s^{-1}	$Q/10^{-28}$ m^2	MHz	
^{1}H	1/2	99.985	4.8371	26.7510	--	100.00	1.0
^{29}Si	1/2	4.70	-0.9609	-5.3141	--	19.867	3.69×10^{-4}
^{27}Al	5/2	100	4.3051	6.9706	0.0149	26.057	0.206
^{13}C	1/2	1.108	1.2162	6.7263	--	25.145	1.76×10^{-4}
^{31}P	1/2	100	1.9581	10.829	--	40.481	0.066
^{15}N	1/2	0.37	-0.4901	-2.7107	--	10.137	3.85×10^{-6}
^{2}H	1	0.015	1.2125	4.1064	2.73×10^{-3}	15.351	1.45×10^{-6}
^{3}H†	1/2	--	5.1594	28.5335		106.663	--
^{17}O	5/2	0.037	-2.2398	-3.36266	-2.6×10^{-2}	13.557	9.25×10^{-2}
^{23}Na	3/2	100	2.8610	7.0760	0.12	26.451	9.25×10^{-2}
^{57}Fe	1/2	2.19	0.1563	0.8644	--	3.231	7.4×10^{-7}
^{119}Sn	1/2	8.58	-1.8029	-9.9707	--	37.29	4.4×10^{-3}
^{207}Pb	1/2	22.6	1.0120	5.5968	--	20.92	2.1×10^{-3}
^{109}Ag	1/2	44.18	-0.2251	-1.2449	--	4.6536	4.9×10^{-5}

† Not present, but useful tracer.

There are three factors that influence the ease by which an element can be detected in the geosphere by NMR. First, the abundance of the *element* is important. Second, the natural abundance of the *isotope* undergoing nuclear magnetic resonance in the element is important. Finally, particular characteristics of the nucleus such as the spin number, the gyromagnetic ratio, and magnetic moment (torque felt when a magnet, i.e. the nucleus, is placed in a magnetic field) are important in determining the sensitivity of the nucleus. Nuclei with small gyromagnetic ratios and magnetic moments are difficult to observe by NMR spectroscopy.

Properties of nuclei in the geosphere, which might be detected by NMR, are listed in Table 10-1 with values of their relative detectability as given by Harris (1978). Protons (^{1}H) are clearly the most easily observed nuclei based on natural isotopic abundance and gyromagnetic ratio, and this is also true in the geosphere. Aluminum and P come next in terms of ease of measurement. The concentration of P is relatively low in the geosphere, but experiments have already been performed which show that this nucleus can, indeed, be studied. The nuclei ^{23}Na, ^{13}C, and ^{29}Si also appear to be good candidates for study by NMR spectroscopy. The literature on ^{13}C-NMR spectroscopy is already extensive for C-rich materials such as coals and humic substances; likewise, ^{29}Si-NMR spectroscopy has been used to study Si-rich materials such as silicate minerals.

10-2 THE ROTATING FRAME

In a real system, i.e. a sample to be studied by NMR spectroscopy, there will be a large number of nuclei, some of which will already be in an excited state due to thermal motion. That is, the population of the two states within the magnetic field is given by a Boltzmann distribution and there will be a net macroscopic magnetization, M, due to the fact that the lower energy state is more occupied than the higher energy excited state.

Let us examine this in more detail. The nuclei in the sample possess a magnetic moment due to their charge and angular momentum. Now, a magnetic moment interacts with a static field in such a way that the applied field tries to force the moment to line up along it. Because of the angular momentum, however, the nuclei experience a torque due to the field acting on the moment. Thus, the nuclei precess about the static magnetic field. The analogy that is frequently quoted in NMR texts is the spinning top, which precesses about the earth's gravitational field. The precession speed ω_0 (radians s^{-1}) of the moment is determined by the product of the gyromagnetic ratio, γ, which is an intrinsic property of the nucleus, and the strength of the magnetic field, H_0,

$$\omega_0 = \gamma H_0. \qquad [2]$$

The quantity ω_0 is called the *Larmor speed*. Since the circumference of a circle is $2\pi r$, the nuclear moment precesses about H_0 with a frequency, ν_0, expressed in cycles s^{-1} (Hz)

$$\nu_0 = \frac{\omega_0}{2\pi} = \frac{\gamma H_0}{2\pi}. \qquad [3]$$

It will be noted that Eq. [1] is the same as Eq. [3].

How then does the magnetic field of the electromagnetic radiation interact with the moment of the nuclei to induce transitions between energy states? It is possible to understand this by very simple vector diagrams, but to do so it is necessary to change to coordinates used for the discussion to rotating frame coordinates. Let us imagine that we wish to measure the distance from New York to London. It would be very complicated to express this mathematically if the coordinates were fixed in space, because both New York and London are moving as the earth turns. Instead, the coordinates are allowed to rotate at the rate of the earth's rotation and this permits one to express the distance as longitude and latitude. The same principle is used in NMR spectroscopy. After the sample is placed in the magnetic field there is a magnetization from each spin (μ) precessing at the Larmor frequency (Fig. 10-1a). In a real sample there are many nuclei all precessing with magnetic moments (μ's). These can be represented by vectors about all points on the ellipse in Fig. 10-1a. After summation they can be described as a net magnetization M, along coordinate z. In a frame rotating at the Larmor frequency the net magnetization is stationary (Fig. 10-1b). At resonance, the field due to the electromagnetic radiation (with magnetic field H_1) also rotates at the frequency of the rotating frame and hence can be regarded as static in the rotating frame at H_1 along y' (Fig. 10-1b). Now, when the field H_1 is applied, it will operate on M so that M will change its orientation with

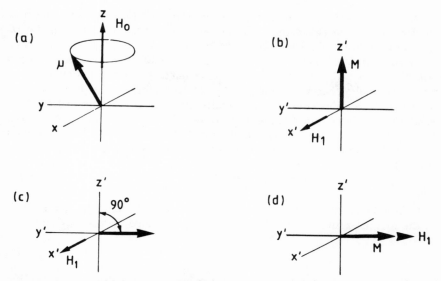

Fig. 10–1. Vector diagrams to describe the behavior of the net magnetization, M, of a collection of nuclear spins in a static magnetic field H_o.

a. μ precesses around laboratory frame coordinates in direction of static magnetic field.

b. Coordinates x', y' rotated at rate of precession.

c. After applying a radiofrequency induced magnetic field, H_1, along rotating coordinate, x', for a period of time until the magnetization is moved by 90°.

d. Shifting the phase of the radiofrequency irradiation so that the magnetic field, H_1, becomes colinear with M. M is said to be "spin locked."

respect to the static field. The amount by which M is moved depends on the intensity and duration of the applied electromagnetic field, H_1. If the applied field H_1 turns M by 90° (Fig. 10–1c) then a 90° pulse is said to have been applied. After H_1 is turned off, the magnetization will return to the z' axis. This is equivalent to the molecules relaxing back to a Boltzmann distribution governed by the static magnetic field, H_o. It should be noted that the phase of the applied electromagnetic irradiation may be changed so that it is colinear with M after a 90° pulse has been applied (Fig. 10–1d). In effect, the magnetization M is now trapped along a second magnetic field H_1 like the static magnetic field, H_o. It will still relax and return to H_o, but at a much slower rate.

10–3 RELAXATION

After H_1 is turned off the situation is that depicted in Fig. 10–2. Because of the natural processes that cause nuclei to exchange energy with each other, the net magnetization represented by M spreads out in the xy plane (Fig. 10–2a). This does not mean there is any loss of energy from the sample as a whole because energy is transferred from some nuclei to other nuclei.

This exchange process is called *spin-spin relaxation* (T_2). At the same time nuclei may lose energy to the surroundings as they relax back to their Boltzmann distribution. This is called *spin-lattice relaxation* (T_1). Both processes occur simultaneously so that the actual magnetization looks like a collapsing cone (Fig. 10-2b, c, d). In actual fact, the magnetic fields are not perfectly homogeneous so that some samples experience slightly different values of H_0 and hence precess at slightly different frequencies, some slower and some faster than the rotating frame. Thus, loss of magnetization in the rotating frame (x', y') is usually a little faster than T_2. This is termed T_2^*, or depending on the nature of the experimental measurement, T_2'. In general,

$$T_2^* < T_2 < T_1.$$

All relaxation processes are exponential, so that after a period of time, τ, the magnetization M is related to the initial magnetization M^0 by

$$M = M^0 \exp(-\tau/T). \qquad [4]$$

Experimentally we measure signal intensity (I) so that

$$I = I^0 \exp(-\tau/T) \qquad [5]$$

where I^0 and I are the initial signal intensity at time $= 0$ and at time τ, respectively, and t is the relaxation constant T_1 or T_2. From Eq. [5] one can

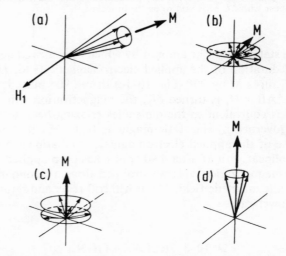

Fig. 10-2. Behavior of spins composing net magnetization, M, during relaxation.

a. Each of the lighter drawn arrows represents the spins of individual nuclei rather than the net magnetization, M. After a 45° pulse, M is aligned at 45° to z'. After a time t, spins fan out due to interchange of energy (T_2 process).
b. Later at time nt, spins fan out further in x, y plane.
c. and d. Spins being to coalesce again in z' directions as they lose energy to surroundings (T_1 process).

measure T from the reciprocal slope of a plot of $\ln I$ vs. time, t, but usually to measure T_1 or T_2 the equation is modified in various ways to reduce experimental errors (Becker, 1980; Shaw, 1984; Wilson, 1987). Finally one should note the situation when the phase of the applied electromagnetic radiation is changed (Fig. 1d). Here, the magnetization, M, also decays but at a different rate, termed *spin-lattice relaxation in the rotating frame* ($T_{1\rho}$). In logarithmic form this is described by Eq. [6].

$$\ln I = \ln I^0 - (\tau/T_{1\rho}).$$ [6]

10-4 FREE INDUCTION DECAYS AND FOURIER TRANSFORMATION

Figure 10-3 shows the exponential decay (termed the *free induction decay*) that results from a radiofrequency pulse applied at the resonance frequency of a nucleus. The decay measures the decrease in magnetization in the xy plane since the detector in a NMR spectrometer is referenced in phase with the radio frequency pulse.

These data can also be expressed in terms of signal intensity as a function of frequency (Fig. 10-4) since signal intensity as a function of time, $I(t)$, and as a function of Larmor frequency (or Larmor speed), $I(f)$, are related by Fourier transformation represented by

$$I(t) = \frac{1}{2\pi} \int_{-\infty}^{\infty} I(f)\, e^{i\omega t}\, d\omega$$ [7]

and

$$I(f) = \int_{-\infty}^{\infty} I(t)\, e^{-i\omega t}\, dt.$$ [8]

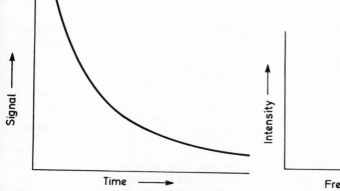

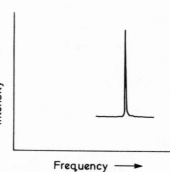

Fig. 10-3. A free induction decay (FID). Variation of signal intensity with time after magnetization in an NMR experiment.

Fig. 10-4. Fourier transform of data in Fig. 10-3.

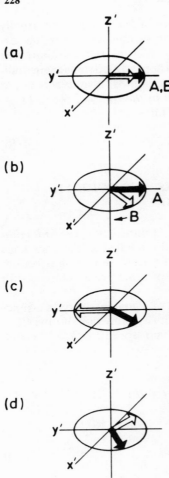

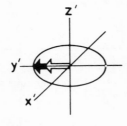

Fig. 10-5. Vector diagrams to show two spins rotating at different rates than the rotating frame and each other. From top to bottom:

a. The magnetizations A,B are colinear.
b. B rotates faster than A.
c. B is now 180° out of phase with position of B in (a) above. Rotation of A can be seen.
d. B almost completed 360° rotation.
e. B at original point in (a) above. A is now at 180° out of phase from (a) above.
f. Both A and B colinear but 180° out of phase.

A magnetization precessing at the rotating frame frequency would be static during the time periods (a) through (f) but some relaxation by T_2 processes may occur so that some nuclei will be rotating slightly faster, and some nuclei slightly slower than the rotating frame.

Suppose that the radiofrequency is slightly different from the Larmor frequency for some of the nuclei. If the reference frame is rotating at the radiofrequency after the 90° pulse (Fig. 10–1c), M lies along the y axis but M now rotates relative to the rotating frame, and hence the magnetization will gradually get out of phase with a signal that arises from nuclei that are exactly in resonance (Fig. 10–5). The free induction decay of this system is complicated and looks somewhat like a sine wave of decreasing amplitude with time. The Fourier transform spectrum of this system shown in Fig. 10–6 has two signals, however, and each nucleus is seen to resonate at different frequencies.

This leads to the most important use of NMR. *Since the intrinsic properties of the molecules govern the precise frequency at which the nuclei resonate, then NMR can be used to distinguish between different types of atoms, e.g. carbon atoms in different chemical environments.* Thus, the two carbons in acetic acid (CH_3COOH) can be distinguished because the CH_3 and COOH carbons have different chemical environments.

Some mention should be made of the methods of processing the free induction decay. Normally, to improve signal-to-noise ratio a large number of free induction decays are added before Fourier transformation, and then a computer is used to weight the data in a desired manner. This is normally done by an exponential weighting function. If the free induction decay is multiplied by a decreasing exponential, the points in the free induction decay at longer time are discriminated against, and since these points have a poorer signal-to-noise ratio in the time domain, the overall peak signal-to-noise ratio in the frequency domain is improved. However, improving the signal-to-noise ratio in this way has its drawbacks. Different parts of the free induction decay contain different information and if the points obtained at longer times are discriminated against, this information is lost.

Information about the existence of magnetization components with large frequency differences are apparent at the start of the free induction decay, but information about smaller frequency differences will not be evident until sufficient time has elapsed. Thus, in samples containing nuclei giving narrow

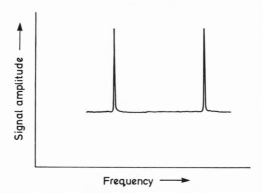

Fig. 10–6. Fourier transform spectrum of signals from two species resonating at slightly different frequencies.

and broad lines, information about the narrow lines is in the tail end of the free induction decay. Thus, truncating the free induction decay to improve signal-to-noise ratio removes this information and causes line broadening. Weighting factors are often termed *line-broadening factors* for this reason. It is necessary, therefore, to use small line broadening factors (e.g., 1 Hz) to look for small signal differences (e.g., 1 Hz). If the lines are broad, however, then the signal-to-noise ratio can be improved greatly by using a large line broadening factor (e.g., 50 Hz).

10-5 CHEMICAL SHIFT

The chemical shift is the principal parameter from which structural information is obtained by NMR in soil science (Wilson, 1981). The primary reason that nuclei resonate at different frequencies is that the electrons around a nucleus alter the magnitude of the magnetic field experienced by the nucleus so that the actual magnetic field is somewhat different from that of the magnet by a shielding factor (σ) (tensor) where $H_{\text{(nucleus)}} = H_o - \sigma H_o$. The shielding factor is small and depends on the orientation of the molecule relative to the applied field. For molecules in rapid motion, such as those in solution, σ averages out so that only one isotropic (σ_{iso}) value is observed. Nevertheless, the isotropic shielding factor, σ, consists of a number of elements (σ_{11}, σ_{22}, σ_{33}) depending on the axes, and these differ for solids.

$$\sigma_{\text{iso}} = 1/3(\sigma_{11} + \sigma_{22} + \sigma_{33}) \qquad [9]$$

From the ^{13}C-NMR spectrum of a solid consisting of only one structural group e.g. polyethylene (Fig. 10-7), the three principal values of the chemical shift tensors can be elucidated. The line shape arises from the weighted average of random orientations of the molecules.

NMR data are measured in frequency units from a chosen reference. The frequency is dependent on the strength of the magnetic field, however, since the chemical shift is induced by the field. Hence, it is desirable to report chemical shifts with respect to some internal standard. These standards differ for different nuclei and are listed in Table 10-2.

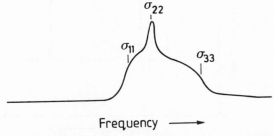

Fig. 10-7. ^{13}C-NMR spectrum of polyethylene without magic angle spinning but with proton decoupling. The chemical shifts of the tensors σ_{11}, σ_{22}, σ_{33} are shown.

Chemical shift, δ, is defined by the equation

$$\delta = \frac{\nu_s - \nu_r}{\nu_r} \times 10^6 \qquad [10]$$

where ν_s is the resonance frequency of the substrate and ν_r is the resonance frequency of the reference. Chemical shift is field independent, and depends only on the magnitude of the shielding factor, σ. One can write

$$\nu_s = \frac{\nu}{2\pi} H_o (1 - \sigma_s) \qquad [11]$$

$$\nu_r = \frac{\nu}{2\pi} H_o (1 - \sigma_r) \qquad [12]$$

where σ_s is the shielding factor for the substrate and σ_r is the shielding factor for the reference

Hence,

$$\frac{\nu_s - \nu_r}{\nu_r} \times 10^6 = \frac{\frac{\nu}{2\pi} H_o [(1 - \sigma_s) - (1 - \sigma_r)] \times 10^6}{\frac{\nu}{2\pi} H_o (1 - \sigma_r)}$$

$$= \frac{(\sigma_r - \sigma_s)}{(1 - \sigma_r)} \times 10^6. \qquad [13]$$

Since $\sigma_r \ll 1$,

$$\delta = 10^6 \times \frac{\nu_s - \nu_r}{\nu_r} = (\sigma_r - \sigma_s) \times 10^6. \qquad [14]$$

There is no field term, H_o, in Eq. [14]. Chemical shifts of various structures containing elements of importance in soil systems are shown in Fig. 10-8 through 10-11.

Table 10-2. Reference standards for chemical shifts.

Nucleus	Standard
^1H	$Si(CH_3)_4$, aqueous solution of $(CH_3)_3Si(CH_2)_3SO_3Na$
^{13}C	$Si(CH_3)_4$, aqueous solution of dioxane
^{31}P	H_3PO_4 (85% in H_2O)
^{15}N, ^{14}N	NH_3

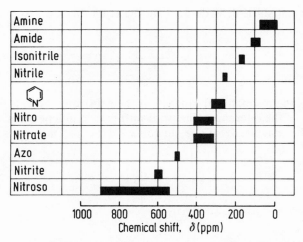

Fig. 10-8. Chemical shifts for nitrogen (^{15}N) in various functional groups relative to ammonia.

10-6 MAGIC ANGLE SPINNING AND HIGH POWER DECOUPLING

As already noted, the NMR spectrum of a solid consists of broad lines because of the different orientations of molecules in the magnetic field. The width of these lines can be greater than the differences in chemical shifts between structural groups. Hence, some method must be employed to reduce the linewidth so that chemical shift information can be obtained.

The technical problems encountered in obtaining solid state spectra in which resonances from discrete C types in the molecule under investigation are to be distinguished from each other (i.e. in obtaining high resolution spectra) are not trivial. In solution, the random motions of liquids average many of the interactions within and between molecules, with the result that only the average interaction is observed. Thus, in solution, the main cause of differences in the resonant irradiation frequencies (chemical shifts) of functional groups is the uneven distributions of electrons in the molecule. The observed chemical shift is the isotropic value brought about by motional averaging. In solids, a range of chemical shifts exists for each structural type because the different orientations in the magnetic field are not averaged. This chemical shift anisotropy is particularly important for structural units that have nonspherical electron distributions such as aromatic and carbonyl carbons. These types of C nuclei experience different shielding in a solid depending on whether the bond axes are parallel or perpendicular to the applied field. In a mixture such as a soil all orientations within and including these extremes are possible.

Another source of line broadening arises because of the interactions between nuclei, which act like charged poles of a magnet. Dipolar interactions between two unlike nuclei such as ^{1}H and ^{13}C can be removed or partly removed by several techniques. Two of these, namely magic angle spinning

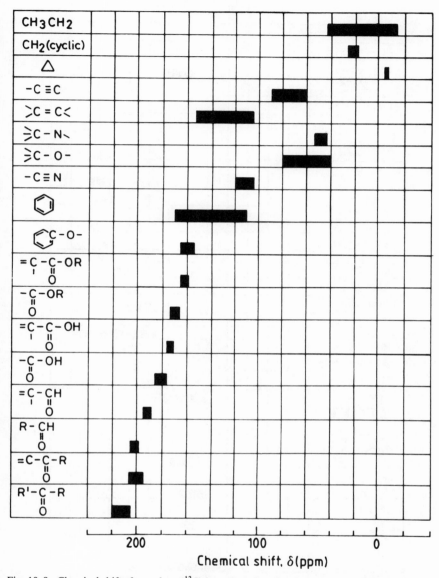

Fig. 10-9. Chemical shifts for carbon (^{13}C) in various functional groups relative to tetramethylsilane.

(MAS) (Lowe, 1959; Andrew, 1971) and high power decoupling (Pines et al., 1973) when used together, can reduce linewidths from kHz to Hz and so allow high resolution spectra to be obtained. These will now be discussed briefly, but the reader is referred to the paper by Pines for a full mathematical treatment.

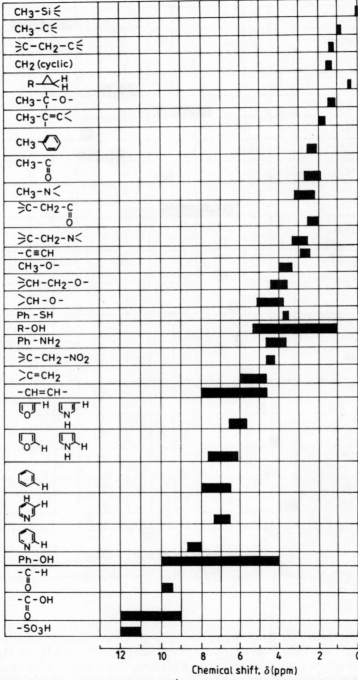

Fig. 10-10. Chemical shifts for protons (^{1}H) in various functional groups relative to tetramethylsilane.

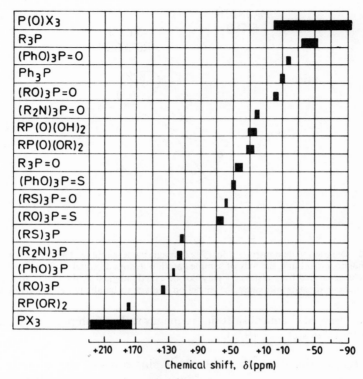

Fig. 10–11. Chemical shifts for phosphorus (^{31}P) in various functional groups relative to phosphoric acid.

As already noted, a magnetic field, having both magnitude and direction, is a vector quantity. Likewise, the potential energy of interaction between two nuclei produced by a dipole can be described as an internuclear vector. The dipolar interaction between two nuclei in an applied magnetic field, and hence the observed chemical shift, depends on the magnitude of the internuclear vector and the angle (θ) between the internuclear vector and the magnetic field vector (Lowe, 1959; Andrew, 1971). More precisely, the interaction depends on ($3 \cos^2 \theta - 1$). This term vanishes in a solid if θ is made equal to 54.74°, (the "magic angle"). In these circumstances the dipolar interaction vanishes and thus the linewidth is drastically reduced. Of course, not all internuclear vectors are oriented at 54.74° to the applied magnetic field vector. However, if one rotates, i.e. "spins", the sample about an axis inclined at 54.74° to the applied field, then the internuclear vector and nuclei exhibit a circular motion about the axis of rotation (Fig. 10–12) and can now be considered, on average, to be at the center of the circle, i.e. at the magic angle. In practice, insufficient spinning speeds are achieved to remove ^{13}C-^{1}H dipolar interactions by this method alone, and high power decoupling is also employed. However, magic angle spinning also removes chemical shift anisotropy as outlined below.

MAGNETIC
FIELD

AXIS OF
ROTATION

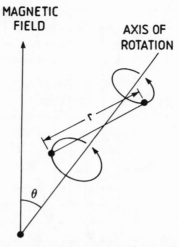

Fig. 10–12. Diagram to demonstrate principles of magic angle spinning. The angle between the static magnetic field and the axis of sample rotation θ is made equal to $54.74°$. Although two nuclei separated by a distance r may not be exactly aligned at the magic angle, rapid rotation about the axis will cause the nuclei on average to be aligned at the magic angle.

The observed chemical shift (in frequency rather than ppm) of a C atom is given by (Pines et al., 1973):

$$\sigma_{obs} = \sin^2 \theta \cos^2 \phi \ \sigma_{11} + \sin^2 \theta \sin^2 \phi \ \sigma_{22} + \cos^2 \theta \ \sigma_{33} \qquad [15]$$

where θ and ϕ are the polar angles relating the principal axes (1, 2, 3) to the laboratory system (x, y, z) and σ_{11}, σ_{22}, σ_{33} are the characteristic values of σ. As already noted, the isotropic chemical shift is given by Eq. [9]. If $\theta = 54.74°$, then $\cos^2 \theta = 1/3$, $\sin^2 \theta = 2/3$, and Eq. [15] reduces to the expression

$$\sigma_{obs} = 1/3(2 \cos^2 \theta \ \sigma_{11} + 2 \sin^2 \phi \ \sigma_{22} + \sigma_{33}). \qquad [16]$$

By placing the sample in the magnetic field, H_o, at the angle of $54.74°$, the magic-angle, and rotating the sample about the axis, $\sin^2 \phi$ and $\cos^2 \phi$ both average to $1/2$ and σ_{obs} in Eq. [16] becomes equivalent to σ_{iso} in Eq. [9].

It is also possible (and essential for obtaining high resolution spectra) to reduce $^{13}C-^1H$ dipolar interactions by high power decoupling. If the proton nuclei are irradiated at close to their resonant frequency, transitions between the energy levels of the protons occur. If these transitions occur fast enough, ^{13}C nuclei will see only an average proton field from the protons. Thus, in the presence of a decoupling field the linewidth of the C resonance is further reduced.

At this point a cautionary note should be expressed. The rate of magic angle spinning needed to remove chemical shift anisotropy is very dependent on the field strength of the magnet used. If the sample is spun at speeds less than that required for maximum line narrowing, the broad line resonance breaks up in a complicated way into a number of "spinning side bands" which may overlap with the isotropic resonances. These sidebands consist of a series of peaks located at integral multiples of the spinner frequency on either side of the true (isotropic) resonance. There are ways to remove sidebands from spectra. The simplest is use a computer to multiply two spectra obtained at different frequencies (Barron & Wilson, 1981). More complex techniques are needed, however, for overlapping broad resonances (Dixon, 1981). Structural information can also be gleaned from sideband signal intensities (Burgar, 1984). It is stressed that these methods are not routine and it is desirable to obtain spectra free from sidebands. At present, this can only be done for ^{13}C nuclei with instruments operating at a C resonant frequency <30 MHz.

Schaefer and Stejskal (1976) were the first to realize the potential of applying magic angle spinning and proton decoupling techniques simultaneously to reduce linewidths in organic solids. In this way they were able to obtain high resolution spectra of organic polymers. A number of researchers have since pioneered the application of these techniques in organic geochemistry (Vander Hart & Retkofsky, 1976; Gerstein, 1979; Maciel et al., 1979; Hatcher et al., 1980; 1981; Wilson et al., 1981a, b, c; Zilm et al., 1981; Havens et al., 1982; Preston & Ripmeester, 1982; Packer et al., 1983).

10-7 CROSS POLARIZATION

Notwithstanding the resolution to be achieved by combining dipolar decoupling and magic angle spinning at sufficient speeds, it is still difficult to obtain a ^{13}C-NMR spectrum of a solid. In the NMR experiment the ^{13}C nuclei under investigation have a Boltzmann distribution among the energy levels produced by the application of the magnetic field. Applying a pulse of radiofrequency radiation alters the distribution of the nuclei among the energy states and the subsequent behavior (free induction decay) of the nuclei can be observed as they return to the Boltzmann distribution. Ideally this would be the end of the experiment. In practice, however, a spectrum can only be obtained by signal-averaging a large number of free induction decays, which must be collected and averaged to improve signal-to-noise ratio. Naturally, it is necessary to allow the system to come to equilibrium, That is, the C nuclei must be able to relax between each individual experiment ("transient" or "scan"). Normally to prevent saturation, it is wise to leave an interval of 5 times the spin lattice relaxation time constant, T_1, between each transient for the nuclei to relax. For pure solids the time lapse necessary between transients is tolerable for protons, but it is intolerable for ^{13}C since it can be of the order of minutes to days. Hence, ^{13}C data are often collected by an alternative process called *cross polarization* (CP) which greatly speeds up the accumulation of data.

This technique is based on energy transfer from protons to carbons (Pines et al., 1973). When a sample containing ^{1}H and ^{13}C nuclei is placed in a magnetic field, the energy levels of these nuclei are split by an amount that depends on the magnetic field and a constant characteristic of each nucleus (the gyromagnetic ratio). In the same magnetic field the energy levels of the protons and carbons, which are different because of the different gyromagnetic ratios, undergo magnetic resonance separately and are excited by different radiofrequencies (Fig. 10–13a). Normally, energy is not transferred between the protons and carbons. If, however, the energy of the two systems can be equalized, energy can be transferred from protons to carbons and vice versa (Fig. 10–13b).

Equalization is accomplished by creating a variable magnetic field on the protons with radiofrequency irradiation. The first step is to apply a 90° pulse to the protons. As noted in section 10–3, the phase of the radiofrequency irradiation can now be changed by 90° so that in the rotating frame the radiofrequency vector is colinear with the proton polarization induced by the initial 90° pulse; this is called *spin locking*. The radiofrequency magnetic field can be adjusted until the energy levels of the protons match those of the carbons (the Hartmann–Hahn condition). Now, any polarization of the proton population from the Boltzmann distribution can be transferred to C and vice versa. It is usual to polarize the C through protons, but they also relax back through the proton population. The important point is that the carbons now relax at the proton relaxation rate rather than the C relaxation rate. The system comes to equilibrium much faster between each individual transient and a spectrum of adequate signal-to-noise ratio can be obtained by averaging a number of free induction decays in a reasonable time.

It should be noted that it takes a finite time for carbons (C) to cross polarize from protons (H) (the cross polarization time T_{CH}). It will be recalled (section 10–3) that the protons relax in a spin locking experiment

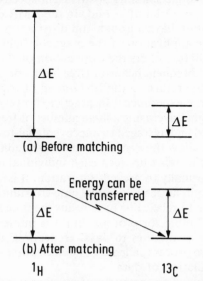

Fig. 10–13. Energy levels in a cross-polarization experiment. (a) Before matching, the energy levels of protons and carbons differ. (b) After matching (spin locking), the energy levels of protons and carbons are similar.

at a rate $T_{1\rho}H$, the spin lattice relaxation time in the rotating frame. It is necessary, therefore, that $T_{CH} \ll T_{1\rho}H$ or the carbons will not polarize and hence will not be observed. When the $T_{1\rho}H$ is long and the T_{CH} is also long it is necessary to allow the protons and carbons to be in contact for longer periods of time, so the carbons will be fully observed. In general, contact times of 1 to 5 ms are used.

The final step in obtaining a cross polarization spectrum involves terminating the ^{13}C pulse and observing the resultant free induction decay while the proton field is irradiated to ensure decoupling. After signal averaging, the free induction decays are Fourier transformed to give a frequency domain spectrum. In practice, to obtain high resolution solid state NMR spectra of soils, cross polarization is combined with magic angle spinning and high power decoupling. The symbols "CPMAS" are used to describe the whole process.

10-8 QUANTIFICATION

At this point it is useful to pause and summarize what has been discussed. NMR spectroscopy is a potentially usefully technique for the analysis of the chemical structure of soil organic matter, and particularly for the elements C, N, and P. Other nuclei bonded to organic matter could also be studied. The most important parameter for organic matter analysis is chemical shift since different structural groups in soil organic matter yield resonances at different chemical shifts. Different structural groups may also have different relaxation times, and could also be characterized in this way. However, relaxation is important for another reason, namely quantification.

It is the nature of the NMR experiment that a large number of signals have to be collected and averaged before an adequate signal is obtained. Although some nuclei may not be observed in the NMR experiment because there are physical limitations of the instrument in observing a nucleus which relaxes extremely fast, it is necessary before repeating each individual transient to ensure that slowly relaxing nuclei have relaxed properly. Because signals are particularly weak from C of organic matter in soils, large numbers of transients are collected. As a result there is a limitation on the time that can be left between each transient; this is about 10 s using current technology if adequate signal-to-noise ratios are to be obtained. Thus, cross polarization is often used on humic substances although there is now some evidence that C nuclei in humic substances and soils relax considerably faster than previously thought (M.A. Wilson, 1988 unpublished work).

Cross polarization, however, introduces new problems. It is necessary that the cross polarization time for each C under observation be shorter than the shortest proton spin lattice relaxation time in the rotating frame ($T_{1\rho}H$) if all the C is to be seen. Moreover, if the relative areas of various structural groups are to be measured directly from a spectrum it is necessary that the sample must have a uniform $T_{1\rho}H$. As shall be seen, both of these conditions are rarely met. The analyst must then consider (i) is the C that is observed representative of the sample as a whole? (ii) if not, is the error significant?, and (iii) what corrections can be made?

To obtain an estimate of the errors it is necessary to (i) study the relaxation properties of the material under investigation, and (ii) study model compounds so that general rules can be formulated. Model compounds are at present being researched extensively. The initial research suggests that carbons that are four or more carbon atoms removed from protons will not cross polarize properly in 1 ms (Alemany et al., 1983). Normally $T_{1\rho}H$ values are much longer than 10 s so that data are quantitative for compounds if a contact time of 5 ms is used. For most substances a contact time of 1 ms is adequate.

Paramagnetic species such as Fe^{3+}, however, are a problem. They cause rapid relaxation (Pfeffer et al., 1984), sometimes to the extent that nuclei cannot be observed even in a conventional noncross polarization experiment. Paramagnetic species are important components of soils, and hence, caution should be taken in interpreting data quantitatively. There has been much discussion of this problem in the literature on coal since paramagnetic species are also components of coals (Davidson, 1986; Wilson, 1987). There is almost general agreement that all C is not observed in a considerable number of samples. Although it is possible to reduce the paramagnetic content by reduction (Wilson, 1987), opinions on the magnitude of errors differ widely. The optimists (to which the author belongs) cite a considerable amount of cross-checked data, which suggest that quantitative data are obtained. Nevertheless, there are samples where paramagnetic species make data nonquantitative, and other samples where a quantitative result cannot be obtained by simply reading off the areas from a single spectrum. The situation for whole-soil NMR is not any simpler. Rather, because a wide range of type and amount of paramagnetic species can be present, more caution should be taken in spectral interpretation.

10-9 STRUCTURE OF ORGANIC MATTER IN SOILS

Quite a number of representations have been proposed to describe the chemical structure of soil organic substances. To be fair, these were not always meant to be accurate descriptions of chemical structure, only generalizations. NMR spectroscopy, however, has shown them to be extremely misleading, and thus they will not be discussed further here, other than to state that they all suggest that soil organic matter is highly aromatic. While this is still true for some soils, more often than not, this has been found not to be the case.

The work that led to these proposals was based on classical degradation techniques (such as oxidation and reduction), pyrolysis/gas chromatography, and functional group analysis. The first two of these methods suffer from the disadvantage that only a small part of the original material is identified, and in the reported works the identified components were mainly aromatic compounds (Kononova, 1966; Schnitzer & Khan, 1972; Schnitzer, 1978; Stevenson, 1982). Functional group analysis is useful, but tells us only about active functional groups, and there has always been doubt concerning the quantitative nature of the results.

It is possible to extract various components from soils using organic solvents but yields are low (<5%). Strong base extracts various amounts of organic C, but the actual yield can depend greatly on the soil under study. Material extracted is divided into two arbitrary fractions based on solubility in base and acid. The base- and acid-soluble fraction is termed the *fulvic acid fraction* [see chapter 2 (Malcom) in this book], and the base-soluble but acid-insoluble fraction is termed the *humic acid fraction*. Material that differs from that found in living organisms is described as humified, and called *humic substances*. This definition is difficult to use in practice, however, since many substances may be only slightly modified from those that are present in the plant. It is necessary, therefore, before an in-depth knowledge of the chemistry of humification can be achieved, to appreciate what types of compounds make up the humification "mix". Here, NMR spectroscopy is becoming invaluable, not just as a technique for use on solutions but also in the analysis of the structure of insoluble materials such as wood and litter fractions.

10-9.1 Wood, Cellulose, and Lignin

Wood is one of the most important precursors of soil humus [See chapter 3 (Stott & Martin) in this book], and in reducing environments is important in the formation of coal. Most of what we know about the chemical structure of wood, like soils, comes from destructive methods. This is especially true of lignin which is easily modified.

There have been several CPMAS ^{13}C-NMR studies of wood and its lignin components (Bartuska et al., 1980; Maciel et al., 1981; Kolodziejski et al., 1982; Haw et al., 1984). Typical spectra of lignins are shown in Fig. 10–14. Differences among spectra may arise because of different extraction efficiencies, because of inherent differences in lignin structure between species, or because of different chemical modifications of lignin during extraction. The resonances in lignin ^{13}C-NMR spectra are assigned as follows: The resonance at 149 ppm corresponds to C_3 and C_4 units in the basic phenyl propane structure of lignin (Fig. 10–15). However, a more detailed analysis than this is possible in solution studies of lignin and should also be possible in the solid state (Nimz et al., 1981; Leary et al., 1987). Other aromatic carbons of lignin resonate in the region 100 to 140 ppm. Different lignin fractions, which are defined by their extraction procedures, have similar spectra in the aromatic region to the original wood. However, Klason lignin, which is prepared by extraction with sulfuric acid, shows some loss of signal intensity around the 130-ppm region compared with wood or other lignins. Brauns native lignin which is prepared solely by extraction with ethanol contains considerably larger amounts of aliphatic material than the other lignins. The sharp resonance at 56.2 ppm in lignin species is assigned as aryl methoxyl C.

Cellulose is an important component of wood. Cellulose can be amorphous or crystalline, and has also been characterized by CPMAS ^{13}C-NMR (Atalla et al., 1980; Earl & Vanderhart, 1980; Maciel et al., 1981). Cellulose consists of long chains of glucose units each of which is combined by a β-glucoside link to the C_4 hydroxyl of another glucose unit (Fig. 10–16) as in the disaccharide cellobiose. Enzymatic hydrolysis of cellulose leads to cellobiose. The molecular weight of cellulose can vary, as can the number of glucose units present. It is not uncommon to find 2000 to 4000 glucose units joined together in cellulose.

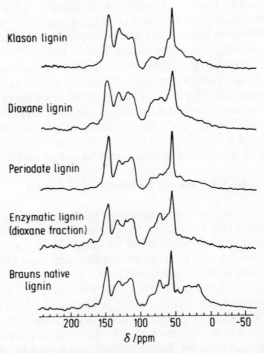

Fig. 10–14. CPMAS ^{13}C-NMR spectra of lignins prepared by different pretreatments of woods.

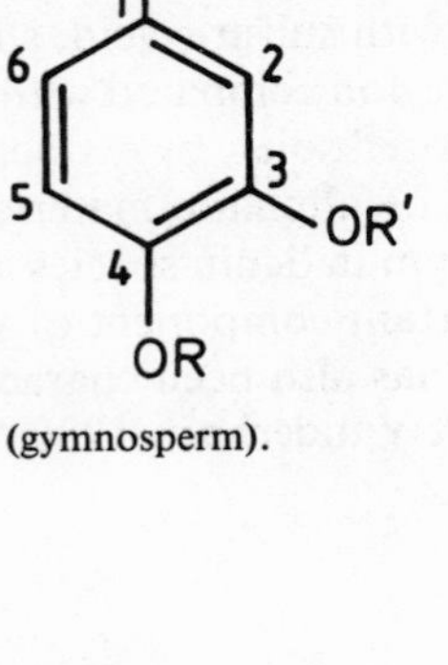

Fig. 10–15. Building block of lignin (gymnosperm).

There are other polysaccharides besides cellulose in the cell wall of plants. These are called hemicelluloses, but the name is unfortunate, as they are made up of pentose units such as xylose. In starch (Fig. 10-17) there are structurally different polysaccharides. Both consist entirely of glucose but the stereochemistry differs from that of cellulose. In the starch amylose the

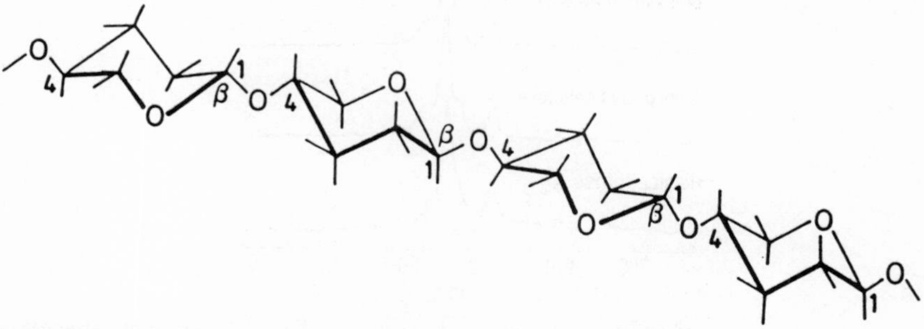

Fig. 10-16. Cellulose.

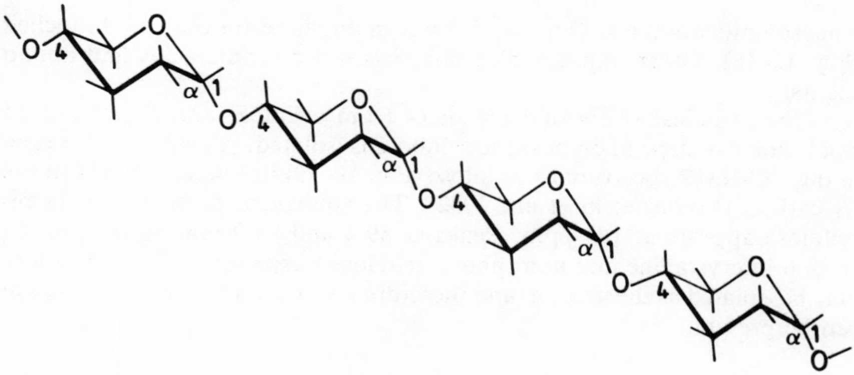

Fig. 10-17. Amylose starch.

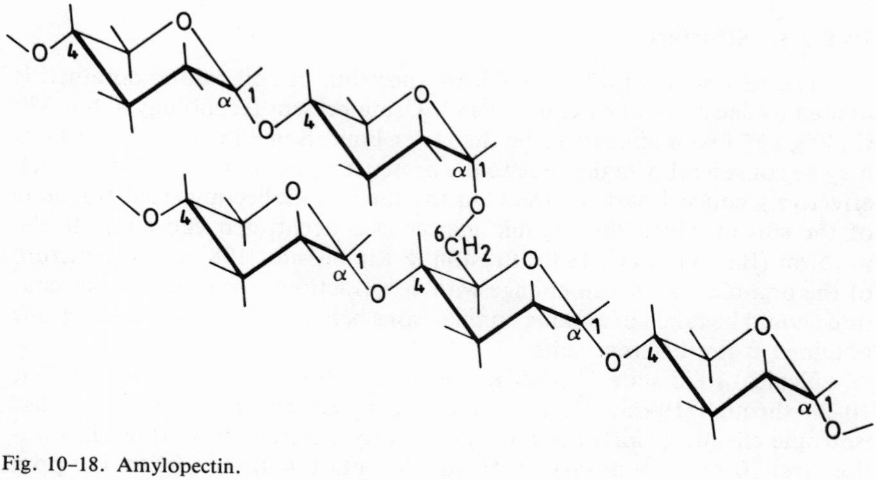

Fig. 10-18. Amylopectin.

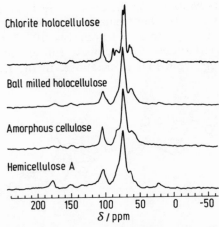

Fig. 10–19. CPMAS ^{13}C-NMR spectra of celluloses prepared by different pretreatments.

glucose units are linear (Fig. 10–17) but in amylopectin they are branched (Fig. 10–18). Other important carbohydrates are uronic acids and amino sugars.

The chemical shifts of resonances from cellulose and the degree of resolution can depend on how the cellulose is isolated. A peak at 105.5 ppm in the ^{13}C-NMR spectrum of cellulose (Fig. 10–19) has been assigned to the C_1 carbon (Kolodziejski et al., 1982). The equivalent resonance in hemicellulose appears at 103 ppm. Peaks at 89.4 and 83.9 ppm arise from C_4 carbon in crystalline and amorphous cellulose, respectively. Hemicellulose may be isolated as the acetate, and therefore resonances at 174 and 21.5 ppm can be present.

10–9.2 Soil Carbon

10–9.2.1 Structure

The ease with which a ^{13}C-NMR spectrum of soil can be obtained is limited by the amount of C in the soil. Using present technology about 3% C (30 g kg^{-1} soil) appears to be the lower limit. Sometimes the lower limit may be considerably higher due to the presence of paramagnetic ions, which effectively conceal parts of the C in the sample. Collecting sized fractions of the soil in which the organic matter is concentrated can alleviate the problem (Barron et al., 1980; Preston & Ripmeester, 1983). The structure of the organic matter can change with size fraction, however, so that caution should be taken in making comparisons between different size fractions obtained from different soils.

Typical solid state ^{13}C-NMR spectra of whole soils are shown in Fig. 10–20 through 10–23. ^{13}C-resonances assigned on the assumption that isotropic chemical shifts are not significantly different from those in solution are: 10 to 50 ppm (alkyl), 50 to 100 ppm (O-alkyl), 100 to 110 ppm

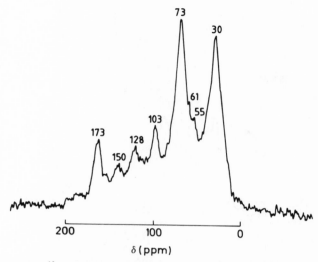

Fig. 10–20. CPMAS ^{13}C-NMR spectrum of Mapourika soil. The spectrum was obtained with a contact time of 1 ms and a recycle time of 0.3 s. 10^5 transients were collected.

(acetal), 110 to 160 ppm (alkene and aromatic), 160 to 200 ppm (carboxyl), 200 to 220 ppm (carbonyl).

The resonances at or around 72 to 75, 84 to 90, and 103 to 106 ppm are usually assigned to carbohydrates. There has been some debate concerning how carbohydrates might be incorporated into humic substances (Wilson et al., 1983b). Nevertheless, CPMAS ^{13}C-NMR studies on whole soils and residues after acid hydrolysis show that almost all of the O-alkyl and dioxygenated alkyl C is removed by hydrolysis (compare Fig. 10–23a and 10–23b). It is noteworthy that much of the COOH-C is also removed by acid hydrolysis. Some of this may be protein and thus forms water-soluble amino acids, but uronic acids also decarboxylate readily.

Typical areas of CPMAS ^{13}C-NMR spectra of soil, and corresponding assignments, are shown in Table 10–3. It is rare to find a soil with an aromaticity, f_a, (fraction of C that is aromatic; see chapter 2) > 50%.

Because the carbon content of soils is usually < 20% and frequently < 5%, large numbers (10^5) of transients must be collected in order to obtain adequate signal-to-noise ratios. Normally, advantages of signal enhancement and rapid relaxation mean that cross-polarization techniques are essential, and recycle times > 1 s are prohibitive. A range of soils have been examined using a contact time of 1 ms (Barron & Wilson, 1981; Wilson et al., 1981a, b; Preston & Ripmeester, 1982; Preston et al., 1984). In some experiments magic angle spinning has not been employed (Barron et al., 1980; Wilson et al., 1981c). There have been several studies of relaxation in soils (Wilson et al., 1983b; Preston et al., 1984; Gillam & Wilson, 1986). For a Maungatua soil sample $T_{1\rho}H$ has been measured directly through protons and through C (Wilson et al., 1983b; Gillam & Wilson, 1986) and shown to be

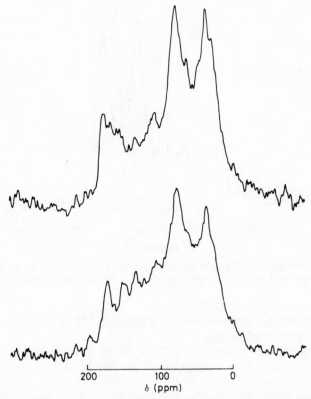

Fig. 10–21. CPMAS ^{13}C-NMR spectra of Franz X soil (upper) and Ikamatu soil (lower). Spectra were obtained with a contact time of 1 ms and a recycle time of 0.3 s. 10^5 transients were collected.

short (~ 10 ms through C and 2.5 ms through protons). The average $T_{1\rho}H$ as measured through C varies from 6 ms to 2.5 ms. For a New Zealand soil sample (Maungatua) there is more than one $T_{1\rho}H$ when measured through protons. The decay of signal intensity with increase in spin locking time can be differentiated into two components with $T_{1\rho}H$ values of 10 and 1.1 ms (Gillam & Wilson, 1986). The fast decaying component constitutes 60% of the signal and the slower decaying component 40% of the signal. It is noteworthy that the $T_{1\rho}H$ of the protons measured through C (4.4 ms) is almost exactly that of the weighted average of the two $T_{1\rho}H$ values measured through protons (4.66 ms).

Preston et al. (1984) studied the effects of contact time on a series of soils differing only in Cu content and degree of decomposition. Initial polarization was found to be rapid with maximum signal intensity reached at or before 0.5 ms contact time. $T_{1\rho}H$ was found to decrease with increase in Cu content. $T_{1\rho}H$ values of the carbohydrate resonance in the 50 to 110-ppm region appear to be less than that of the aryl resonance (Table 10–4). These

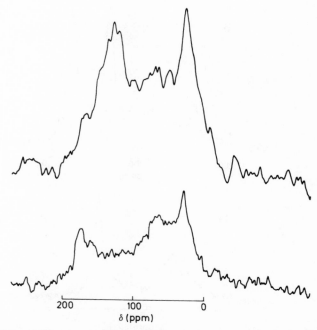

Fig. 10–22. CPMAS ^{13}C-NMR spectra of Hokitika soil (upper) and Tirau soil (lower). Spectra were obtained with a contact time of 1 ms and a recycle time of 0.3 s. 10^5 transients were collected.

results suggest a preferential localization of Cu in the carbohydrate portion of the organic matter since the Cu may assist in relaxing the carbohydrate structures in the soil.

It is clear that there is not spin diffusion between all protons (i.e. $T_{1\rho}H$'s are not all the same) in cross polarization experiments on some soils. Hence, the question arises whether or not the results obtained can be interpreted in a quantitative way. Clearly, it is necessary to extrapolate back the signal intensity from each structural group to an initial signal intensity at which spin lattice rotating frame relaxation has not occurred. Thus the initial carboxyl C signal intensity is given from Eq. [5] as:

$$I^0_{COOH} = \frac{I_{COOH}}{\exp - (\tau/T_{1\rho}H \, COOH)} \qquad [17]$$

and the fraction of signal from carboxyl, f_{COOH}, is given by

$$f_{COOH} = I^0_{COOH}/\sum_s I^0_s \qquad [18]$$

where I^0_s is the initial signal intensity from each structural group including carboxyl.

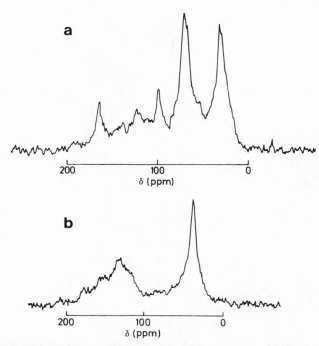

Fig. 10-23. CPMAS ^{13}C-NMR spectrum of (a) Okarito soil and (b) Okarito soil after hydrolysis with 6M HCl. Spectra were obtained with a contact time of 1 ms and a recycle time of
0.3 s. 10^5 scans were collected.

Even after this correction is made, there is no reason to believe that all
of the C in a soil is being observed by CPMAS ^{13}C-NMR. Indeed, it is clear
that the ease of obtaining a signal as measured by signal-to-noise ratio is
not proportional to C content (Wilson et al., 1983 a, b; Wilson et al., 1986,
1987; Vassallo et al., 1987). This problem can be investigated, however, by
spin-counting experiments in which known amounts of a simple organic compound are mixed with soil of known weight. Experiments of this sort have
been carried out on coals and humic acids and it is clear that all of the C
is not observed in a number of cases (Wilson, 1987).

Notwithstanding these limitations, relaxation experiments have the
potential for determining which carbons in the structure are close to protons. In 1979, Opella and Frey were able to show that if the spectrometer
decoupler is turned off (dephasing time) for a period before the free induction decay is collected, the signal intensity from CH or CH_2 carbon is lost
preferentially over that from quaternary or CH_3 carbon. This is because the
T_2' relaxation time of quaternary and CH_3 carbon is longer than that of CH_2
and CH under these conditions. If the spectrometer decoupler is turned off
for 40 μs or longer almost all the CH_2 and CH carbon is removed from the
spectrum, so that the technique (coined "dipolar dephasing") is useful for
identifying these structural groups in soils. Attempts have been made to estimate the fraction of aromatic C that is protonated in soils and humic acids

Table 10-3. Estimates of fractions of carbon types in soils (from Wilson et al., 1981a).

Soil	Alkyl 30	O-Alkyl 73	Acetal 103	Arly-H† 128	Arly-R‡,§ 150	f_a	COOR¶ 173	CO 198
				Assignment of chemical shift peak, ppm				
Mapourika	0.30	0.35	0.09	0.10	0.05	0.15	0.10	--
Ikamatu	0.26	0.33	0.11	0.11	0.08	0.19	0.08	0.03
Hokitika	0.27	↕0.30	↕0.37			0.37	0.06	--
Franz X	0.33	↕0.33	↕0.07	↕0.17		0.17	0.10	--
Tirau	0.30	↕0.37	↕0.23			0.23	0.10	--

† Some substituted carbon of aryl carboxylic acids or phenolic aryl carboxylic acids may occur in this region.
‡ R = 0.
§ Largely aromatic oxygenated carbon.
¶ R = H, metal ion, alkyl.

Table 10-4. Relaxation data for soils.

| Soil | $T_{1\rho}H$ | | | |
	COOH	Ar	O-Alkyl	Alkyl
		ms		
New Brunswick Terric Mesisol†	4.2	6.0	4.5	4.1
New Brunswick Terric Mesisol†	4.5	4.5	4.0	3.9
New Brunswick Terric Humisol†	2.1	4.2	3.5	4.0
New Brunswick Terric Humisol†	3.1	4.2	2.2	3.9
Maungatua humic loam	9.0	4.2	4.8	4.4

† Values estimated from plots in Preston et al. (1984).

by this method. It must be assumed that the relaxation rates of all CH carbons are the same. Likewise, all nonprotonated aromatic carbons must be assumed to have the same relaxation rate. The data are then fitted to a double exponential or similar equation and the contributions of signal from CH and nonprotonated C are evaluated (Wilson et al., 1983b). For humic substances and coals, the signals from protonated and nonprotonated C can be recognized (Wilson et al., 1983b; Wilson, 1987). The T_2 values for protonated aromatic C are short ($< 30 \ \mu s$) and for nonprotonated aromatic C they are long ($\geq 75 \ \mu s$).

The overall loss in signal intensity of a Maungatua soil sample with dephasing time is shown in Fig. 10-24. Clearly, loss of signal intensity of the aryl C can be described by a single exponential with $T'_{2B} = 339 \ \mu s$ (Wilson et al., 1983b). Thus, almost all of the observed aromatic C is nonprotonated. Likewise, the decay constant for the aliphatic C is short ($T'_{2C} = 28 \ \mu s$), which clearly shows that most of the aliphatic C is protonated.

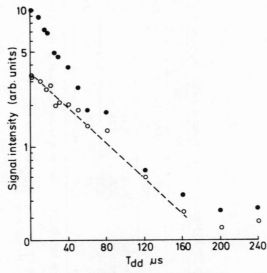

Fig. 10-24. Effect of varying dipolar dephasing time, T_{dd}, on signal intensity (• = total signal; o = aryl plus carboxyl signal).

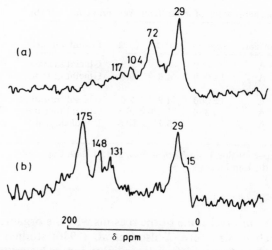

Fig. 10–25. Dipolar dephased spectra of Maungatua soil. (a) Spectrum of protonated carbon other than methyl. Difference spectrum between spectrum obtained in a conventional cross polarization experiment and that obtained at a dipolar dephasing time of 40 μs. (b) Spectrum obtained at a dipolar dephasing time of not greater than 40 μs.

Subspectra measured at 40 μs (Fig. 10–25b) and as the difference between spectra obtained at dephasing times of 0 and 40 μs are shown in Fig. 10–25a. It is clear that such spectra reveal that much of the C resonating at 29 ppm arises from polymethylene units. Likewise, a resonance at 15 ppm is present from methyl carbons at the end of alkyl chains.

In the aromatic region, nonprotonated resonances at 131 and 148 ppm are observed. The former is probably from C-substituted C including bridge-head C and the latter from O-aryl C. It is noteworthy that there is no evidence for methoxy C in these spectra and hence most of the O-aryl C is probably phenolic although one cannot rule out the presence of aryl ethers other than methoxy. Dipolar dephasing has also been used to produce two-dimensional NMR spectra of the same Maungatua soil sample (Wilson, 1984).

10–9.2.2 Geochemistry

Soils of different age (Table 10–5) or developed on contrasting parent material have quite different functional groups as revealed by CPMAS ^{13}C-NMR spectra. The variation almost certainly is not because of inherent effects of paramagnetic species on the amount of "invisible carbon". In the Reefton chronosequence from the South Island of New Zealand the oldest soil (14 000 yr) was found to be the most aliphatic, and the youngest Hokitika (300 yr) the most aromatic (Wilson et al., 1981c). Soils formed under cold and extremely wet conditions, which retard the breakdown and mineralization of newly formed organic matter, contain large amounts of carbohydrate (Wilson et al., 1981a). In contrast, the young Hokitika soil is almost devoid of carbohydrate C.

Table 10–5. Characteristics of soils (from Wilson et al., 1981b).†

Soil	Horizon	Age, yr	pH‡	C, %	Parent material	New Zealand classification
Hokitika	A	300	5.5	2.8	Glacial gravels	Recent
Ikamatu	A	14 000	4.6	9.8	Glacial gravels	Lowland yellow brown earth
Mapourika	A	12 000	4.2	7.9	Glacial moraines	Gley podzol
Franz X	A	350	5.2	6.6	Glacial moraines	Gley podzol
Tirau	A	--	5.7	10.5	Rhyolitic ash	Central yellow brown loam

† Reprinted by permission from *Nature*, Vol. 294, p. 649. Copyright © 1981, Macmillan Magazines Ltd., London.
‡ 1:25 with H_2O.

In order to unravel some of the reasons why the organic matter in soils varies so much in structure, Wilson et al. (1986) studied soils from the Antarctic. In these samples there is no lignin input. Nevertheless, aromatic C was found to be present showing that lignin is not a necessary precursor for aromatic components in soils. These soils also contained considerable amounts of carbohydrate C as measured by NMR and thus it would appear that under the extremely cold conditions in Antarctica, the transformation of fresh organic residues in humus is slow and incomplete. Hence low temperature is an important environmental factor affecting humus formation (Alexander, 1965; Dormaar, 1975; Duchaufour, 1976). Most of the cellulose is likely to have originated from the residual plant components that have resisted decomposition although it could also be derived from direct microbial synthesis and transformation.

In the warmer climates of New Zealand, metabolism of carbohydrates can still be fairly slow. Wilson et al. (1983a) have shown that the ratio of lignin to cellulose content of pine (*Pinus radiata*) and beech (*Nothafogus truncata*) leaves increases with duration of deposition on the forest floor. ^{13}C-NMR spectroscopic studies show also that the leaves lose carbohydrates. Dipolar dephasing studies, either by two-dimensional NMR or conventional CPMAS ^{13}C-NMR spectroscopy, clearly indicate that much of the dioxygenated C is nonprotonated (Wilson et al., 1983b; Wilson, 1984) due to the presence of tannins (Wilson & Hatcher, 1988). However, the most surprising result is how small the changes in the NMR spectra are. There is also a remarkable resemblance between the decayed leaf spectra and some of the spectra of whole soils.

High temperatures and abundant precipitation provide conditions for intensive microbiological activity leading to intensive mineralization of organic matter to simple compounds and its final product CO_2. Humic extracts from these soils have low aromaticities (Skjemstad et al., 1983). On the other hand, good drainage but relatively low microbiological activity can produce highly aromatic soils (M.A. Wilson, 1989, unpublished work). Thus, the nature and content of humus in soils is determined largely by the balance of two processes, the synthesis and the decomposition of humic substances expressed through the activity of soil microorganisms, which in turn is influenced by climate, including temperature and precipitation and vegetation.

The decomposition of ^{13}C-labelled compounds in soil can also be traced by NMR spectroscopy. Preston and Ripmeester (1983) have shown that ^{13}C-labelled acetate becomes incorporated in a number of functional groups in soils including carboxyl and carbohydrate.

10-10 SOIL PHOSPHORUS

Although present in low concentration in soil samples the observation of P by NMR spectroscopy is quite easy. Hence, there is a vast literature concerning solution spectra of P compounds. There are some difficulties since a suitable internal standard for referencing spectra is not available. Normally H_3PO_4 is used, but the resonance from this substance is broad, and hence chemical shifts can be in error by several parts per million. A book has been published on ^{31}P-NMR spectroscopy (Crutchfield et al., 1967).

^{31}P chemical shifts cover a range of about 700 ppm (Fig. 10–11) and are highly dependent on oxidation state. P^{5+} compounds give resonances between -50 and 100 ppm, although P^{3+} chemical shifts can extend from -460 to $+250$ ppm. Coupling is found with protons and can extend from 200 to 1100 Hz depending on the oxidation state of the P and the electronegativity of substituents on the P. There is little relevant work on relaxation in P compounds, although in solution T_1 can be quite long. Values as long as 55 s have been reported for 2-membered cyclic metaphosphates but this decreases when the size of the ring increases. Little relevant solid state work has been reported on compounds of geochemical interest although there is a fairly rapidly growing literature on phospholipid chemistry.

^{31}P-NMR spectroscopy can distinguish a range of inorganic and organic species in extracts from soils including inositol phosphates (Fig. 10–26), orthophosphate monoesters, and polyphosphates (Table 10–6). Newman and Tate (1980) were also able to detect previously unreported phosphates in soils (Fig. 10–27). They also measured T_1 values of the P resonances in soil extracts. The T_1 values were found to be short (<4 s). The amount of orthophosphate diesters measured in extracts from a climosequence of soils showed a positive correlation with precipitation (Tate & Newman, 1981). Williams et al. (1981) have obtained solid-state ^{31}P spectra of a soil and

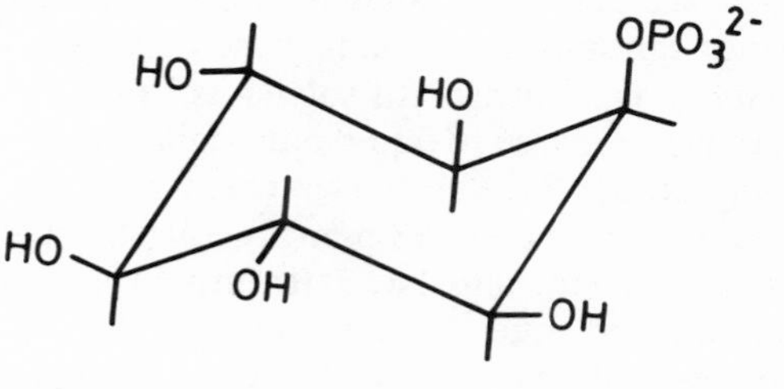

Fig. 10–26. Inositol phosphate structure.

Table 10-6. Types of compounds expected or observed by ^{31}P-NMR in soils or soil extracts.

Compound	Chemical shift†
Hydroxyapatite $Ca_5(PO_4)_3OH$	38.2‡
Crandallite $CaAl_3(PO_4)_2(OH)_2H_2O$	33.8‡
Alkylphosphonic esters $RCH_2PO_3R'R'$	19.8
Alkylphosphonic acid $RCH_2PO_3^{2-}$	18.3
Orthophosphate PO_4^{3-}	5.3
Choline phosphate $(CH_3)_3N(CH_2)_2PO_4^{2-}$	3.56
Orthophosphate monoesters $ROPO_3^{2-}$	3.5–5.0
Inositol phosphates	5.26, 4.37, 4.02, 3.85
Orthophosphate diester $(RO)(RO')PO_2^-$	−0.8
Pyrophosphate $P_2O_7^{4-}$	−5.5
Polyphosphate, trimetaphosphate $P_3O_9^{3-}$	−21.4

† From 85% H_3PO_4.
‡ Corrected to 85% H_3PO_4, assuming the isotopic chemical shift of $AlPO_4$ in solution is the same in the solid state.

various P containing minerals including hydroxyapatite $Ca_5(PO_4)_3OH$, aluminium phosphate ($AlPO_4$), and crandallite $CaAl_3(PO_4)_2(OH)_2(H_2O)$. Although spectral interpretation is hindered by spinning side bands because spinning is slow compared to the shift anisotropy, the isotropic chemical shifts of the minerals were found to vary and differed from that of the soil.

10-11   SOIL NITROGEN

Nitrogen possesses two isotopes that can be used for NMR spectroscopy. ^{14}N is 99.6% abundant but possesses a quadrupole moment ($I = 1$). Although easily observed, lines are broad with linewidths sometimes as great as 1000 Hz (e.g., $\sim$150 ppm). Hence, little resolution is observed. ^{15}N has a spin quantum number I of 1/2, but is only 0.37% abundant and is thus difficult to observe. Nevertheless, it is possible with high field instruments to obtain spectra of naturally occurring ^{15}N. For complex macromolecules, however, studies at present are limited to enriched samples. A book has been published on ^{15}N-NMR spectroscopy (Levy & Lichter, 1979).

^{15}N and ^{14}N chemical shifts are very similar, and hence it is possible to use data from either type of measurement. Chemical shifts range over about 900 ppm but in solution measurements a range of internal standards has been used, some of which are solvent dependent. Solvent effects in ^{15}N-NMR spectroscopy vary substantially more than those of C. For $CH_3\,^{15}NO_2$ the chemical shift difference as a function of solvent is 7 ppm. Typical chemical shifts for N compounds referenced to liquid ammonia are shown in Fig. 10–8. Other internal standards that have been used extensively, and referenced to ammonia, are CH_3NO_2 (neat) at 380.23 ppm, NH_4^+ (saturated aqueous ammonium nitrate) at 20.68 ppm and NO_3^- (saturated aqueous ammonium nitrate) at 376.25 ppm. Solid state ^{15}N spectra are readily obtained, and can be free of sidebands because chemical shift anisotropy is small (Schaefer et al., 1981).

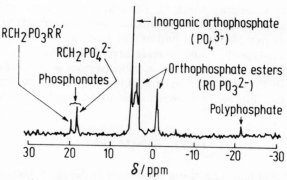

Fig. 10–27. Solution ^{31}P-NMR spectrum of an extract from McKerrow soil (Newman & Tate, 1980). Reprinted from *Communications in Soil Science and Plant Analysis*, Vol. 11, p. 835–842, by courtesy of Marcel Dekker, Inc., New York.

Preston et al. (1982) have used ^{15}N solution NMR spectroscopy to study the structure of synthetic humic acids prepared from *p*-benzoquinone and ammonium chloride. They observed peaks at 30, 55, 82, 116, 132, 148, and 153 ppm. Bearing in mind that the chemical shift of ammonium is solvent and counter ion dependent, the resonance at 30 ppm can be assigned to ammonium. The peak at 55 ppm was assigned to aromatic amines like aniline. The other peaks were not assigned. In a solid-state study (Benzing-Purdie et al., 1983), ^{15}N-labelled melanoidins were synthesized and the NMR spectra obtained. A soil was also incubated with $Na^{15}NO_3$. Both the melanoidins and soil spectra showed a prominent peak at 97 ppm, which they claim as

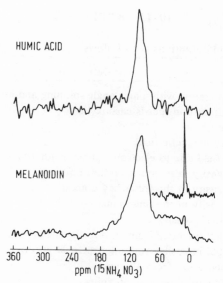

Fig. 10–28. CPMAS ^{15}N-NMR spectra of ^{15}N-labelled materials. The internal standard $^{15}NH_4$, NO_3 at 0 ppm is also shown.

characteristic of secondary amide N, although tertiary amides also resonate in this region (Fig. 10–28). Small resonances were also observed at 155 ppm and assigned as pyrrole-type N and at 128 ppm attributed to a secondary amide or pyrrole-type N. Some increased resolution was obtained by varying the cross-polarization time. They ascribe the poor acid hydrolysis properties of N in soils to the formation of secondary amides, and draw attention to the similarities in spectra between melanoidins and humic substances.

10–12 SUMMARY

It is clear that high resolution solid-state NMR spectroscopy is a useful new technique for the analysis of organic matter in soils. Not only is it useful for analysis of C types in soils but there are a number of applications for analysis of other elements as well. Two major problems exist.

1. The presence of paramagnetic materials severely limits the applications of the technique, but it is often possible to remove these materials by reduction (Wilson, 1987).
2. The low concentration of C in some soils is also limiting, but higher field magnets will overcome this limitation provided sufficient spinning speeds can be achieved in order not to introduce a different set of problems.

It is also clear that researchers must understand the problems of nuclear relaxation so as not to misinterpret their own data or those of other workers.

10–13 APPENDIX

A list of symbols and definitions is as follows:

Symbol	Definition
CPMAS	Cross polarization with magic angle spinning and high power decoupling
f_{COOH}	Fraction of carbon that is carboxylic
h	Plank's constant
H_o	Laboratory magnetic field
H_1	Magnetic field due to electromagnetic irradiation
I	Signal intensity; also spin quantum number
$I(f)$	Signal intensity in the frequency domain
$I(t)$	Signal intensity in the time domain
I^0	Initial signal intensity
I^0_{COOH}	Initial signal intensity from carboxylic carbon.
I^0_s	Initial signal intensity from all structural groups
M	Net magnetisation from all spins
M^0	Initial magnetisation from all spins

(continued on next page)

Appendix continued.

Symbol	Definition
T	Relaxation time
T_{CH}	Cross polarization time constant
T_1	Spin lattice relaxation time
$T_{1\rho}H$	Spin lattice relaxation time in the rotating frame of protons
T_2	Spin-spin relaxation time
T_2^*	Observed spin-spin relaxation time measured from linewidths
T_2'	Spin-spin relaxation time measured by dipolar dephasing
T_{2A}'	Spin-spin relaxation time of protonated aromatic carbons
T_{2B}'	Spin-spin relaxation time of nonprotonated aromatic carbons
T_{2C}'	Spin-spin relaxation time of aliphatic carbons
t	Time period
x, y, z	Laboratory frame coordinates
$x', y', z',$	Rotating frame coordinates
γ	Gyromagnetic ratio
δ	Chemical shift
θ	Angle between H_o and internuclear vector
ϕ	Polar angle
μ	Magnetic moment
ν	Frequency
ν_r	Reference frequency
ν_s	Sample frequency
σ	Shielding factor
σ_r	Shielding factor for reference
σ_s	Shielding factor for sample
$\sigma_{11}, \sigma_{22}, \sigma_{33}$	Chemical shift tensors
σ_{iso}	Isotropic shielding tensor
π	Pi
τ	Time period
ω_o	Precessional speed

REFERENCES

Alemany, L.B., D.M. Grant, R.J. Pugmire, T.D. Alger, and K.W. Zilm. 1983. Cross polarization and magic angle spinning NMR spectra of model organic compounds. II. Molecules of low or remote protonation. J. Am. Chem. Soc. 105:2142–2147.

Alexander, M. 1965. Biodegradation: Problems of molecular recalcitration and microbial fallibility. Adv. Appl. Microbiol. 7:35–76.

Andrew, E.R. 1971. The narrowing of NMR spectra of solids by high speed specimen rotation and the resolution of chemical shift and spin multiplet structure for solids. Prog. Nucl. Magn. Reson. Spectros. 8:1–39.

Attalla, R.H., J.C. Gast, D.W. Sindorf, V.J. Bartuska, and G.E. Maciel. 1980. ^{13}C NMR spectra of cellulose polymorphs. J. Am. Chem. Soc. 102:3249–3251.

Barron, P.F., and M.A. Wilson. 1981. Humic soil and coal structure study with magic angle spinning ^{13}C-CP-NMR. Nature (London) 289:275–276.

Barron, P.F., M.A. Wilson, J.F. Stephens, B.A. Cornell, and K.R. Tate. 1980. Cross polarization ^{13}C-n.m.r. spectroscopy of whole soils. Nature (London) 286:585–586.

Bartuska, V.J., G.E. Maciel, H.I. Bolker, and B.I. Fleming. 1980. Structural studies of lignin isolation procedures by [13]C N.M.R. Holzforschung. 34:214–217.

Becker, E.D. 1980. High resolution NMR. Theory and chemical applications. 2nd ed. Academic Press, New York.

Benzing-Purdie, L., J.A. Ripmeester, and C.M. Preston. 1983. Elucidation of the nitrogen forms in melanoidins and humic acid by N-15 cross polarization-magic angle spinning nuclear magnetic resonance. J. Agric. Food Chem. 31:913–915.

Burgar, M.I. 1984. New approach to the analysis of the CP/MAS NMR spectra of coal. Fuel 63:1621–1623.

Crutchfield, M.M., C.H. Dungan, J.H. Letcher, V. Mark, and J.R. Van Wazer. 1967. Topics in phosphorus chemistry. Vol. 5. M. Grayson and F.J. Griffith (ed.) Interscience Publ., New York.

Davidson, R.M. 1986. Nuclear magnetic resonance studies of coal. In International Energy Authority Review. Rep. ICTIS/IEAR. London.

Dixon, W.T. 1981. Spinning sideband free NMR spectra. J. Mag. Reson. 44:220–223.

Dormaar, J.F. 1975. Susceptibility of organic matter of chernozemic A_h horizons to biological decoposition. Can. J. Soil Sci. 55:473–480.

Duchaufour, P. 1976. Dynamics of organic matter in soils of temperate regions: Its actions on pedogenesis. Geoderma 15:31–40.

Earl, W.L., and D.L. Vander Hart. 1980. High resolution magic angle sample spinning ^{13}C NMR of solid cellulose. I.J. Am. Chem. Soc. 102:3251–3252.

Fukushima, E., and S.B.W. Roeder. 1981. Experimental pulse NMR—A nuts and bolts approach. Addison Wesley Publ. Co., Reading MA.

Fyfe, C.A. 1983. Solid state NMR for chemists. CFG Press, Guelph, Ontario, Canada.

Gerstein, B.C. 1979. Fingerprinting solid coals using pulse and multiple pulse nuclear magnetic resonance. p. 425–444. In C. Kass (ed.) Analytical methods for coal and coal products. Vol. III. Academic Press, New York.

Gillam, A.J., and M.A. Wilson. 1986. Structural analysis of aquatic substances by NMR spectroscopy. p. 128–141. In M. Soon (ed.) Organic marine geochemistry. ASC Symp. Ser. no. 305. Am. Chem. Soc., Washington, DC.

Harris, R.K. 1978. Introduction. p. 1–19. In R.K. Harris, and B.E. Mann (ed.) NMR and the periodic table. Academic Press, New York.

Hatcher, P.G., G.E. Maciel, and L.W. Dennis. 1981. Aliphatic structures of humic acids. A clue to their origin. Org. Geochem. 3:43–48.

Hatcher, P.G., R. Rowan, and M.A. Mattingly. 1980. H-1 and C-13 NMR of marine humic acids. Org. Geochem. 2:77–85.

Havens, J.R., J.L. Koenig, and P.C. Painter. 1982. Chemical characterization of solid coal through magic angle spinning ^{13}C NMR. Fuel 61:393–396.

Haw, J.F., G.E. Maciel, and H.A. Schroeder. 1984. Carbon-13 nuclear magnetic resonance spectrometric study of wood and wood pulping with cross polarization and magic angle spinning. Anal. Chem. 56:1323–1329.

Kolodziejski, W., J.S. Frye, and G.E. Maciel. 1982. Carbon-13 nuclear magnetic resonance spectrometry with cross polarization and magic angle spinning for analysis of lodgepole pine woods. Anal. Chem. 54:1419–1424.

Kononova, M.M. 1966. Soil organic matter. Pergamon Press, Oxford.

Leary, G.J., K.R. Morgan, and R.H. Newman. 1987. Solid state carbon-13 nuclear magnetic resonance study of Pinus radiata wood. Appita 40:181–184.

Levy, G.C., and R.L. Lichter. 1979. Nitrogen-15 nuclear magnetic resonance. John Wiley & Sons, New York.

Lowe, I.J. 1959. Free induction decay of rotating solids. Phys. Rev. Lett. 2:285–287.

Maciel, G.E., V.J. Bartuska, and F.P. Miknis. 1979. Characterization of solid coals through magic angle ^{13}C nuclear magnetic resonance. Fuel 58:391–394.

Maciel, G.E., D.J. O'Donnell, J.J.H. Ackerman, B.H. Hawkins, and V.J. Bartuska. 1981. A ^{13}C NMR study of four lignins in the solid and solution states. Makromol. Chem. 182:2297–2304.

Nimz, H.H., D. Robert, O. Faix, and M. Nemr. 1981. Carbon-13 NMR spectra of lignins. 8. Structural differences between lignins of hardwoods, softwoods, grasses and compression wood. Holzforschung 35:16–26.

Newman, R.H., and K.R. Tate. 1980. Soil phosphorus characterization by ^{31}P nuclear magnetic resonance. Commun. Soil Sci. Plant Anal. 11:835–842.

Opella, S.J., and M.H. Frey. 1979. Selection of non-protonated carbon resonances in solid state NMR. J. Am. Chem. Soc. 101:5854–5856.

Packer, R.J., R.K. Harris, A.M. Kenwright, and C.E. Snape. 1983. Quantitative aspects of solid state ^{13}C n.m.r. coals and related materials. Fuel 62:999–1002.

Pfeffer, P.E., W.V. Gerasimowicz, and E.G. Piotrowski. 1984. Effect of paramagnetic iron on quantitation in carbon-13 cross polarization magic angle spinning nuclear magnetic resonance spectrometry of heterogeneous environmental matrices. Anal. Chem. 56:734–741.

Pines, A., M.G. Gibby, and J.S. Waugh. 1973. Proton enhanced NMR of dilute spins in solids. J. Chem. Phys. 59:569–590.

Preston, C.M., R.L. Dudley, C.A. Fyfe, and S.P. Mathur. 1984. Effects of variations in contact times and copper contents in a ^{13}C CP MAS study of samples of four organic soils. Geoderma 33:245–253.

Preston, C.M., B.S. Rauthan, C. Rodger, and J.A. Ripmeester. 1982. A hydrogen-1, carbon-13 and nitrogen-15 nuclear magnetic resonance study of p-benzoquinone polymers incorporating aminonitrogen compounds ("synthetic humic acids"). Soil Sci. 134:277–293.

Preston, C.M., and J.A. Ripmeester. 1982. Application of solution and solid state ^{13}C NMR to four organic soils, their humic acids, fulvic acids, humins and hydrolysis residues. Can. J. Spectros. 27:99–105.

Preston, C.M., and J.A. Ripmeester. 1983. ^{13}C-labelling for NMR studies of soils: CP MAS NMR observation of ^{13}C acetate transformation in a mineral soil. Can. J. Soil Sci. 63:495–500.

Schaefer, J., and E.O. Stejskal. 1976. Carbon-13 nuclear magnetic resonance of polymers spinning at the magic angle. J. Am. Chem. Soc. 98:1031–1032.

Schaefer, J., E.O. Stejskal, G.S. Jacob, and R.A. McKay. 1981. Natural abundance N-15 NMR of the solids from reaction of HCN and ammonia. Appl. Spectros. 36:179–182.

Schnitzer, M. 1978. Humic substances: Chemistry and reactions. p. 1–58. In M. Schnitzer and S.U. Khan. Soil organic matter. Elsevier, New York.

Schnitzer, M., and S.U. Khan. 1972. Humic substances in the environment. Marcel Dekker, New York.

Shaw, D. 1984. Fourier transform NMR spectroscopy. 2nd ed. Elsevier, Amsterdam.

Skjemstad, J.O., R.L. Frost, and P.F. Barron. 1983. Structural units in humic acids from South-Eastern Queensland soils as detected by ^{13}C NMR spectroscopy. Aust. J. Soil Res. 21:539–547.

Stevenson, F.J. 1982. Humus chemistry: Genesis, composition, reactions, John Wiley & Sons, New York.

Tate, K.R., and R.H. Newman. 1981. Phosphorus fractions of a climosequence of soils in New Zealand Tussock grassland. Soil Biol. Biochem. 14:191–196.

Vander Hart, D.L., and H.L. Retkofsky. 1976. Estimation of coal aromaticities by proton decoupled carbon-13 magnetic resonance spectra of whole coals. Fuel 55:202–206.

Vassallo, A.M., M.A. Wilson, P.J. Collin, J.M. Oades, and A.G. Waters. 1987. Structural analysis of geochemical samples by solid state nuclear magnetic resonance spectrometry. Role of paramagnetic material. Anal. Chem. 59:558–562.

Williams, R.J.P., R.G.F. Giles, and A.M. Posner. 1981. Solid state phosphorus NMR spectroscopy of minerals and soils, J. Chem. Soc., Chem. Commun. 1051.

Wilson, M.A. 1981. Applications of nuclear magnetic resonance spectroscopy to the study of the structure of soil organic matter. J. Soil Sci. 32:167–186.

Wilson, M.A. 1984. Soil organic matter maps by NMR. J. Soil Sci. 35:209–215.

Wilson, M.A. 1987. NMR techniques and applications in geochemistry and soil chemistry. Pergamon Press, Oxford.

Wilson, M.A., P.F. Barron, and K.M. Goh. 1981a. Differences in structure of organic matter in two soils as demonstrated by ^{13}C cross polarization nuclear magnetic resonance spectroscopy with magic angle spinning. Geoderma 26:323–327.

Wilson, M.A., P.R. Barron, and K.M. Goh. 1981c. Cross polarization ^{13}C-N.M.R. spectroscopy of some genetically related New Zealand soils. J. Soil Sci. 32:419–425.

Wilson, M.A., K.M. Goh, P.J. Collin, and L.G. Greenfield. 1986. Origins of humus variation. Org. Geochem. 9:225–231.

Wilson, M.A., and P.G. Hatcher. 1988. Detection of tannins in modern and fossil barks and in plant residues by high resolution solid state ^{13}C n.m.r. Org. Geochem. 12:539–546.

Wilson, M.A., S. Heng, K.M. Goh, R.J. Pugmire, and D.M. Grant. 1983a. Studies of litter and acid insoluble organic matter fractions using ^{13}C-CP NMR with magic angle spinning. J. Soil Sci. 34:83–97.

Wilson, M.A., M. Perdue, A.M. Vassallo, and J.H. Reuter. 1987. A compositional and solid
state nuclear magnetic resonance study of humic and fulvic acid fractions of soil organic
matter. Anal. Chem. 59:551–558.

Wilson, M.A., R.J. Pugmire, and D.M. Grant. 1983b. Nuclear magnetic resonance spectroscopy
of soils and related materials. Relaxation of ^{13}C nuclei in cross polarization nuclear magnetic resonance experiments. Org. Geochem. 5:121–129.

Wilson, M.A., R.J. Pugmire, K.M. Zilm, K.M. Goh, S. Heng, and D.M. Grant. 1981b. Cross
polarization ^{13}C NMR spectroscopy characterizes organic matter in whole soils. Nature
(London) 294:648–650.

Zilm, K.W., R.J. Pugmire, S.R. Larter, J. Allan, and D.M. Grant. 1981. Carbon-13 CP/MAS
spectroscopy of coal macerals. Fuel 60:717–723.

Chapter 11

Humic Substances in Soil and Crop Sciences: An Overview

P. MacCARTHY, *Colorado School of Mines, Golden, Colorado*

P. R. BLOOM, *University of Minnesota, St. Paul, Minnesota*

C. E. CLAPP, *Agricultural Research Service, USDA,
University of Minnesota, St. Paul, Minnesota*

R. L. MALCOLM, *U.S. Geological Survey, Denver, Colorado*

ABSTRACT

The roles of humic substances in soil and crop sciences, discussed in detail in the preceding chapters of this book, are summarized here. The topics addressed include: the extraction, composition, characterization, and formation of soil humic substances. Particular attention is focused on the application of ^{13}C-NMR spectroscopy to these materials. Other subjects addressed are the stabilization of soil organic matter through humification, sorption of nonionic compounds by soil organic matter, and sewage sludge humic materials. The influences of humic substances on soil fertility and plant growth are also discussed. The chapter concludes with an ecological view of humic substances that provides a rationale for their highly intractable nature.

The chapters in this book cover a wide range of topics related to humic substances in soils. The general topics discussed include (i) composition, (ii) formation, and (iii) functions of humic substances as related to crop production.

11-1 COMPOSITION OF HUMIC SUBSTANCES

The structures of humic substances are not known. That is, these materials cannot be represented by discrete molecular or structural formulas. Even the term *structure* itself is difficult to define in the context of humic substances. These materials are comprised of complex mixtures, which to date, have defied all attempts at separation into even reasonably pure fractions. Because of this heterogeneity, chemical and physical probes into the structure of humic materials generally yield data that are difficult to interpret at best, and perhaps impossible to decipher in many cases. Limits on the interpretability of data obtained from complex mixtures have recently been discussed (MacCarthy & Rice, 1985). A recent treatise provides a thorough and critical review of what is known about the chemistry and structure of humic substances, and the reader is referred to that text for a comprehensive coverage of this subject (Hayes et al., 1989). Even though the structures of humic substances are not known (and the question of the existence of an ultimate structure, as such, may in fact be moot), much is known about the major structural components and functional groups that determine many of the properties of humic substances. In other words, extensive information on the *composition* of humic substances is available; but it appears that the structural components in humic substances are linked together in a random manner resulting in a material of extraordinary complexity and heterogeneity. Consequently, the term *structure* has an elusive meaning in the context of humic substances.

In this book some of the more recent physical methods that have added to our knowledge of the chemical composition of humic substances are discussed. Carbon-13 nuclear magnetic resonance (^{13}C-NMR) spectroscopy, in particular, stands out in terms of the contribution it has made to our understanding of humic substances. It has provided more refined information on the composition of humic substances than any other single technique. ^{13}C-NMR spectroscopy can be used to study aqueous solutions of humic substances as described in chapters 2 and 4. Solid-state NMR spectroscopy, using the techniques of cross polarization and magic angle spinning (CPMAS), has been applied to extracted humic substances, as well as to whole soils (chapters 2, 4, and 10). The detailed discussion of the theory and practice of solid-state NMR spectroscopy in chapter 10 indicates the difficulty of obtaining solid-state spectra, which can be used for estimating relative quantities of C atoms in different molecular environments. For example, C atoms which are four carbons away from a proton cannot be detected by CPMAS NMR spectroscopy, and paramagnetic ions, such as Fe^{3+} and Cu^{2+}, can decrease signal intensity for nearby C atoms. Paramagnetic ions, which may be a problem for whole-soil samples, are not likely to pose a problem for spectra of extracted and purified humic substances. With proper attention to detail, the areas under spectral bands are essentially proportional to the number of C atoms in the corresponding chemical environment. This quantitative aspect is one of the features that makes NMR spectroscopy particularly useful in the study of humic substances.

Perhaps the most significant result from the ^{13}C-NMR investigations is that, contrary to previous opinions based on data from chemical degradation studies, humic substances rarely contain >50% aromatic C (chapters 2 and 10). Unfortunately, in the literature of this subject, and as indicated in chapter 2, the fraction of aromatic C, fa, has been reported in two different ways. The aromatic content is expressed either as a fraction of the total C (fa_1) or as a fraction of the total C less the C in carboxyl, ester, and ketone groups (fa_2). The data in chapter 2 suggest that fa_2 is larger than fa_1 by a factor of 1.2 to 1.3; however, both methods of calculation give aromaticity values of <50%. The author of chapter 2 clearly favors the use of fa_1, while the author of chapter 4 favors the use of fa_2.

An example of the special utility of NMR spectroscopy in the study of humic substances is its ability to clearly identify polysaccharide residues in these materials. Data presented in chapter 2 show that traditionally prepared soil fulvic acid fractions contain very significant quantities of polysaccharide C that can be readily removed by adsorption of fulvic acid on an XAD-8 resin column; in this procedure the polysaccharide molecules are not adsorbed, and they pass through the column. As a result of these considerations it is important to distinguish between *fulvic acid* and the *fulvic acid fraction* which contains the fulvic acid in addition to identifiable constituents (see chapter 2). Such distinctions were not generally appreciated prior to the development of resin technology for separating the fulvic acid from the polysaccharides. Thus, many of the samples referred to as "fulvic acid" in the literature correspond to what would be classified as "fulvic acid fractions" today. Nevertheless, many workers still do not make this distinction in reporting their contemporary results. Humic and fulvic acids also contain polysaccharide residues which are covalently bonded into the structure. These residues, however, may be partially degraded saccharide units (chapter 4). Data presented in chapter 9 show that neutral sugars comprise 10 to 20% of "fulvic acids" extracted from sewage sludges.

Soil humic acids prepared using similar extraction techniques are similar in composition to each other. Fulvic acid fractions, however, vary with source; but, the data in chapter 2 suggest that fulvic acids per se, after separation from the saccharides in the fulvic acid fraction, may actually be quite similar to each other. However, more data are needed to further evaluate these generalizations. It does appear that the variation in composition of humic acids from soils and streams is less than that of fulvic acids from different soils. Soil humic substances have a higher N content than those from streams. Soil fulvic acids contain older C than aquatic fulvic acids, which are generally of recent origin (see chapter 2).

NMR spectroscopy has also been used to study P and N in humic substances (chapter 10). Phosphorus-31 NMR spectroscopy can distinguish a range of inorganic and organic P in soil extracts, including inositol phosphates, orthophosphate monoesters, orthophosphate diesters, and polyphosphates. The technique has not yet been applied to the study of P in extracted humic or fulvic acids.

Nitrogen-15 NMR spectroscopy is more difficult than ^{13}C-NMR spectroscopy because of the low natural abundance of ^{15}N. This technique has been applied to the study of synthetic polymers, labeled fungal melanins, and short-term (7 mo) labeled soil organic matter. In the studies of melanins and soil, secondary (and possibly also tertiary) amines were found as well as some evidence for the presence of pyrrole-type N. This is a potentially promising area for future research.

Pyrolysis mass spectrometry, discussed in chapters 4 and 8, provides structural data for humic substances that are quite different from those provided by NMR spectroscopy. From the observed fragmentation patterns, the analyst can learn more about the nature of the structural components observed by NMR spectroscopy. The mass spectrum of a pyrolyzed humic acid shows an abundance of signals due to polysaccharides, as well as pyridine, methylpyridine, and indole units (chapter 4). One of the problems with pyrolysis techniques (as with some chemical degradative methods), however, is the difficulty of distinguishing between products that may have been an integral part of the humic structure and those that are merely "artifacts" of the degradative process. Even such artifacts may be of value in structural studies if sufficient information is available to allow intelligent speculation on possible mechanistic pathways by which precursor molecules could have led to the formation of the observed products. The contribution of lignin to the aromatic character of humic acids is indicated by signals from lignin monomers and dimers in the pyrolysis–mass spectra. The mass spectral data, however, are selective for the more volatile components, e.g., carbohydrates and N-containing constituents. For example, in one study only 31% of humic acid was volatilized. This may account for the low intensity of the signal for lignin fragments in humic acids (chapters 2 and 3). With whole soils, however, the lignin signals were prominent (chapter 8).

Electron spin resonance (ESR) spectrometry has been used to study the free radicals that are generally attributed to quinone structures present in low concentration in humic substances (chapter 4). These structures are considered to be important in the redox behavior of humic materials, but definitive evidence for this conclusion is lacking. ESR spectrometry, however, cannot reveal refined details about the molecular structures that give rise to the free radicals in humic substances.

Over the years, many methods have been used to extract humic substances from soils and from other substrates. The major goals of extraction are to isolate the materials in high yields with minimal alteration of the substances from their natural form in the environment. One of the more recent approaches in this area has been the use of supercritical fluids as extractants. Supercritical gas extraction can be used with different solvents for extracting fractions from soil organic matter or for further fractionating humic substances already isolated by conventional base extraction procedures (chapter

4). The data so far, however, indicate poor extraction yields (<5% using n-pentane as the extractant) suggesting that the method may be useful only for studying certain minor fractions in soil organic matter.

Discussion of the molecular structure of humic substances in this book clearly indicates that there is much yet to learn. Knowledge has now progressed to the stage where we have a general idea of the quantities of various molecular subunits. In particular, functional group composition is known reasonably well. We also now know that aromatic C makes up much less of the structure than once thought. Little is known, however, about the arrangement of the molecular subunits. The heterogeneity results in an inability to form crystals that can be subjected to x-ray diffraction analysis, and except for lignin dimers and trimers the molecular structure decomposes into small units in pyrolysis mass spectral analysis. Also, pyrolysis mass spectroscopy determines only the most volatile fractions. Thus, the determination of structural arrangement in humic substances will continue to present a very difficult challenge.

11-2 FORMATION OF HUMIC SUBSTANCES IN SOILS

When plant residues are added to soils, much of the C is rapidly oxidized to CO_2. For crop residues, typically 70% of the C is converted to CO_2 in 1 yr. Carbon added in the form of water-soluble compounds that are assimilated by microbes is mineralized even more rapidly, with as much as 90% evolved as CO_2 in 1 yr (chapter 3). Polysaccharides and proteins in soil are also readily oxidized to CO_2, with 70 to 85% of these materials being mineralized in a period of 6 to 12 mo. The remainder of this nonaromatic C is incorporated into the soil organic matter and microbial biomass.

Lignin and some aromatic compounds are much more resistant to decomposition than the nonaromatic materials mentioned above (chapter 3). Only 30% of the aromatic C from lignin or catechol in soil is oxidized in 1 yr. Very little of the lignin C is incorporated into the microbial biomass, and lignin degradation products constitute one of the sources for the aromatic components in humic substances.

Stabilization of nonaromatic C in soil organic matter is largely due to the incorporation of microbial decay products into humic structures (chapter 3). Peptide and polysaccharide fragments can be covalently coupled to humic structures forming part of the nonaromatic fraction of humic substances. Nonaromatic C can also be transformed into the aromatic C in humic materials. For example, soils in Antarctica, where there are no sources of lignin, contain aromatic C. Fungi and actinomycetes can form melanins, which have some similarity to humic substances. These dark-colored, phenolic, macromolecular materials are very resistant to degradation and contain aromatic units in their structures.

The stability of the aromatic carbons in lignin and melanins suggests that aromaticity may be an important factor in the stability of humic substances. Whole soil ^{13}C-NMR spectroscopy of associated soils under different agricultural and agroforestry systems in Yucatan, Mexico (chapter 8) demonstrates that under cultivation, polysaccharide C content is decreased relative to aromatic carbons. Cultivation and addition of plant nutrients result in a relative increase in the oxidation of polysaccharide carbons. The stability of the man-made *Terra Preta* soils in the Amazon basin is also correlated with a much higher aromaticity compared to adjacent soils. The *Terra Preta* soils have three times the organic C content as adjacent soils. Pyrolysis-mass spectral data suggest that much of the aromaticity in both the *Terra Preta* and the adjacent oxisol is due to lignin fragments. The signals due to lignin fragments in those soils were much greater than those for humic acids reported in chapters 3 and 4. Perhaps lignin fragments are more important in humic substances in soils of the humid tropics. The evidence suggests that the stability of the *Terra Preta* might be due to long-term additions of lignin-rich plant material.

Carbon-14 data for soil organic matter provides further evidence for the stability of humic substances. Typical mean ages for soil C are in excess of 1000 yr. The fraction remaining after $6M$ HCl hydrolysis, which removes peptides and polysaccharides, is older than the hydrolyzed fraction (chapter 3). The intrinsic resistance to microbial degradation of humic substances contributes to stability. Another factor contributing to stability is the formation of clay–humus and clay–humus–metal complexes in the soil. The authors of chapter 3 speculate that much of the nonaromatic fraction in humic substances is due to ring cleavage during microbial degradation.

The origin of humic substances is understood only in very general terms. It is impossible to describe in detail formation reactions for substances whose structure is not known in detail. Progress in structural and composition investigations will ultimately lead to progress in studies of the reactions that form humic substances. It does seem, however, that both lignin degradation products and melanins contribute to aromaticity and that coupling reactions incorporating amino acids, peptides, and polysaccharides into aromatic fragments are important.

11–3 FUNCTIONS OF HUMIC SUBSTANCES RELATED TO CROP PRODUCTION

Humic substances, which are naturally present in the soil, are very important in crop production. They represent slow release sources of N, P, and S for plant nutrition. In addition, they contribute to cation retention because of their considerable cation exchange capacity (CEC) and ability to form complexes with many metals ions. In a typical agricultural soil, the organic matter constitutes a much smaller fraction by weight than the mineral components. Nevertheless, because of the greater CEC (per kilogram) of the organic matter, both fractions contribute about equally to the CEC of many soils. This exchange capacity is important for retaining macro- and micronutrients that otherwise would be leached. The nutrient cations retained can subsequently be released for plant uptake as necessary. Humic substances may influence plant growth beyond the contribution of supplying mineral nutrients. For example, they seem to have a direct stimulatory effect on plants (chapter 7). Also, humic substances influence the effects and mobility of non-ionic organic compounds, both pesticides and pollutants, by removing these compounds from aqueous solution (chapter 6).

Nitrogen in humic substances is quite stable compared to N in the unhumified portion of soil organic matter (chapter 5). Humic substances, however, contribute to both mineralization and immobilization of N. When fertilizer N is added to soil, 20 to 35% is retained in organic forms after the first year. This is similar to the retention of C added in the form of metabolically active C compounds (chapter 4). Eventually, a portion of this N will become incorporated into the well-humified fraction of soil. This well-humified N has a mean residence time similar to that of old C in soils.

Presumably the most biologically active fraction of N in humic substances can be represented by acid hydrolysis products. Acid hydrolysis can remove 50 to 65% of the N from humic acids of which the largest identifiable fraction consists of amino acids. Acid hydrolysis also releases amino sugars and NH_3, the latter resulting from the hydrolysis of amines. A portion of the amino acid fraction also comes from loosely bound proteins and peptides that can be removed from humic acid by phenol extraction or cation exchange. This peptide N is subject to cleavage by proteolytic enzymes. However, much of the amino acid N seems to be covalently bonded in the humic acid structure.

The acid-insoluble and unidentifiable, hydrolyzable N accounts for about 50% of the N in humic substances. There is much speculation about the nature of these fractions of N, but little hard evidence. The authors of chapter 5 suggest that bridge constituents linking quinone groups may be important in the acid-insoluble fraction. Quinones, however, account for only a small fraction of C in soils. Amino acids bound to phenols also are not released by hydrolysis. The ^{15}N-NMR spectroscopic evidence suggests that secondary and possibly tertiary amines may be components of the nonhydrolyzable N. Heterocyclic N (e.g., purine and pyrimidine) may also contribute to the unidentifiable, hydrolyzable N.

Studies of the processes involved in N immobilization and mineralization suffer from the meager understanding of the structure of humic materials. Nitrogen-15 NMR spectroscopy could contribute greatly to the understanding, but the natural abundance of ^{15}N is low, and the time for incorporation of tagged fertilizer N into humic substances is long.

Sewage sludges are often applied to soils both as a means of disposal and as a fertilizer amendment for crop growth (chapter 9). These materials, which are the organic-rich remains after treatment of sewage residues in anaerobic or aerobic digestors, contain materials that conform to the operational definitions of humic substances. The humic acids and fulvic acids from sewage sludge fractions have higher N and a lower C/N ratio than soil humic and fulvic acids. As a result, they are good sources of plant-available N. Humic substances from sewage sludge have high H/C ratios suggesting a high aliphatic content. Unfortunately, ^{13}C-NMR spectroscopic data are not yet available for these materials so aromaticity values are not available. These humic and fulvic acids also have a very high S content because of the incorporation of sulfonate detergents. In fact, the fulvic acid fractions, prepared without a resin separation technique, are contaminated with xenobiotic surfactant materials as well as containing indigenous polysaccharides, amino sugars, and amino acids.

The fulvic acids in sewage sludges are of particular interest because of the possible role of these materials in the mobilization of trace metals contained in the sludge. Because of the high N content of these materials, the ESR spectra of Cu(II) complexes have a very significant contribution due to electron donation from N in the complexes. This fact suggests stronger bonding of Cu(II) by sewage sludge fulvic acids than by normal soil fulvic acids.

Addition of sewage sludges to soil strongly influences the native soil humic and fulvic acids. After a period of a few years, however, the influence of sludge on the composition of humic substances is expected to be minimal even after large additions of sludges (chapter 9). Interestingly, other studies have shown that even though the applied sewage sludges contain significant concentrations of toxic metal ions, which accumulate in the upper layer of the soil, there is no evidence that these metal ions are appreciably assimilated into plants grown on sludge-amended soils (Knuteson et al., 1988).

There are reports that addition of humic substances can have a positive effect on plant growth even when the supply of mineral nutrients is not limiting (chapter 7). In some cases, as with alleviation of high-lime chlorosis by adding Fe(III)-organic matter, complexation with humic substances can increase the supply of a micronutrient to plant roots. Humic substances also can have a direct effect on plant growth. The mechanisms involved in this direct type of response are not clearly understood but there is some evidence that humic substances can increase membrane permeability and macronutrient uptake. Also, there is evidence that humic acids have growth hormone-like activity (chapter 7).

Foliar sprays of humic substances, as well as application in nutrient solutions, can increase both root and top growth. This observation suggests that humic materials are sorbed and translocated within the plants. Studies using tagged materials, however, show that only a small fraction of applied materials is active in plant metabolism and growth.

The application of commercial products containing humic substances to soil to increase plant production is generally not economical because of the cost of the materials and the large quantity needed. On the other hand, foliar sprays containing humic substances may have some promise as a growth promoter because of the smaller quantity required in this mode of application (chapter 7).

Much more work is needed to determine the cause and effect relationship involved in plant response to humic substances. The first question that needs to be addressed is the nature of the fraction of humic substance preparations that is biologically active. Are humic type structures involved in stimulation of plant growth or are there other separate components that are biologically active? It seems unlikely that the large molecular weight core in humic acid could be taken up by plants. Fulvic acids are also quite large. Once the active component or components are identified, the mode of action must be established. Until these questions are answered, reports of humic substances as growth promoters will be viewed with skepticism by most scientists who study plant nutrition (chapter 7).

Soil organic matter is the fraction in surface soils that is most active in the sorption of nonionic pesticides and pollutant molecules from aqueous systems (chapter 6). Dissolved humic substances also can increase the water solubility of nonionic organic compounds by forming complexes. The competition from water adsorption on mineral surfaces greatly inhibits adsorption of nonionic organic compounds on mineral surfaces at humidities normally found in soils. The author of chapter 6 suggests that drying of mineral soils may increase adsorption and decrease vapor phase transport of nonionic compounds, but the data presented refer to soils in equilibrium with air having humidities $< 90\%$. Such low humidities rarely occur in soil except possibly in the top few millimeters under very dry conditions.

The sorption of nonionic organic compounds by soil organic matter can be treated as a solvent-partitioning process whereby the organic compound is distributed between the soil organic matter and water (chapter 6). This partitioning can be correlated with the octanol water partition coefficient of the nonionic solutes. The ability of soil organic matter to function as a partitioning medium in this manner is apparently due to the presence of hydrophobic moieties within the organic matter. Humic substances actually exhibit a duality of character in terms of their hydrophilic/hydrophobic properties. Because of their high content of polar functional groups, such as carboxyl and hydroxyl, these materials are quite hydrophilic, as manifested by their moisture-holding ability. The hydrophobic nature of these materials accounts for the retention of nonionic solutes, and is also manifested in the difficulty of rewetting these materials once they have been dried (in the hydrogen form).

Much more work needs to be done to establish the nature of the partitioning of nonionic solutes into humic and fulvic acids and to determine if other fractions, such as lipids, are important in the process. This may help determine why partition coefficients for organic matter vary from soil to soil.

11-4 CONCLUDING COMMENTS: AN ECOLOGICAL VIEW OF HUMIC SUBSTANCES

It is evident that humic substances play a very important role in soils and in plant growth. Many of the effects of humic substances can be explained in terms of the known properties of these materials. However, there are still many "unknowns" at the fundamental level. Among the most dramatic of the unanswered questions are queries relating to the structure and formation of these materials, and the mechanisms of their interactions with plants. As already mentioned in the introduction to this chapter, the question of the ultimate structure of humic substances may be moot because these materials apparently are constructed from a random combination of the constituents in the "witch's brew" that results from the undirected decay of plant and animal residues. MacCarthy and Rice (1990) have recently proposed that this highly random nature actually constitutes the very essence of humic substances, and have stated that if these materials were comprised of anything other than a highly complex mixture of intractible molecules we would be faced with a serious dilemma.

As early as 1962, Swaby and Ladd (1962, 1966) suggested that the resistance of humic substances to microbial decay could be explained in terms of the random structure of these materials. The disordered structure of humic substances would not serve as a receptive substrate for enzymatic decomposition, in contrast to the facile enzymatic degradation of proteins, polysaccharides, and other ordered biopolymers. Likewise, the irregular humic molecules would not serve as templates for guiding the evolution of future generations of organisms capable of rapidly decomposing these materials.

According to MacCarthy and Rice (1990), the generation of a highly complex mixture of irregular molecules is ecologically desirable because it results in a material having the many essential properties for soil, and consequently for plant growth, *and which is also persistent in the environment.* The organic molecules that confer the desirable properties on soils must be sufficiently long-lived to have the opportunity to exercise these characteristics. While proteins, polysaccharides, and other biopolymers may, in fact, have many general characteristics in common with humic substances, such as hydrophilic nature, metal-complexing properties, ability for sorption on mineral surfaces, acid/base buffering capacity, and so on, these materials generally undergo rapid microbial decomposition in the soil, in contrast to the persistent humic substances.

From this perspective, humic substances can be considered as a genuine class of natural products where the very essence of these materials is their heterogeneous and irregular nature. In all other areas of chemistry we are accustomed to working with materials that are pure or that can be purified. The exceedingly complex nature of humic substances has not allowed their separation into pure components, and consequently we cannot apply the conventional tools of chemistry to these materials with the degree of success that we are accustomed to in other areas.

The difficulty in establishing mechanisms for the formation of humic substances is also consistent with the above discussion. There is no need for a controlled process in the formation of a random mixture, and it appears that these materials are not formed by a single pathway or by a small number of pathways. This is further consistent with the fact that humic substances can form from a diversity of precursor materials. Thus, the many "failed" attempts to determine the structure of humic substances and the mode(s) of their formation should, in fact, not be regarded as failures but rather as providing overwhelming evidence for the highly random nature of these materials. This is not to say that researchers should abandon all attempts at fractionating humic substances or at trying to establish structural features or mechanistic aspects of their formation. Any success in achieving fractions of decreased heterogeneity will facilitate further advances in this area. The ecological viewpoint discussed here, however, should help us to better understand the true nature of these materials. The multicomponent and random character of humic substances imposes severe restrictions on our ability to interpret experimental data measured on these materials. An appreciation of this fact would help to minimize the frequency of over-interpretation of data that is common to this area.

In normal chemical parlance, understanding the chemical nature of a substance means knowing its chemical structure, or the structures of its constituent molecules if it is a mixture. For pure substances, and for simple mixtures that can be separated into pure components, this is an achievable goal because pure substances are composed of an assemblage of identical molecules. This simplifies experimental data recorded on these materials whether it be the degradation products of a hydrolytic procedure or absorption data from spectroscopic measurements. For an intractible mixture, such as humic substances, this level of microscopic detail is not currently an achievable goal, and it appears that this situation will not change in the foreseeable future. The term *structure of humic substances* must not be interpreted in the conventional chemical context because such microscopic detail is simply beyond our reach. The ecological viewpoint discussed above does provide a framework for understanding the nature of humic substances.

Finally, it is interesting to note that the functions of humic substances in the soil depend only on the gross, or average, properties of the material, and are not dependent on any unique molecule. Accordingly, most of the properties and functions of humic substances can be rationalized in terms of what we do know about these substances.

REFERENCES

Knuteson, J., C.E. Clapp, R.H. Dowdy, and W.E. Larson. 1988. Sewage sludge management. Land application of municipal sewage sludge to a terraced watershed in Minnesota. Minn. Agric. Exp. Stn. Publ. 56-1988, St. Paul, MN.

Hayes, M.H.B., P. MacCarthy, R.L. Malcolm, and R.S. Swift (ed.). 1989. Humic substances II. In search of structure. John Wiley & Sons, Ltd., Chichester, UK.

MacCarthy, P., and J.A. Rice. 1985. Spectroscopic methods (other than NMR) for determining functionality in humic substances. p. 527–559. In G.R. Aiken et al. (ed.) Humic substances in soil, sediment, and water: Geochemistry, isolation, and characterization. Wiley-Interscience, New York.

MacCarthy, P., and J.A. Rice. 1990. An ecological rationale for the heterogeneity of humic substances: A holistic perspective of humus. In S.H. Schneider and P.J. Boston (ed.) Proc. of Chapman Conference on the Gaia Hypothesis. (San Diego, CA) 7–11 March 1988. MIT Press, Cambridge, MA. (In press.)

Swaby, R.J., and J.N. Ladd. 1962. Chemical nature, microbial resistance, and origin of soil humus. p. 197–202. In G.J. Neale (ed.) Trans. Jt. Meet. Comm. 4, 5, Int. Soc. Soil Sci. 13–22 Nov. 1962, Palmerston North, New Zealand. Soil Bureau, P.B., Lower Hutt, New Zealand.

Swaby, R.J., and J.N. Ladd. 1966. Stability and origin of soil humus. p. 153–159. In The use of isotopes in soil organic matter studies. Report FAO/IAEA in cooperation with the Int. Soc. Soil Sci. Brunswick-Völkenrode, Sept. 1963. (Supplement to J. Appl. Rad. Isot.) Pergamon Press, New York.

SUBJECT INDEX

Acetamide, soil, 164
Acid hydrolysis, soil humic acids, 50–51
Acid insoluble-N
  description, 92
  soil, 267
  soil humic substances, 95–99
Acid soil, 10
Acidity
  humic substances, 9–10
  sewage sludge humic substances, 207–208, 210–213
  sludge-amended soil humic substances, 215–216
  tropical soils, 193
Adsorbate, 113
Adsorbent, 113
Adsorption, by soil, 112–113, 121–128
Adsorption capacity, 126
Adsorption equilibrium, 124
Adsorption isotherm, 130
Aggregates, soil, 49, 57
Alcohol groups, 8, 10
Aliphatic acids, degradation in soil, 41
Aliphatic carbon
  groundwater humic substances, 31–33
  sewage sludge humic substances, 208, 214
  sludge-amended soil humic substances, 216
  soil, 250
  soil humic substances, 17–19, 21–22, 25–26, 32, 66, 69
  stream humic substances, 27, 29, 31
  tropical soils, 195–200
Alkanes, soil humic substances, 66, 79–81, 83–84
Alkanoic acids, soil humic substances, 66, 80–82, 84
Alkyl benzene sulfonate, sewage sludge, 207
Alkylphosphonic acids, soil, 254
Alkylphosphonic esters, soil, 254
Amalgam reductive degradation, soil humic substances, 52–53, 95
Amazon soil. *See* Brazilian Amazon region soil
Amide groups
  sewage sludge humic substances, 208
  tropical soils, 195, 199
Amino acids
  degradation in soil, 39, 41, 43, 48
  description, 92
  incorporation into humic substances, 45
  sewage sludge humic substances, 210–211, 215, 268
  sludge-amended soil humic substances, 215
  soil, 4–5, 49, 267
  soil humic substances, 7, 38, 50, 84, 92–99, 101–102
  tropical soils, 196
Amino sugars

degradation in soil, 48
description, 93
incorporation into humic substances, 45
sewage sludge humic substances, 210–211, 215, 268
sludge-amended soil humic substances, 215
soil, 48–49
soil humic substances, 38, 93–98, 101, 267
structure, NMR spectroscopy, 244
Ammonia-N
  description, 92
  sewage sludge humic substances, 210–211, 215
  sludge-amended soil humic substances, 215
  soil, 267
  soil humic substances, 93–94, 96–98, 100–102
Amylopectin, 243–244
Amylose, 243–244
Anaerobically digested sludge, 204–205
Aniline, 124
Anionic surfactant, 204, 210–211
Anisole, 137
Anomeric carbon
  soil humic substances, 17, 19, 24
  stream humic substances, 27, 29
Antarctic soil, 252, 265
Anthracene, 124
Anthropogenic soil. *See Terra Preta* soil
Arachis hypogea L. *See* Peanut
Arctic soil, 10
Aromatic carbon, 263, 265
  groundwater humic substances, 31–33
  lignin, 241
  soil, 248–252
  soil humic substances, 17–22, 25–26, 33, 52–53, 66–69, 84
  stream humic substances, 27, 29, 31, 33
  tropical soils, 195–196, 199–200, 266
Aromatic compounds
  degradation in soil, 41–42
  incorporation into humic substances, 45
Aromaticity, 15, 18–21, 26–29, 67–69, 85–86, 187, 199, 240, 245, 263, 266
Auximones, 163

Barley, growth response to humic substances, 165
Bean, growth response to humic substances, 167–168
Beech, degradation in soil, 252
Beet, growth response to humic substances, 172, 174–176
Begonia, growth response to humic substances, 167–168, 175–176
Begonia semperflorens L. *See* Begonia
Benzene, 124, 126, 128–129, 134–137, 150–151